STUDENT SUPPLEMENT
VOLUME 1: CHAPTERS 1–11
TO ACCOMPANY

SWOKOWSKI'S

CALCULUS WITH ANALYTIC GEOMETRY
FOURTH
EDITION

THOMAS A. BRONIKOWSKI
MARQUETTE UNIVERSITY

PWS-KENT PUBLISHING COMPANY
BOSTON

PWS-KENT
Publishing Company

20 Park Plaza
Boston, Massachusetts 02116

PWS-KENT Publishing Company is a division of Wadsworth, Inc.

Portions of this book previously appeared in the Student Supplement Volume 1: Chapters 1-11, to accompany Swokowski's Calculus with Analytic Geometry, Alternate Edition, copyright © 1983 by PWS Publishers, and in the Student Supplement Volume 1: Chapters 1-11, to accompany Swokowski's Calculus with Analytic Geometry, Third Edition, copyright © 1984 by PWS Publishers.

Printed in the United States of America.

91 92 -- 10 9 8 7 6 5 4

ISBN 0-87150-131-7

INTRODUCTION

This supplement to the Fourth Edition of Earl W. Swokowski's
Calculus with Analytic Geometry has been written to help you, the student,
develop the skill and art of solving problems in calculus. I have included the
solution of every third problem (numbers 1, 4, 7, 10, etc.) from every section
of the text. Occasionally this sequence is modified slightly to provide a
greater variety of solved exercises. However, theoretical problems involving
proofs are sometimes not included because many instructors use such problems as
special assignments and prefer to have no more than a hint given. In sections
of the text concerned with applications of the definite integral I have included,
in addition to solutions of exercises in the normal sequence, integral answers
to the remaining odd-numbered exercises. In this way you can quickly check to
see if a wrong answer is due to setting up the problem incorrectly or to merely
making an error in evaluating the correct integral.

Clearly, not every arithmetic and algebraic detail can be included in every
solution if the size of this book is to remain reasonable. You will find enough
detail in all solutions to see how the correct answers are obtained. The
symbols "$\Rightarrow$" and "$\Leftrightarrow$" are used frequently and are to be read "implies" and
"if and only if" respectively. All other symbols will be familiar to you from
the text or from other courses. The author would appreciate being informed of
errors or misprints in this work. A note to me at the address below will be
promptly acknowledged with thanks.

I would like to express my deepest appreciation to David Geggis of PWS-KENT
and to Kathi Townes for their untiring efforts to expedite and smooth the path
of this project. My thanks too to my friend and colleague, Earl W. Swokowski,
for his thoughtful recommendations. To my wife, Irene, I owe a special debt of
gratitude for her good-natured, generous, unflagging participation in all phases
of the production of this work.

Prof. Thomas A. Bronikowski
Department of Mathematics
Marquette University
Milwaukee, Wisconsin 53233

TABLE OF CONTENTS

CHAPTER 1

FUNCTIONS AND GRAPHS

EXERCISES 1.1

1. (a) $-2 > -5$ since $-2 - (-5) = 3 > 0$. (b) $-2 < 5$ since $5 - (-2) = 7 > 0$.
(c) $(6-1) = (2+3)$ since $(6-1) - (2+3) = 0$. (d) $2/3 > 0.66$ since $2/3 - 0.66 = 0.666... - 0.66 = 0.006... > 0$. (e) $2 = \sqrt{4}$ by definition of the $\sqrt{}$ symbol.
(f) $\pi < 22/7$ since $22/7 - \pi = 3.14285... - 3.14159... = 0.00126... > 0$.

4. (a) $|4-8| = |-4| = -(-4) = 4$. (b) $|3-\pi| = -(3-\pi) = \pi-3$. (c) $|-4| - |-8| = -(-4) - (-(-8)) = 4-8 = -4$. (d) $|-4+8| = |4| = 4$. (e) $|-3|^2 = (-(-3))^2 = 3^2 = 9$. (f) $|2-\sqrt{4}| = |2-2| = |0| = 0$. (g) $|-0.67| = -(-0.67) = 0.67$.
(h) $-|-3| = -[-(-3)] = -3$. (i) $|x^2+1| = x^2+1$ since $x^2+1 \geq 1 > 0$ for all x.
(j) $|-4-x^2| = |-(4+x^2)| = -[-(4+x^2)] = 4+x^2$ since $-4-x^2 \leq -4 < 0$ for all x.

7. $5x-6 > 11 \iff 5x > 17 \iff x > 17/5$. $\therefore$ $(17/5,\infty)$ is the solution set.

10. $7-2x \geq -3 \iff 10 \geq 2x \iff 5 \geq x$. $\therefore$ $(-\infty,5]$ is the solution set.

13. $3x+2 < 5x-8 \iff 10 < 2x \iff 5 < x$. $\therefore$ $(5,\infty)$ is the solution set.

16. $-4 < 2-9x < 5 \iff -6 < -9x < 3 \iff 2/3 > x > -1/3$. $\therefore$ $(-1/3,2/3)$ is the solution set.

19. $5/(7-2x) > 0 \iff 7-2x > 0$ (since $5 > 0$) $\iff 7 > 2x \iff 7/2 > x$.
$\therefore$ $(-\infty,7/2)$ is the solution set.

22. $|(2x+3)/5| < 2 \iff |2x+3| < 10 \iff -10 < 2x+3 < 10 \iff -13 < 2x < 7 \iff -13/2 < x < 7/2$ or $(-13/2,7/2)$.

25. $|25x-8| > 7 \iff 25x-8 > 7$ or $25x-8 < -7$. The first yields $x > 3/5$; the second yields $x < 1/25$. Thus the solution set is $(-\infty,1/25) \cup (3/5,\infty)$.

28. $2x^2 - 9x + 7 < 0 \iff (2x-7)(x-1) < 0$. For the product to be negative, the factors must differ in sign. The points where one or the other changes sign are $x = 1$ and $x = 7/2$. Schematically we have

```
(2x-7)        -           -           +
         ———————————————+———————————+—————————  x
(x-1)         -          1    +    7/2     +
```

Thus $(1,7/2)$ is the desired solution set.

31. $1/x^2 < 100 \iff 1/|x| < 10 \iff |x| > 1/10 \iff x > 1/10$ or $x < -1/10$. Thus $(-\infty,-1/10) \cup (1/10,\infty)$ is the solution set.

34. It is convenient to look at the inequality separately on $(-\infty,-2)$, $(-2,9)$, $(9,\infty)$. On the 1st, both denominators are negative and $3/(x-9) > 2/(x+2) \iff (3x+6)/(x-9) < 2 \iff 3x+6 > 2x-18 \iff x > -24$. Since $x \in (-\infty,-2)$, we get $(-24,-2)$. On the 2nd interval the left side is < 0, the right side is > 0, and, so, no number here can satisfy the given inequality. On the 3rd, $(9,\infty)$, both denominators are positive and, as above, we obtain $x > -24$. Since every

x ε (9,∞) satisfies this, we get the entire interval (9,∞). Combining results we get (-24,-2) ∪ (9,∞) as the solution set.

37. F = 4.5x = $\frac{9}{2}$x, 10 ≤ F ≤ 18 ⟹ 10 ≤ $\frac{9}{2}$x ≤ 18. Now multiply through by $\frac{2}{9}$ to
     obtain $\frac{20}{9}$ ≤ x ≤ 4.

40. pν = 200 ⟹ p = 200/ν. 25 ≤ ν ≤ 50 ⟹ 200/50 ≤ 200/ν ≤ 200/25,
     or 4 ≤ p ≤ 8.

EXERCISES 1.2

1. (a) d(A,B) = $\sqrt{(2-6)^2 + (1-(-2))^2}$ = $\sqrt{(-4)^2 + 3^2}$ = $\sqrt{16+9}$ = $\sqrt{25}$ = 5. (b) Mid-
    point is ((6+2)/2,(-2+1)/2) = (4,-1/2).

4. (a) d(A,B) = $\sqrt{(4-4)^2 + (5-(-4))^2}$ = $\sqrt{0+9^2}$ = 9. (b) Midpoint is
    ((4+4)/2,(5-4)/2) = (4,1/2).

7. Since d(A,B)^2 = 5^2 + 5^2 = 50, d(B,C)^2 = 7^2 + 7^2 = 98, d(A,C)^2 = 12^2 + 2^2 =
    148, it is a right triangle with legs AB and BC and hypotenuse AC. The area
    is $\frac{1}{2}$d(A,B)d(B,C) = $\frac{1}{2}$ $\sqrt{50}$ $\sqrt{98}$ = (5$\sqrt{2}$)(7$\sqrt{2}$)/2 = 35.

10. A point (x,y) satisfies y = -3 if and only if it is 3 units below the x-axis.
     Thus the graph is a horizontal line 3 units below and parallel to the x-axis.

13. |x| < 2 ⟺ -2 < x < 2 and |y| > 1 ⟺ y > 1 or y < -1. Thus the graph
     consists of the two vertical strips shown between the vertical lines x = ±2.
     The upper strip corresponds to y > 1; the lower to y < -1.

NOTE: In graphing the equations in #15-32 of the text, all you can do now is
tabulate some points, plot them and connect them smoothly. Only after Chapters 3,
4 and 12 will you have the mathematical background to sketch graphs accurately
with only a few significant points plotted. The graphs for the odd-numbered
exercises are in the text.

16. For y = 4x-3, we obtain the table of points and then
     plot to obtain the line shown.

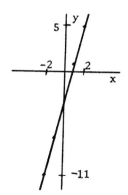

x	-2	-1	0	1	2
y	-11	-7	-3	1	5

There are no symmetries in this graph.

19. For $y = 2x^2 - 1$, we plot

x	± 2	± 1	0
y	7	1	-1

to obtain the graph in the text, a parabola. If we replace x by -x and use $(-x)^2 = x^2$, we obtain the same equation. Hence, the graph is symmetric with respect to the y-axis.

22. Rewrite the equation as $y = -x^2/3$ and tabulate as above.

x	± 3	± 2	± 1	0
y	-3	-4/3	-1/3	0

As above, the graph is symmetric with respect to the y-axis.

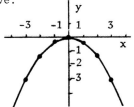

25. For $y = x^3 - 2$, we obtain

x	-2	-1	0	1	2
y	-10	-3	-2	-1	6

There are no symmetries.

28. For $y = \sqrt{x} - 1$, x must be ≥ 0 and we get

x	0	1	4	9
y	-1	0	1	2

There are no symmetries.

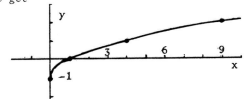

31. The graph is a circle of radius 4 with center at (0,0), symmetric with respect to both axes and the origin.

34. $x = \sqrt{4 - y^2} \implies x \geq 0$, and $x^2 = 4 - y^2$, or $x^2 + y^2 = 4$. Since $x \geq 0$, the graph is the right half of the circle with C(0,0) and r = 2.

37. Directly we obtain $(x - 3)^2 + (y - (-2))^2 = 4^2$ or $(x - 3)^2 + (y + 2)^2 = 16$.

40. $r = d(C,P) = ((-4 - 1)^2 + (6 - 2)^2)^{1/2} = (5^2 + 4^2)^{1/2} = 41^{1/2}$. Thus: $(x + 4)^2 + (y - 6)^2 = 41$.

43. Since AB is a diameter, C is the midpoint, $(1,2)$, and
     $r = d(A,B)/2 = (6^2 + 10^2)^{1/2}/2 = (136/4)^{1/2} = 34^{1/2}$. Thus,
     $(x - 1)^2 + (y - 2)^2 = 34$.

46. We solve by completing the squares:
$$(x^2 - 10x \qquad) + (y^2 + 2y \qquad) = -22$$
$$(x^2 - 10x + 25) + (y^2 + 2y + 1) = -22 + 26$$
$$(x - 5)^2 \qquad\qquad + (y + 1)^2 \qquad = 4$$
     Thus, $C(5,-1)$, $r = 2$.

49. We begin as above, but here we divide by 2 to convert the
     coefficients of x^2 and y^2 to 1:
$$(2x^2 - x \qquad\qquad) + (2y^2 + y \qquad\qquad) = 3$$
$$(\ x^2 - x/2 \qquad\qquad) + (\ y^2 + y/2 \qquad\qquad) = 3/2$$
$$(\ x^2 - x/2 + 1/16) + (\ y^2 + y/2 + 1/16) = 3/2 + 2/16$$
$$(x - 1/4)^2 + (y + 1/4)^2 = 26/16.$$
     Thus, $C(1/4, -1/4)$, $r = \sqrt{26}/4$.

EXERCISES 1.3

NOTE: In some of the solutions below, m_{AB} will denote the slope of the line
through points A and B.

 1. $m_{AB} = (18-6)/(-1-(-4)) = 12/(-1+4) = 4$.

 4. $m_{AB} = (4-4)/(2+3) = 0$.

 7. $m_{AB} = (12-15)/(11-6) = -3/5$ and $m_{CD} = (-5+8)/(-6+1) = -3/5$. Thus AB $\parallel$ CD.
     Also, $m_{AD} = 20/12 = 5/3$ and $m_{BC} = (-20)/(-12) = 5/3$ and AD $\parallel$ BC. Thus the
     figure is a parallelogram and since the slopes of adjacent sides are nega-
     tive reciprocals, these sides are $\perp$ and the figure is a rectangle.

10. Let E, F, G, H be the midpoints of the segments AB, BC, CD, DA, respectively.
     Then
$$m_{EF} = (\frac{y_2+y_3}{2} - \frac{y_1+y_2}{2}) \Big/ (\frac{x_2+x_3}{2} - \frac{x_1+x_2}{2}) = (y_3-y_1)/(x_3-x_1) \quad (\text{if } x_1 \neq x_3).$$
$$m_{HG} = (\frac{y_3+y_4}{2} - \frac{y_1+y_4}{2}) \Big/ (\frac{x_3+x_4}{2} - \frac{x_1+x_4}{2}) = (y_3-y_1)/(x_3-x_1).$$
     Thus EF $\parallel$ HG since, if $x_1 \neq x_3$, the slopes are equal and if $x_1 = x_3$ both
     lines are vertical. Similarly $m_{EH} = m_{FG} = (y_2-y_4)/(x_2-x_4)$, if $x_2 \neq x_4$.

13. $m_{AB} = \frac{(-4-(-7))}{(3-(-5))} = \frac{3}{8}$. Using this slope and the point $A(-5,-7)$ in the point-

slope formula (1.15), we obtain: $y-(-7) = (3/8)(x-(-5))$ or $8(y+7) = 3(x+5)$
or $3x - 8y - 41 = 0$.

16. Using $m = 6$, and the point $(-2,0)$, we get $(y-0) = 6(x-(-2))$ or $y = 6(x+2)$.

19. The given line, $2x - 5y = 8$, has slope $2/5$ (obtained by writing it as
     $y = (2/5)x - (8/5)$). Thus the desired slope is $-5/2$ and the equation is
     $y + 3 = (-5/2)(x-7)$.

22. The desired line must pass through the origin and must have inclination $135°$.
     Since $\tan 135° = -1$, the equation is $y = -x$.

25. Rewriting the equation in the form $y = -\frac{1}{2}x + 0$, we see that
     $m = -1/2$, $b = 0$.

28. Rewriting as $y = 2x + 4$, we obtain $m = 2$, $b = 4$.

31. Substituting the coordinates of $P(-1,2)$ into the equation, we
     obtain: $k(-1) + 2(2) - 7 = 0$, or $-k - 3 = 0$. Thus, $k = -3$.

34. If $x_1 \neq x_2$, then $m = (y_2 - y_1)/(x_2 - x_1)$. Using this together
     with P_1 in the point-slope formula, the two-point formula is
     obtained after multiplying by $x_2 - x_1$. If $x_1 = x_2$, then the
     line is vertical with equation $x = x_1$. In this case too, the
     two-point formula is satisfied since each side is zero
     $(x_2 - x_1 = 0$ and $x - x_1 = 0)$. Using $P_1 = A$ and $P_2 = B$, we obtain:
     $(y + 1)(4 - 7) = (6 + 1)(x - 7)$ or $-3(y + 1) = 7(x - 7)$.

37. (a) $R_0 =$ resistance at $T = 0°C$.

     (b) Substituting $T = -273$ and $R = 0$, we obtain $R_0(1 - 273\alpha) = 0$
     $\Longrightarrow 273\alpha = 1$ or $\alpha = 1/273$.

     (c) With $R_0 = 1.25$ and $\alpha = 1/273$, we have $R = 1.25(1 + T/273)$.
     Setting $R = 2R_0 = 2.5$, we solve $2.50 = 1.25(1 + T/273)$
     $\Longrightarrow 1 + T/273 = 2 \Rightarrow T = 273°C$.

40. (a) The radius from $(0,0)$ to $(-4,-3)$ has slope $3/4$. Thus, the
     tangent line has slope $-4/3$ and equation $y + 3 = (-4/3)(x + 4)$.
     When $y = -50$, we obtain $x + 3 = (3/4)47 = 141/4$. Thus, $x = 129/4$.

     (b) Let (a,b) be the unknown release point. The radial slope is
     b/a, and the tangent line equation is $(y - b) = (-a/b)(x - a)$ or
     $ax + by = a^2 + b^2 = 25$ (since (a,b) is on the circle of $r = 5$).

Since $(0,-50)$ is also on this line we obtain $-50b = 25$ or $b = -1/2$ and $a = -(5^2 - 1/4)^{1/2} = -\sqrt{99}/2$.

EXERCISES 1.4

1. $f(x) = x^3 + 4x - 3$. (a) $f(1) = 1^3 + 4 \cdot 1 - 3 = 2$. (b) $f(-1) = (-1)^3 + 4(-1) - 3 = -1 - 4 - 3 = -8$. (c) $f(0) = 0^3 + 4 \cdot 0 - 3 = -3$. (d) $f(\sqrt{2}) = (\sqrt{2})^3 + 4\sqrt{2} - 3 = 2\sqrt{2} + 4\sqrt{2} - 3 = 6\sqrt{2} - 3$.

4. $f(x) = 1/(x^2+1)$. (a) $f(a) = 1/(a^2+1)$. (b) $f(-a) = 1/((-a)^2+1) = 1/(a^2+1)$. (c) $-f(a) = -1/(a^2+1)$. (d) $f(a+h) = 1/((a+h)^2+1)$. (e) $f(a) + f(h) = (1/(a^2+1)) + (1/(h^2+1))$.

(f) $\dfrac{f(a+h) - f(a)}{h} = \dfrac{1}{h}\left(\dfrac{1}{(a+h)^2+1} - \dfrac{1}{a^2+1}\right) = \dfrac{1}{h}\dfrac{(a^2+1) - ((a+h)^2+1)}{((a+h)^2 + 1)(a^2+1)} =$

$\dfrac{1}{h}\dfrac{a^2+1-a^2-2ah-h^2-1}{((a+h)^2+1)(a^2+1)} = -\dfrac{(2a+h)}{((a+h)^2+1)(a^2+1)}$.

7. The domain consists of all real numbers x for which $3x-5 \geq 0$. Solving, we get $3x \geq 5$ or $x \geq 5/3$.

10. Here, the domain $= \{x: x^2 - 9 \geq 0\}$. Since $x^2 - 9 \geq 0 \Longleftrightarrow x^2 \geq 9 \Longleftrightarrow |x| \geq 3$, the domain is $(-\infty,3] \cup [3,\infty)$.

13. f is one-to-one since $a \neq b \Longrightarrow 2a \neq 2b \Longrightarrow 2a+9 \neq 2b+9$, or $f(a) \neq f(b)$.

16. f is not one-to-one. The graph of f is a parabola opening upward. This suggests that different x values will yield the same functional value. In particular $f(0) = f(.5) = -3$.

19. f is not one-to-one. If $a > 0$ then $a \neq -a$, but $f(a) = a$, $f(-a) = |-a| = -(-a) = a$.

22. $f(-a) = 7(-a)^4 - (-a)^2 + 7 = 7a^4 - a^2 + 7 = f(a)$ $\therefore$ even.

25. f is even since $f(a) = f(-a) = 2$.

28. f is even since $f(-a) = \sqrt{(-a)^2+1} = \sqrt{a^2+1} = f(a)$.

31. As a polynomial, the domain is all of $\mathbb{R}$. The range is also $\mathbb{R}$ since, if $y \in \mathbb{R}$, $f((3-y)/4) = y$.

34. Domain is $\mathbb{R}$; range is $\{3\}$; graph is the horizontal line $y = 3$.

37. Domain $= \{x: 4-x^2 \geq 0\} = [-2,2]$. The graph is that of the equation $y = \sqrt{4-x^2}$, i.e., the upper half of the circle $y^2 = 4-x^2$ or $x^2 + y^2 = 4$. Thus the range is the set of y-values corresponding to this graph, namely $[0,2]$.

40. Domain $= \{x:x \neq 4\}$, range $= \{y:y < 0\}$ since for any $y < 0$, $f(4 \pm \sqrt{-1/y}) = y$. See graph below.

43. Domain = $\{x:4 - x \geq 0\}$ = $(-\infty,4]$, range = $\{y:y \geq 0\}$ since, by
     definition, $\sqrt{4 - x} \geq 0$ and if $y \geq 0$, $f(4 - y^2)$ = y.

46.   40. (Graph)

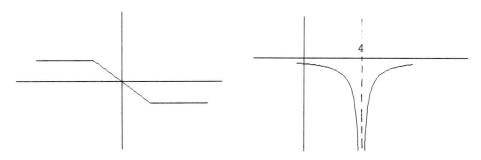

49. If $x \neq 2$, $f(x)$ = $(x + 2)(x - 2)/(x - 2)$ = x + 2, and the graph
     is a straight line for such x.

52. Let f be one-to-one. If some horizontal line intersected the
     graph in 2 points (a, f(a)) and (b, f(b)), then a $\neq$ b but
     f(a) = f(b) contradicting f being one-to-one. Conversely, if
     every horizontal line intersects the graph of f in one point, if
     f were not one-to-one, we could find a,b with a $\neq$ b but f(a) =
     f(b). But this means that y = f(a) cuts the graph in 2 points.

55. Let h = height above the ground at time t. Then h = 2t and
     d = $\sqrt{100^2 + h^2}$ = $\sqrt{10,000 + 4t^2}$ = $2\sqrt{2500 + t^2}$.

58. Let d_r = distance rowed, d_w = distance walked, T_r = time rowing,
     T_w = time walking. Then, in terms of x, d_r = $\sqrt{4 + (6 - x)^2}$,
     d_w = x. Since time = distance/rate, T_r = $\sqrt{4 + (6 - x)^2}/3$,
     T_w = x/5. Finally, T = T_r + T_w to combine both.

EXERCISES 1.5

  4. (Graph)   10. (Graph)

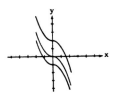
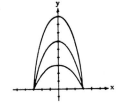

13. $(f{\pm}g)(x) = f(x) \pm g(x) = 3x^2 \pm 1/(2x-3)$. $(fg)(x) = f(x)g(x) = 3x^2(1/(2x-3)) =$ $3x^2/(2x-3)$. $(f/g)(x) = f(x)/g(x) = 3x^2/(1/(2x-3)) = 3x^2(2x-3)$ (if $x \neq 3/2$).

16. $(f{\pm}g)(x) = (x^3+3x) \pm (3x^2+1)$. $(fg)(x) = (x^3+3x)(3x^2+1) = 3x^5 + 10x^3 + 3x$. $(f/g)(x) = (x^3+3x)/(3x^2+1)$.

19. If $f(x) = 2x^2 + 5$, $g(x) = 4 - 7x$ then $(f{\circ}g)(x) = 2g(x)^2 + 5 = 2(4-7x)^2 + 5 =$ $2(16-56x+49x^2) + 5 = 98x^2 - 112x + 37$; and $(g{\circ}f)(x) = 4 - 7f(x) = 4-7(2x^2+5)$ $= 4 - 14x^2 - 35 = -14x^2 - 31$.

22. If $f(x) = \sqrt{x^2+4}$, $g(x) = 7x^2+1$ then $(f{\circ}g)(x) = \sqrt{g(x)^2+4} = \sqrt{(7x^2+1)^2 + 4} =$ $\sqrt{49x^4 + 14x^2 + 5}$, and $(g{\circ}f)(x) = 7f(x)^2 + 1 = 7(x^2+4) + 1 = 7x^2 + 29$.

25. Here, $(f{\circ}g)(x) = \sqrt{2g(x) + 1} = \sqrt{2x^2+7}$ and $(g{\circ}f)(x) = f(x)^2 + 3 = (2x+1) + 3 =$ $2x+4$, for $x \geq -1/2$ (the domain of f).

28. $(f{\circ}g)(x) = \sqrt[3]{g(x)^2+1} = \sqrt[3]{(x^3+1)^2+1} = \sqrt[3]{x^6+2x^3+2}$. $(g{\circ}f)(x) = f(x)^3+1 = (\sqrt[3]{x^2+1})^3$ $+ 1 = (x^2+1) + 1 = x^2 + 2$.

31. $(f{\circ}g)(x) = 2g(x) - 3 = 2(\frac{x+3}{2}) - 3 = (x+3) - 3 = x$. $(g{\circ}f)(x) = \frac{f(x)+3}{2} =$ $\frac{(2x-3)+3}{2} = \frac{2x}{2} = x$.

34. Let $f(x) = a_m x^m + \cdots + a_0$, $g(x) = b_n x^n + \cdots + b_0$, where $a_m \neq 0$ and $b_n \neq 0$. Then $(f{\circ}g)(x) = a_m g(x)^m + \cdots + a_0 =$ $a_m(b_n^m x^{mn} + \cdots b_0^m) + \cdots + a_0$, which is of degree mn since $a_m b_n^m \neq 0$.

37. Let h = height of balloon and L = length of rope paid out t seconds after release. Then $L^2 = h^2 + 20^2$ and $h = \sqrt{L^2 - 400}$. At t = 0, L = 20 since the rope is flat on the ground between pulley and pad. Since rope pays out at 5 ft/sec, L = 20 + 5t and $L^2 = 400 + 200t + 25t^2$. Thus, $h = \sqrt{200t + 25t^2} = 5\sqrt{8t + t^2}$.

40. Since he walks at the rate of
1 ft/sec, d = t. We construct
the triangle shown with base
2' above the ground, height
the upper 28' of the pole, and
hypoteneuse the length L of
the rope. We see that
$L = \sqrt{50^2 + 28^2} = \sqrt{3284} = 2\sqrt{821}$.
Then h = y + 2, and by similar
triangle, y/d = 28/L. Thus,
y = 28d/L = 28t/L, and h = 28t/L + 2 = $14t/\sqrt{821}$ + 2.

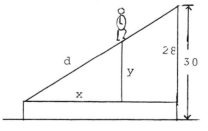

EXERCISES 1.6 (Review)

1. $4-3x > 7+2x \iff -3 > 5x \iff -3/5 > x \iff x \epsilon (-\infty,-3/5)$.

4. Observing that $|6x-7| \leq 1 \iff -1 \leq 6x-7 \leq 1 \iff 6 \leq 6x \leq 8 \iff 1 \leq x \leq 4/3$, the desired solution set of $|6x-7| > 1$ is the complement of $[1,4/3]$, namely $(-\infty,1) \cup (4/3,\infty)$. (This is equivalent to the method of #25, Sec. 1.1.)

7. If $x > 1/3$, both denominators are positive and $1/(3x-1) < 2/(x+5) \iff x+5 < 6x-2 \iff 7 < 5x \iff 7/5 < x$, yielding the partial solution $(7/5,\infty)$. If $-5 < x < 1/3$, $3x-1 < 0$, $x+5 > 0$ and every number in $(-5,1/3)$ satisfies the inequality. If $x < -5$, both denominators are negative and, as in the first part, the given inequality is equivalent to $7/5 < x$ which is not true for any $x < -5$. Hence $(-5,1/3) \cup (7/5,\infty)$ is the solution set.

10. $3x-5y = 10$ is a line with slope $3/5$ and y-intercept -2. There are no sym-metries since replacement of x by -x and/or y by -y changes the equation and the solution set.

13. $|x+y| = 1$ is satisfied if $x+y = 1$ or $x+y = -1$. Thus the graph consists of these 2 parallel lines. The graph is symmetric with respect to the origin since replacing x and y by -x and -y, respectively, we get $|-x-y| = |-(x+y)| = |x+y| = 1$, the same equation.

16. A point (x,y) satisfies $x^2 + y^2 < 1$ if, and only if, its distance from $(0,0)$ is < 1. Thus the graph of W is the set of points inside, but not <u>on</u>, the circle $x^2 + y^2 = 1$.

19. Since $C(-4,-3)$ is 9 units from the vertical line $x = 5$, the radius is 9 and the equation is $(x+4)^2 + (y+3)^2 = 81$.

22. $m_{AC} = (-5-2)/(2-(-4)) = -7/6$ and the equation is $y-6 = (-7/6)(x-3)$ or $7x + 6y - 57 = 0$.

25. Any line parallel to the y-axis has equation x = constant. Since it passes through $A(-4,2)$, its equation is $x = -4$.

28. Because of the radical in the denominator, the domain is $\{x: 16 - x^2 > 0\} = (-4,4)$.

31. If $f(x) = 1/\sqrt{x+1}$, (a) $f(1) = 1/\sqrt{2}$, (b) $f(3) = 1/\sqrt{4} = 1/2$, (c) $f(0) = 1/\sqrt{1} = 1$, (d) $f(\sqrt{2}-1) = 1/\sqrt{(\sqrt{2}-1)+1} = 1/\sqrt{\sqrt{2}} = 1/\sqrt[4]{2}$, (e) $f(-x) = 1/\sqrt{-x+1}$, (f) $-f(x) = -1/\sqrt{x+1}$, (g) $f(x^2) = 1/\sqrt{x^2+1}$, (h) $(f(x))^2 = (1/\sqrt{x+1})^2 = 1/(x+1)$.

34. If x is close to -1, the numbers -1/(x+1) are very
large, positive if x < -1, negative if x > -1. If
|x| is very large, then -1/(x+1) is nearly 0.

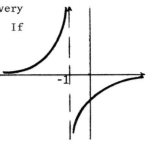

40. If $f(x) = x^2 + 3x + 1$, $g(x) = 2x - 1$, then $(f + g)(x) = x^2 + 5x$,
$(f - g)(x) = x^2 + x + 2$, $(fg)(x) = (x^2 + 3x + 1)(2x - 1) =$
$2x^3 + 5x^2 - x - 1$, $(f/g)(x) = (x^2 + 3x + 1)/(2x - 1)$,
$(f \circ g)(x) = (2x - 1)^2 + 3(2x - 1) + 1 = 4x^2 + 2x - 1$,
$(g \circ f)(x) = 2(x^2 + 3x + 1) - 1$.

CHAPTER 2
LIMITS OF FUNCTIONS

EXERCISES 2.1

1. (a) With $P(a,f(a)) = P(a,5a^2-4)$ and $Q(x,f(x)) = Q(x,5x^2-4)$, we

   have $m_{PQ} = \dfrac{(5x^2 - 4) - (5a^2 - 4)}{x - a} = \dfrac{5(x^2 - a^2)}{x - a} = \dfrac{5(x - a)(x + a)}{x - a}$

   $= 5(x + a)$. Thus, $m = \lim\limits_{x \to a} m_{PQ} = \lim\limits_{x \to a} 5(x + a) = 5(2a) = 10a$.

   (b) $a = 2 \Rightarrow P(2,16)$, $m = 20$, and the tangent line is $y - 16 = 20(x - 2)$.

4. (a) Here, $P(a,f(a)) = P(a,a^4)$ and $Q(x,f(x)) = (x,x^4)$. Thus,

   $m_{PQ} = \dfrac{x^4 - a^4}{x - a} = \dfrac{(x - a)(x^3 + x^2a + xa^2 + a^3)}{x - a} = x^3 + x^2a + xa^2 + a^3$,

   and $m = \lim\limits_{x \to a} m_{PQ} = 4a^3$.

   (b) $a = 2 \Rightarrow P(2,16)$, $m = 32$, and the tangent line is $y - 16 = 32(x - 2)$.

7. Here we have $P(a,1/a)$, $Q(x,1/x)$, $m_{PQ} = \dfrac{1/x - 1/a}{x - a} = \dfrac{(a - x)/ax}{(x - a)} =$

   $-\dfrac{1}{ax}$. Thus, $m = \lim\limits_{x \to a} m_{PQ} = -\dfrac{1}{a^2}$. If $a = 2$, $m = -1/4$, $P(2,1/2)$ and

   the tangent line is $y - \dfrac{1}{2} = -\dfrac{1}{4}(x - 2)$.

10. If $1 < x < 2$, then the secant line PQ is <u>horizontal</u> since both P and Q lie on the horizontal line $y = 1$. If, however, $0 < x < 1$, then PQ has positive slope and PQ approaches a <u>vertical</u> line as x approaches 1. Thus, there is no unique limiting position for the secant lines and, thus, no tangent line.

13. In mathematical terms, the question is if the tangent line to $y = 1 + \dfrac{1}{x}$ at $P(1,2)$ contains a point of the form $(n,0)$ where $n = 1, 2, 3, 4,$ or 5. As in #7 above, we find that $m = -1/a^2$. Thus, at P, $a = 1$, $m = -1$, and the tangent line is $y - 2 = -(x - 1)$ or $x + y = 3$. When $y = 0$, we see that $x = 3$ and the creature at $(3,0)$ is hit. If the point of release is $(3/2,5/3)$, then $a = 3/2$, $m = -1/(3/2)^2 = -4/9$, and the tangent line is $y - \dfrac{5}{3} = -\dfrac{4}{9}(x - \dfrac{3}{2})$ or $4x + 9y = 21$. Now when $y = 0$, $x = 21/4$, so it misses all creatures.

16. (a) The average velocity during the time interval from a to t, v_{ave}, is $v_{ave} = \dfrac{s(t) - s(a)}{t - a} = \dfrac{t^3 - a^3}{t - a} = t^2 + at + a^2$. Setting $a = 1$ and $t = 1.2, 1.1, 1.01,$ and 1.001, we obtain $v_{ave} = 3.64,$ $3.31, 3.0301,$ and 3.003001 cm/sec.

   (b) (1) $= \lim\limits_{t \to 1} v_{ave} = 3$.

(c) $v(a) = \lim_{t \to a} v_{ave} = \lim_{t \to a} (t^2 + at + a^2) = 3a^2$.   Motion is in the positive direction when $v(a) > 0$ or $3a^2 > 0$.   This is true for all $a \neq 0$.

(d) $v(a)$ is never negative.   Thus, there is never any motion in the negative direction.

19. $v(a) = \lim_{t \to a} \dfrac{(t^2/5 + 8t) - (a^2/5 + 8a)}{t - a}$

$= \lim_{t \to a} \dfrac{(t^2 - a^2)/5 + 8(t - a)}{t - a} = \lim_{t \to a} \dfrac{(t + a)}{5} + 8$

$= \dfrac{2a}{5} + 8$ ft/sec.

(a) $v(0) = 2(0)/5 + 8 = 8$.

(b) $v(5) = 10/5 + 8 = 10$.

(c) To find when he crosses the finish line, we seek the time, a, at which $s(a) = 100 \Rightarrow a^2/5 + 8a = 100 \Rightarrow a^2 + 40a - 500 = (a + 50)(a - 10) = 0$.   We use only the positive solution, $a = 10$, and $v(10) = 2(10)/5 + 8 = 12$.

EXERCISES 2.2

1.   (a) We examine functional values $f(x)$ when x is close to 2 and $x < 2$.   We see that for such x, points on the graph of f are close to the open circle at $(2,3)$.   Thus, $f(x)$ is nearly 3 for such x, and $\lim_{x \to 2^-} f(x) = 3$.

(b) Now, we examine $f(x)$ if x is close to 2 and $x > 2$.   Again, from the graph, we see that for such x, $f(x)$ is close to 1. Thus, $\lim_{x \to 2^+} f(x) = 1$.

(c) Because the left- and right-hand limits differ, $\lim_{x \to 2} f(x)$ does not exist.

(d)-(f) All answers are 2 since, if x is close to 0, $f(x)$ is close to 2 on both sides of $x = 0$.

4.   (a) If x is close to 2 and $x < 2$, then from the graph, $f(x)$ is close to 1.   Thus, $\lim_{x \to 2^-} f(x) = 1$.

(b) If x is close to 2 and $x > 2$, then $f(x)$ is close to 2.   Thus, $\lim_{x \to 2^+} f(x) = 2$.

(c) Because the left- and right-hand limits differ, $\lim_{x \to 2} f(x)$ does not exist.

(d)-(f) All answers are -1 since $f(x)$ is close to -1 when x is close to 0.

7. (a) $\lim_{x \to 1^-} f(x) = \lim_{x \to 1^-} (x^2 + 1) = 2.$

    (b) $\lim_{x \to 1^+} f(x) = \lim_{x \to 1^+} (x + 1) = 2.$

    (c) Since the left- and right-hand limits are both equal to 2, $\lim_{x \to 1} f(x) = 2.$

10. (a) If $x < 1$, then $x - 1 < 0$, and $f(x) = -(x - 1)$. $\lim_{x \to 1^-} f(x) = \lim_{x \to 1^-} (1 - x) = 0.$

    (b) If $x > 1$, then $x - 1 > 0$ and $f(x) = |x - 1| = x - 1$. $\lim_{x \to 1^+} f(x) = \lim_{x \to 1^+} (x - 1) = 0.$

    (c) $\lim_{x \to 1} f(x) = 0$ since both one-sided limits are 0. (Note: The fact that $f(1) = 1$ has absolutely nothing to do with the discussion of limits as x approaches 1. The only x's one works with here are those for which $x \neq 1$.)

13. (a) $x < 4 \Rightarrow x - 4 < 0 \Rightarrow |x - 4| = -(x - 4)$. $\lim_{x \to 4^-} f(x) = \lim_{x \to 4^-} \frac{-(x - 4)}{x - 4} = \lim_{x \to 4^-} (-1) = -1.$

    (b) $x > 4 \Rightarrow x - 4 > 0 \Rightarrow |x - 4| = x - 4$. $\lim_{x \to 4^+} f(x) = \lim_{x \to 4^+} \frac{(x - 4)}{(x - 4)} = 1.$

    (c) $\lim_{x \to 4} f(x)$ does not exist because the one-sided limits differ.

16. Note that as $x \to 5/2^-$, $x < 5/2$, $5 - 2x > 0$ and such x are in the domain of f.

    $$\lim_{x \to 5/2^-} (\sqrt{5-2x} - x^2) = \sqrt{5-2(5/2)} - (5/2)^2 = 0 - 25/4.$$ (The right-hand limit, however, does not exist for if $x > 5/2$ then $5 - 2x < 0$, and f is not defined for such x.)

19. $\lim_{x \to 2} \frac{x^2 - 4}{x - 2} = \lim_{x \to 2} \frac{(x+2)(x-2)}{x-2} = \lim_{x \to 2} x + 2 = 4.$

22. $\lim_{r \to -3} \frac{r^2 + 2r - 3}{r^2 + 7r + 12} = \lim_{r \to -3} \frac{(r+3)(r-1)}{(r+3)(r+4)} = \lim_{r \to -3} \frac{r-1}{r+4} = -4.$

25. $\lim_{k \to 4} \frac{k^2 - 16}{\sqrt{k} - 2} = \lim_{k \to 4} \frac{(k+4)(k-4)}{\sqrt{k} - 2} = \lim_{k \to 4} \frac{(k+4)(\sqrt{k}+2)(\sqrt{k}-2)}{\sqrt{k} - 2} = \lim_{k \to 4} (k+4)(\sqrt{k}+2)$

    $= (4+4)(\sqrt{4}+2) = 8 \cdot 4 = 32.$

28. $\lim\limits_{h\to 2} \dfrac{h^3-8}{h^2-4} = \lim\limits_{h\to 2} \dfrac{(h-2)(h^2+2h+4)}{(h-2)(h+2)} = \lim\limits_{h\to 2} \dfrac{h^2+2h+4}{h+2} = \dfrac{4+4+4}{4} = 3.$

31.   See the discussion in the answer section of the text.

34,37 The calculations for both exercises are listed in the table below. The functional values appear as they would on a calculator which displays numbers to 8 decimal place accuracy.

x	#34 $(1+2x)^{3/x}$	#37 $\left(\dfrac{4^{\lvert x\rvert}+9^{\lvert x\rvert}}{2}\right)^{1/\lvert x\rvert}$
1.00000	27.00000000	6.50000000
0.10000	237.37631380	6.04951023
0.01000	380.23450806	6.00493407
0.00100	401.01867128	6.00049323
0.00010	403.18684107	6.00004932
0.00001	403.40458881	6.00000493

EXERCISES 2.3

1.   With $f(x) = 3x$, $a = 4$, $L = 12$ we have $\lvert f(x) - L\rvert = \lvert 3x - 12\rvert = 3\lvert x - 4\rvert$. Thus, $\lvert f(x) - L\rvert < \varepsilon$ if and only if $3\lvert x - 4\rvert < \varepsilon$ or $\lvert x - 4\rvert < \varepsilon/3$. Thus, choosing $\delta = \varepsilon/3$ we find that if x satisfies $0 < \lvert x - 4\rvert < \delta$ then $\lvert f(x) - L\rvert = 3\lvert x - 4\rvert < 3\delta = 3(\varepsilon/3) = \varepsilon$.

4.   With $f(x) = 2x + 1$, $a = -3$, $L = -5$, we have $\lvert f(x) - L\rvert = \lvert(2x + 1) - (-5)\rvert = \lvert 2x + 6\rvert = 2\lvert x + 3\rvert = 2\lvert x - (-3)\rvert$. Thus, $\lvert f(x) - L\rvert < \varepsilon$ if and only if $2\lvert x - a\rvert = 2\lvert x - (-3)\rvert < \varepsilon$ or $\lvert x + 3\rvert < \varepsilon/2$. Choosing $\delta = \varepsilon/2$, we find that if $\lvert x - (-3)\rvert < \delta$ then $\lvert f(x) - L\rvert = 2\lvert x - (-3)\rvert < 2\delta = 2(\varepsilon/2) = \varepsilon$.

7. $\lvert f(x) - L\rvert = \lvert 5 - 5\rvert = 0$ which is less than any $\varepsilon > 0$ for any x. Thus, δ may be any positive number whatever.

10. With $f(x) = 9 - x/6$, $a = 6$, $L = 8$, we have $\lvert f(x) - L\rvert = \lvert(9 - \frac{x}{6}) - 8\rvert = \lvert 1 - \frac{x}{6}\rvert = \frac{1}{6}\lvert 6 - x\rvert = \frac{1}{6}\lvert x - 6\rvert < \varepsilon$ if and only if $\lvert x - 6\rvert < 6\varepsilon$. With $\delta = 6\varepsilon$, we find that $\lvert f(x) - L\rvert < \dfrac{\delta}{6} = \varepsilon$.

13. We consider the case where $a > 0$. Here, $L = a^2$. The horizontal line $y = a^2 - \varepsilon$ intersects the graph of f at the point with abscissa $x_1 = \sqrt{a^2 - \varepsilon}$ (obtained by setting $y = a^2 - \varepsilon$ in $y = x^2$, and solving for x). The line $y = a^2 + \varepsilon$ intersects the graph when $x = x_2 = \sqrt{a^2 + \varepsilon}$.

$\lim\limits_{x\to -a} x^2 = a^2$

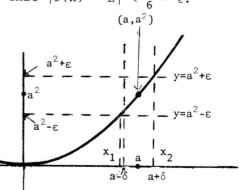

Then any $x \varepsilon (x_1, x_2)$ satisfies the condition that the point (x, x^2) on the graph of f lies between the horizontal lines $y = a^2 \pm \varepsilon$, i.e. $f(x) \varepsilon (a^2 - \varepsilon, a^2 + \varepsilon)$. To obtain an x-interval of the proper form, $(a - \delta, a + \delta)$, let δ be any number which is positive and less than or equal to the smaller of $x_2 - a$ and $a - x_1$. (It can be shown that $x_2 - a$ is the smaller). Then $(a - \delta, a + \delta) \varepsilon (x_1, x_2)$ and satisfies the definition.

16. The solution and figure are essentially the same as in #13 if you change a^2 to a^4, x^2 to x^4, and $\sqrt{}$ to $\sqrt[4]{}$.

19. $x < 3 \Rightarrow f(x) = -(x - 3)/(x - 3) = -1$ and $x > 3 \Rightarrow f(x) = (x - 3)/(x - 3) = +1$. Thus, $\lim_{x \to 3^-} f(x) = -1$, $\lim_{x \to 3^+} f(x) = +1$.

Thus the limit does not exist.

22. $x > 5 \Rightarrow f(x) = (2x - 10)/(x - 5) = 2$ and $x < 5 \Rightarrow f(x) = -2$. Thus left- and right-hand limits are -2 and 2, respectively, and the limit does not exist.

25. If $\lim_{x \to -5} \frac{1}{x+5} = L$ existed, then given any $\varepsilon > 0$ we could find $\delta > 0$ such that if $-5 - \delta < x < -5 + \delta$, $x \neq -5$ (i.e., $-\delta < x+5 < \delta$, $x+5 \neq 0$) then $L - \varepsilon < \frac{1}{x+5} < L + \varepsilon$ by (2.10). But this is impossible since $1/(x+5)$ can be made larger than $L + \varepsilon$ by taking $x+5$ small enough. (For example, if $L > 0$ then $L + \varepsilon > 0$ and $1/(x+5) > L + \varepsilon$ if $0 < x+5 < 1/(L+\varepsilon)$.) Thus, the limit does not exist.

28. Informally, with $f(x) = [x]$ if $a < x_1 < a+1$ then $f(x_1) = a$ whereas if $a-1 < x_2 < a$, then $f(x_2) = a-1$, even if x_1 and x_2 are very close to a. Thus there is no single number which all functional values are close to if x is close to a.

More formally, from the comments made above, the left- and right-hand limits of $f(x)$ as x approaches a are $a-1$ and a, respectively.

Since these are unequal, the limit does not exist.

EXERCISES 2.4

1. $\lim_{x \to -2} (3x^3 - 2x + 7) = 3(-2)^3 - 2(-2) + 7 = -13.$

4. $\lim_{t \to -3} (3t+4)(7t-9) = \lim_{t \to -3} (3t+4) \cdot \lim_{t \to -3} (7t-9) = (3(-3)+4)(7(-3)-9) = (-5)(-30) = 150.$

7. $\lim\limits_{x\to 7} 0 = 0$ and 10. $\lim\limits_{x\to 15} \sqrt{2} = \sqrt{2}$.

13. $\lim\limits_{x\to 2} \dfrac{x-2}{x^3-8} = \lim\limits_{x\to 2} \dfrac{x-2}{(x-2)(x^2+2x+4)} = \lim\limits_{x\to 2} \dfrac{1}{x^2+2x+4} = \dfrac{1}{4+4+4} = \dfrac{1}{12}$.

16. $\lim\limits_{x\to -2} \dfrac{x^3+8}{x^4-16} = \lim\limits_{x\to -2} \dfrac{(x+2)(x^2-2x+4)}{(x+2)(x-2)(x^2+4)} = \lim\limits_{x\to -2} \dfrac{(x^2-2x+4)}{(x-2)(x^2+4)} = \dfrac{12}{(-4)(8)} = -\dfrac{3}{8}$.

19. $\lim\limits_{x\to 1} \left(\dfrac{x^2}{x-1} - \dfrac{1}{x-1}\right) = \lim\limits_{x\to 1} \dfrac{x^2-1}{x-1} = \lim\limits_{x\to 1} \dfrac{(x+1)(x-1)}{x-1} = \lim\limits_{x\to 1} x+1 = 2$.

22. $\lim\limits_{x\to -8} \dfrac{16x^{2/3}}{4-x^{4/3}} = \dfrac{16(-8)^{2/3}}{4-(-8)^{4/3}} = \dfrac{16(-2)^2}{4-(-2)^4} = \dfrac{64}{-12} = -\dfrac{16}{3}$.

25. $\lim\limits_{h\to 0} \dfrac{4-\sqrt{16+h}}{h} = \lim\limits_{h\to 0} \dfrac{4-\sqrt{16+h}}{h} \dfrac{(4+\sqrt{16+h})}{(4+\sqrt{16+h})} = \lim\limits_{h\to 0} \dfrac{16-(16+h)}{h(4+\sqrt{16+h})} = \lim\limits_{h\to 0} \dfrac{-1}{4+\sqrt{16+h}} = -\dfrac{1}{8}$.

28. $\lim\limits_{x\to 6} (x+4)^3(x-6)^2 = (6+4)^3(6-6)^2 = (1000)(0) = 0$.

31. (a) As $x \to 5^-$, $x < 5$ and $5-x > 0$. Thus such x are in the domain of f.
     $\lim\limits_{x\to 5^-} \sqrt{5-x} = \sqrt{5-5} = 0$. (b) As $x \to 5^+$, $x > 5$ and $5-x < 0$. Thus such x are not
     in the domain of f and $\lim\limits_{x\to 5^+} \sqrt{5-x}$ does not exist. (c) This limit does not
     exist since the right-hand limit does not exist.

34. All three limits exist and equal $(-8)^{2/3} = (-2)^2 = 4$.

37. $x > 3 \Rightarrow x - 3 > 0 \Rightarrow |x - 3| = x - 3$. Thus $\lim\limits_{x\to 3^+} \dfrac{|x - 3|}{x - 3} =$
     $\lim\limits_{x\to 3^+} \dfrac{x - 3}{x - 3} = \lim\limits_{x\to 3^+} 1 = 1$.

40. $\lim\limits_{x\to 4^+} \dfrac{\sqrt[4]{x^2 - 16}}{x + 4} = \dfrac{\sqrt[4]{16 - 16}}{8} = \dfrac{0}{8} = 0$

43. If $n < x < n + 1$, then $f(x) = 0$ and $\lim\limits_{x\to n^+} f(x) = 0$. If
     $n - 1 < x < n$, then $f(x) = 0$ and $\lim\limits_{x\to n^-} f(x) = 0$ also. Thus,
     $\lim\limits_{x\to n} f(x) = 0$. (The value $f(n) = n$ has <u>absolutely nothing</u> to do
     with these limits since in all cases $x \neq n$.)

49. $0 \leq f(x) \leq c \Rightarrow 0 \leq x^2 f(x) \leq cx^2$. Since $\lim\limits_{x\to 0} 0 = 0$ and $\lim\limits_{x\to 0} cx^2 = 0$,
     it follows from the Sandwich Theorem that $\lim\limits_{x\to 0} x^2 f(x) = 0$ also.

52. The limit does not exist by #46 since the denominator,
$\sqrt{1 - v^2/c^2}$ approaches 0 as $v \to c^-$. Thus, m becomes larger and
larger as v approaches c. Thus, c appears to be the ultimate
speed in the universe because the force required to accelerate
the object to the speed of light increases without bound as v
approaches c. (Recall, force = (mass)(acceleration).) Note also
that if $v > c$, then m would be a purely imaginary complex number!

EXERCISES 2.5

1. $\lim_{x \to 4} (\sqrt{2x-5} + 3x) = \lim_{x \to 4} \sqrt{2x-5} + \lim_{x \to 4} 3x = \sqrt{3} + 12 = f(4).$

4. $f(x) = 1/x$ is continuous at every number $a \neq 0$ since it is the
quotient of the continuous functions 1 and x, and if $a \neq 0$, the denominator
is nonzero at a.

7. f is defined on [4,8] since $x-4 \geq 0$ for all x there. If $c \varepsilon (4,8)$ then
$\lim_{x \to c} f(x) = \lim_{x \to c} \sqrt{x-4} = \sqrt{c-4} = f(c)$ by limit theorems.   Similarly $\lim_{x \to 4^+} \sqrt{x-4}$
$= 0 = f(4)$ and $\lim_{x \to 8^-} \sqrt{x-4} = 2 = f(8)$. Thus f is continuous on [4,8].

10. If $c \varepsilon (1,3)$ then $\lim_{x \to c} f(x) = \lim_{x \to c} \frac{1}{x-1} = \frac{1}{c-1} = f(c)$,
valid since $1 < c < 3 \Longrightarrow 0 < c-1$.

13. The domain of f is $\{x: 2x-3 \geq 0\} = [3/2, \infty)$. If $a \varepsilon (3/2, \infty)$, $\lim_{x \to a} f(x) =$
$\sqrt{2a-3} + a^2 = f(a)$. Also, $\lim_{x \to 3/2^+} f(x) = \sqrt{2(3/2)-3} + (3/2)^2 = 9/4 = f(3/2)$.
Thus f is continuous throughout its domain.

16. The domain of f is (-1,1), and f is continuous there by the quotient theorem
for continuous functions, valid since the numerator is a polynomial and the
denominator is continuous.

19. As a rational function, f is continuous at every number in its domain.
Since the denominator $x^3-x^2 = x^2(x-1)$ is 0 when $x = 0$ and $x = 1$, the domain
of f is $\{x: x \neq 0,1\} = (-\infty,0) \cup (0,1) \cup (1,\infty)$.

22. The numerator, $\sqrt{9-x}$, is continuous throughout its domain which is
$\{x: 9-x \geq 0\} = \{x: 9 \geq x\} = (-\infty,9]$. Similarly the denominator, $\sqrt{x-6}$, is continuous on $[6,\infty)$ but is 0 at $x = 6$. Thus both are continuous and the denominator is nonzero on $(-\infty,9] \cap (6,\infty) = (6,9]$ where f is continuous by the
quotient theorem.

25. The function of #9, Sec. 2.2 is continuous everywhere. At $x = 1$,
$\lim_{x \to 1^+} f(x) = \lim_{x \to 1^+} (3x - 1) = 2$, $\lim_{x \to 1^-} f(x) = \lim_{x \to 1^-} (3 - x) = 2$.
Thus, $\lim_{x \to 1} f(x) = 2$ and $f(1) = 3 - 1 = 2$, and f is continuous at
$x = 1$. On each of the intervals $(-\infty,1)$ and $(1,\infty)$, f is defined
to be a polynomial function. Thus, f is also continuous on each
of these intervals as well as at $x = 1$.

28. If $x > 1$, the function of #12, Sec. 2.2, is simply x^2 (since
$|x - 1| = x - 1$ for such x). Thus, $\lim_{x \to 1^+} f(x) = \lim_{x \to 1^+} x^2 = 1$.
If $x < 1$, $f(x) = -x^2$ (since $|x - 1| = -(x - 1)$ for such x).
Thus, $\lim_{x \to 1^-} f(x) = \lim_{x \to 1^-} (-x^2) = -1$. Thus, f has a jump discontinuity at $x = 1$. Elsewhere, f is continuous since it is a poly-
nomial on $(-\infty,1)$ and $(1,\infty)$.

31. f is continuous on $(-3,3)$ by the quotient theorem since the denominator is
non-zero there. (It is 0 only at $x = \pm3$.) For continuity of f at $x = 3$, we
must have $\lim_{x \to 3^-} f(x) = f(3)$. Since $f(3) = d$, we find d by computing this
limit. $\lim_{x \to 3^-} f(x) = \lim_{x \to 3^-} \frac{9-x^2}{4-\sqrt{x^2+7}} = \lim_{x \to 3^-} \frac{9-x^2}{4-\sqrt{x^2+7}} \cdot \frac{4+\sqrt{x^2+7}}{4+\sqrt{x^2+7}} =$
$\lim_{x \to 3^-} \frac{(9-x^2)(4+\sqrt{x^2+7})}{16-(x^2+7)} = \lim_{x \to 3^-} (4 + \sqrt{x^2+7}) = (4 + \sqrt{9+7}) = 8$. Hence, $d = 8$.
For continuity at $x = -3$, we need $\lim_{x \to -3^+} f(x) = f(-3) = c$. As above, $\lim_{x \to -3^+}$
$f(x) = 8$ and, thus, $c = 8$ also.

34. (a) $x \in [n,n+1) \Rightarrow [x] = n \Rightarrow f(x) =$
$(x-[x])^2 = (x-n)^2$. As a polynomial, f
is continuous on the open interval
$(n,n+1)$. Since $\lim_{x \to n^+} f(x) = \lim_{x \to n^+} (x-n)^2$
$= (n-n)^2 = 0 = f(n)$, f is continuous at
$x = n$ and, thus, on $[n,n+1)$.
(b) f is not continuous on $[n,n+1]$
since it is not continuous at $x =$
$n+1$. We see that $f(n+1) =$
$((n+1)-[n+1])^2 = ((n+1)-(n+1))^2 = 0$, but $\lim_{x \to n+1^-} f(x) = \lim_{x \to n+1^-} (x-n)^2 =$
$((n+1)-n)^2 = 1 \neq f(n+1)$.

$y = (x-[x])^2$

37. No. $\lim_{x \to 3^+} \frac{|x-3|}{x-3} = \lim_{x \to 3^+} \frac{x-3}{x-3} = 1$ since, as $x \to 3^+$, $x > 3$, $x-3 > 0$, and $|x-3|$

= x-3. However, $\lim\limits_{x\to 3^-} \frac{|x-3|}{x-3} = \lim\limits_{x\to 3^-} \frac{-(x-3)}{x-3} = -1$ since as $x \to 3^-$, $x < 3$, x-3 <

0, and $|x-3| = -(x-3)$. Since the left and right-hand limits differ,

$\lim\limits_{x\to 3} f(x)$ does not exist.

40. HINT: Use the fact that every interval $(a-\delta, a+\delta)$ contains rational numbers, at which f(x) = 0, and irrational numbers, at which f(x) = 1, to prove that $\lim\limits_{x\to a} f(x)$ does not exist for every real number, a.

43. Let w be given between f(-1) = 0 and f(2) = 9. We must show that there is a number c between -1 and 2 such that f(c) = w. Substituting, we get $c^3+1=w \Rightarrow$ $c = \sqrt[3]{w-1}$. Note that $0 < w < 9 \Rightarrow -1 < w - 1 < 8 \Rightarrow -1 < \sqrt[3]{w-1} < 2$, or -1 < c < 2, as required.

46. Let w be given between f(-1) = 2 and f(3) = 6. Then $f(c) = w \Rightarrow c^2-c = w$ $\Rightarrow c^2-c-w = 0 \Rightarrow c = \frac{1 \pm \sqrt{1+4w}}{2}$. To decide which sign to choose, we note that $2 < w < 6 \Rightarrow 9 < 1+4w < 25 \Rightarrow 3 < \sqrt{1+4w} < 5$. Thus $2 < (1+\sqrt{1+4w})/2 <$ 3, but $-2 < (1-\sqrt{1+4w})/2 < -1$. Since c must be in the interval (-1,3), we must choose the + sign. Thus, $c = (1 + \sqrt{1+4w})/2$.

49. (a) T(t) = 0 if t = 0, 12, 24. The test value k = 6 in (0, 12) yields T(6) = 32.4, and the test value k = 18 in (12, 24) yields T(18) = -32.4. Thus T(t) > 0 on (0, 12), and T(t) < 0 on (12, 24).

52. Setting each factor equal to zero, we see that f(x) = $(x+1)^2 (x-3)(x-5) = 0$ if x = -1, 0, 3, and 5. We tabulate as follows:

Interval	Test Value k	f(k)	f(x)
(-∞, -1)	-2	-70	-
(-1, 0)	-1/2	-77/32	-
(0, 3)	1	32	+
(3, 5)	4	-100	-
(5, ∞)	6	882	+

EXERCISES 2.6 (Review)

1. $\lim\limits_{x\to 3} \frac{5x+11}{\sqrt{x+1}} = \frac{5(3)+11}{\sqrt{3+1}} = \frac{26}{2} = 13.$

4. $\lim\limits_{x\to 4^-} (x - \sqrt{16-x^2}) = 4 - \sqrt{16-4^2} = 4.$ (The right-hand limit, however, does not exist.)

7. $\lim\limits_{x\to2} \dfrac{x^4-16}{x^2-x-2} = \lim\limits_{x\to2} \dfrac{(x+2)(x-2)(x^2+4)}{(x-2)(x+1)} = \lim\limits_{x\to2} \dfrac{(x+2)(x^2+4)}{x+1} = \dfrac{(4)(8)}{3} = \dfrac{32}{3}.$

10. $\lim\limits_{x\to5} \dfrac{(1/x)-(1/5)}{x-5} = \lim\limits_{x\to5} \dfrac{(5-x)/5x}{x-5} = \lim\limits_{x\to5} \dfrac{-1}{5x} = -\dfrac{1}{25}.$

13. $\lim\limits_{x\to3^+} \dfrac{3-x}{|3-x|} = \lim\limits_{x\to3^+} \dfrac{3-x}{-(3-x)} = -1$ since $x > 3 \Longrightarrow 3-x < 0 \Longrightarrow |3-x| = -(3-x).$

16. $\lim\limits_{x\to-3} \sqrt[3]{\dfrac{x+3}{x^3+27}} = \lim\limits_{x\to-3} \sqrt[3]{\dfrac{x+3}{(x+3)(x^2-3x+9)}} = \lim\limits_{x\to-3} \sqrt[3]{\dfrac{1}{x^2-3x+9}} = \sqrt[3]{\dfrac{1}{9+9+9}} = \sqrt[3]{\dfrac{1}{27}} = \dfrac{1}{3}.$

19. $\lim\limits_{x\to2^+} \dfrac{\sqrt{(x-2)^2}}{2-x} = \lim\limits_{x\to2^+} \dfrac{|x-2|}{2-x} = \lim\limits_{x\to2^+} \dfrac{x-2}{2-x} = \lim\limits_{x\to2^+} (-1) = -1$ since as $x \to 2^+,$

$x > 2,$ $x-2 > 0$ and $|x-2| = x-2.$

22. $\lim\limits_{x\to2^-} f(x) = \lim\limits_{x\to2^-} x^3 = 8,$ $\lim\limits_{x\to2^+} f(x) = \lim\limits_{x\to2^+} (4-2x) = 0,$ and

$\lim\limits_{x\to2} f(x)$ does not exist.

25. $\lim\limits_{x\to1^-} f(x) = \lim\limits_{x\to1^-} x^2 = 1,$ $\lim\limits_{x\to1^+} f(x) = \lim\limits_{x\to1^+} (4-x^2) = 3,$ and

$\lim\limits_{x\to1} f(x)$ does not exist.

28. If $2 < x < 3,$ then $[x] = 2.$ Thus $\lim\limits_{x\to3^-} [x] = 2$ and $\lim\limits_{x\to3^-} [x] - x^2 = 2-9 = -7.$

(However, in #27, $\lim\limits_{x\to3^+} [x] = 3$ and $\lim\limits_{x\to3^+} [x] - x^2 = 3-9 = -6.$)

31. If c is any real number, $\lim\limits_{x\to c} f(x) = \lim\limits_{x\to c} (2x^4 - \sqrt[3]{x} + 1) = 2c^4 - \sqrt[3]{c} + 1 = f(c)$

by limit theorems of Section 4. Thus f is continuous.

34. $\sqrt{x}$ requires $x \geq 0,$ and x^2-1 in the denominator means $x = 1$ must be excluded. Thus the domain of f is $[0,1) \cup (1,\infty)$ where f is continuous.

37. f is a rational function which is continuous everywhere except at the points where the denominator is $0.$ Here $x^2-2x = x(x-2) = 0$ at $x = 0$ and $x = 2$ where f is discontinuous.

EXERCISES 3.1

1. (a) (b) $f'(x) = \lim\limits_{h \to 0} \dfrac{f(x+h) - f(x)}{h} = \lim\limits_{h \to 0} \dfrac{37-37}{h} = \lim\limits_{h \to 0} 0 = 0$ for all x. (c) $y - 37 = 0(x - 1) = 0$, or $y = 37$.

4. (a) (b) $f'(x) = \lim\limits_{h \to 0} \dfrac{(7(x+h)^2 - 5) - (7x^2 - 5)}{h} = \lim\limits_{h \to 0} \dfrac{7(x^2 + 2xh + h^2) - 7x^2}{h} =$

   $\lim\limits_{h \to 0} \dfrac{14xh + 7h^2}{h} = \lim\limits_{h \to 0} 14x + 7h = 14x$ for all x.

   (c) $f(1) = 2$, $f'(1) = 14 \Rightarrow$ tangent line is $y - 2 = 14(x - 1)$.

7. (a) (b) $f'(x) = \lim\limits_{h \to 0} \dfrac{1}{h}[\dfrac{1}{x+h-2} - \dfrac{1}{x-2}] = \lim\limits_{h \to 0} \dfrac{1}{h}[\dfrac{(x-2)-(x+h-2)}{(x+h-2)(x-2)}] =$

   $\lim\limits_{h \to 0} \dfrac{1}{h} \dfrac{-h}{(x+h-2)(x-2)} = \lim\limits_{h \to 0} \dfrac{-1}{(x+h-2)(x-2)} = \dfrac{-1}{(x-2)^2}$, for $x \neq 2$.

   (c) $f(1) = -1$, $f'(1) = -1 \Rightarrow$ tangent line is $y + 1 = -(x - 1)$.

10. (a) (b) $f'(x) = \lim\limits_{h \to 0} \dfrac{1}{h}[\dfrac{1}{2(x+h)} - \dfrac{1}{2x}] = \dfrac{1}{2} \lim\limits_{h \to 0} \dfrac{1}{h}[\dfrac{x-(x+h)}{(x+h)x}] = \dfrac{1}{x+h} - \dfrac{1}{x}$

    $\dfrac{1}{2} \lim\limits_{h \to 0} \dfrac{1}{h}[\dfrac{-h}{(x+h)x}] = \dfrac{1}{2} \lim\limits_{h \to 0} \dfrac{-1}{(x+h)x} = -\dfrac{1}{2x^2}$, for $x \neq 0$.

    (c) $f(1) = 1$, $f'(1) = -1/2 \Rightarrow$ tangent line is $y - 1 = (-1/2)(x - 1)$.

13. If $y = f(x)$, then $D_x y$ is just $f'(x)$. Thus,

    $D_x y = \lim\limits_{h \to 0} \dfrac{[2(x+h)^3 - 4(x+h) + 1] - [2x^3 - 4x + 1]}{h}$

    $= \lim\limits_{h \to 0} \dfrac{[(2x^3 + 6x^2 h + 6xh^2 + 2h^3) - (4x + 4h) + 1] - [2x^3 - 4x + 1]}{h}$

    $= \lim\limits_{h \to 0} \dfrac{6x^2 h + 6xh^2 + 2h^3 - 4h}{h} = \lim\limits_{h \to 0} 6x^2 + 6xh + 2h^2 - 4 = 6x^2 - 4$.

$f(x) = \sqrt{2}x$

16. $f'(a) = \lim\limits_{x \to a} \dfrac{f(x) - f(a)}{x - a} = \lim\limits_{x \to a} \dfrac{\sqrt{2}\, x - \sqrt{2}\, a}{x - a} = \lim\limits_{x \to a} \dfrac{\sqrt{2}(x-a)}{x - a} = \lim\limits_{x \to a} \sqrt{2} = \sqrt{2}$.

19. $f'(a) = \lim\limits_{x \to a} \dfrac{[1/(x+5) - 1/(a+5)]}{x - a} = \lim\limits_{x \to a} \dfrac{(a+5) - (x+5)}{(x-a)(x+5)(a+5)} = \lim\limits_{x \to a} \dfrac{a-x}{(x-a)(x+5)(a+5)}$

    $= \lim\limits_{x \to a} \dfrac{-1}{(x+5)(a+5)} = -\dfrac{1}{(a+5)^2}$.

22. First note that $\lim\limits_{h \to 0^+} [5+h] = 5$ and $\lim\limits_{h \to 0^-} [5+h] = 4$. Then, $\lim\limits_{h \to 0^+} \dfrac{f(5+h) - f(5)}{h}$

    $= \lim\limits_{h \to 0^+} \dfrac{[5+h] - [5]}{h} = \lim\limits_{h \to 0^+} \dfrac{5-5}{h} = 0$, but $\lim\limits_{h \to 0^-} \dfrac{[5+h] - [5]}{h} = \lim\limits_{h \to 0^-} \dfrac{4-5}{h} = \lim\limits_{h \to 0^-} \dfrac{-1}{h}$

    which does not exist. Both limits must exist and be equal in order for f to be differentiable at 5. Thus f is not differentiable there.

25. The graph of f in the answer section of the text is readily obtained by replacing $|x|$ by x if x > 0, -x if x < 0. The slopes of the components of the graph are either +1 or -1. The sharp "corners" in the graph at x = 0, ±1 occur when the slope abruptly changes from +1 to -1 or -1 to +1. At these points, f'(x) does not exist.

28. By an example of Sec. 2.1, $D_x(x^2) = 2x$, and on any interval where y = 1, $D_x y = 0$. Thus, we obtain the graph of f' shown. f' fails to exist at x = ±1.

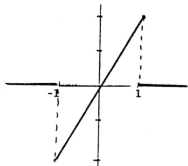

EXERCISES 3.2

1. $f'(x) = D_x(10x^2 + 9x - 4) = D_x(10x^2) + D_x(9x) - D_x(4) = 10\, D_x(x^2) + 9\, D_x(x)$
   $- D_x(4) = 10(2x) + 9(1) - 0 = 20x + 9.$

4. $f'(t) = D_t(12 - 3t^4 + 4t^6) = D_t(12) - D_t(3t^4) + D_t(4t^6) = D_t(12) - 3\, D_t(t^4)$
   $+ 4\, D_t(t^6) = 0 - 3(4t^3) + 4(6t^5) = -12t^3 + 24t^5.$

7. $h'(r) = D_r(r^2(3r^4 - 7r + 2)) = D_r(3r^6 - 7r^3 + 2r^2) = 3(6r^5) - 7(3r^2) + 2(2r)$
   $= 18r^5 - 21r^2 + 4r.$

10. $h'(x) = D_x(\frac{8x^2 - 6x + 11}{x-1}) = \dfrac{(x-1)D_x(8x^2 - 6x + 11) - (8x^2 - 6x + 11)D_x(x-1)}{(x-1)^2}$
    $= [(x-1)(16x-6) - (8x^2 - 6x + 11)(1)]/(x-1)^2$
    $= (16x^2 - 6x - 16x + 6 - 8x^2 + 6x - 11)/(x-1)^2$
    $= (8x^2 - 16x - 5)/(x-1)^2$

13. $D_x(3x^3 - 2x^2 + 4x - 7) = D_x(3x^3) + D_x(-2x^2) + D_x(4x) + D_x(-7) = 3(3x^2) +$
    $(-2)(2x) + 4(1) + 0 = 9x^2 - 4x + 4.$

16. $D_x(2x + 1/2x) = D_x(2x) + D_x((1/2)x^{-1}) = 2 + (1/2)(-1)x^{-2} = 2 - \dfrac{1}{2x^2}.$

19. $D_v\dfrac{(v^3-1)}{(v^3+1)} = \dfrac{(v^3+1)D_v(v^3-1) - (v^3-1)D_v(v^3+1)}{(v^3+1)^2} = \dfrac{(v^3+1)(3v^2) - (v^3-1)(3v^2)}{(v^3+1)^2}$
    $= \dfrac{6v^2}{(v^3+1)^2}.$

22. $D_x(1 + 1/x + 1/x^2 + 1/x^3) = D_x(1) + D_x(x^{-1}) + D_x(x^{-2}) + D_x(x^{-3})$
    $= 0 - x^{-2} - 2x^{-3} - 3x^{-4} = -(1/x^2 + 2/x^3 + 3/x^4).$

25. $D_s((3s)^{-4}) = D_s(3^{-4}s^{-4}) = 3^{-4}D_s(s^{-4}) = 3^{-4}(-4)s^{-5} = -4/(81s^5) = -(4/81)s^{-5}.$

let $g(r) = u$ $D_r(u^{-3}) = -2u^{-3}$ =

28. First note that if $f = g$ in the product rule we get $(f^2)' = 2ff'$. Here

$f(r) = (5r-4)^{-1} = \dfrac{1}{5r-4}$ and $f'(r) = \dfrac{(5r-4)D_r(1) - 1\,D_r(5r-4)}{(5r-4)^2} = \dfrac{-5}{(5r-4)^2}$.

Combining these calculations gives us $D_r(5r-4)^{-2} = 2(5r-4)^{-1}(-5(5r-4)^{-2})$

$= -10(5r-4)^{-3}$.

31. $D_x(\dfrac{2x^3 - 7x^2 + 4x + 3}{x^2}) = D_x(2x - 7 + 4x^{-1} + 3x^{-2}) = 2 - 4x^{-2} - 6x^{-3}$.

34. (a) $D_x(\dfrac{x^2+1}{x^4}) = \dfrac{x^4(2x) - (x^2+1)(4x^3)}{x^8} = -\dfrac{2x^2 + 4}{x^5}$.

(b) $D_x((x^2+1)x^{-4}) = (x^2+1)(-4x^{-5}) + x^{-4}(2x) = -2x^{-3} - 4x^{-5} = -(2x^2 + 4)x^{-5}$, the same as (a).

37. Here $y' = 3x^2 + 4x - 4$. (a) The tangent line is horizontal when $y' = 0$. $3x^2 + 4x - 4 = (3x-2)(x+2) = 0$ if $x = 2/3$ and $x = -2$. (b) The given line, $2y + 8x - 5 = 0$ has slope -4. The tangent line is parallel to it when $y' = -4$. $3x^2 + 4x - 4 = -4 \iff 3x^2 + 4x = x(3x+4) = 0 \implies x = 0$ and $x = -4/3$.

40. $y' = \dfrac{(a^2 + x^2)(ab) - (abx)(2x)}{(a^2 + x^2)^2} = \dfrac{a^3b - abx^2}{(a^2 + x^2)^2}$. When $x = a$, $y' = 0$, and the tangent line is the horizontal line $y = b/2$.

43. (a) $(f + g)'(2) = f'(2) + g'(2) = -1 + 2 = 1$
(b) $(f - g)'(2) = f'(2) - g'(2) = -1 - 2 = -3$
(c) $(4f)'(2) = 4f'(2) = -4$
(d) $(fg)'(2) = f(2)g'(2) + g(2)f'(2) = (3)(2) + (-5)(-1) = 11$
(e) $(f/g)'(2) = (g(2)f'(2) - f(2)g'(2))/g(2)^2 = ((-5)(-1) - (3)(2))/25 = -1/25$.

46. (a) $(3f - 2g)'(2) = 3f'(2) - 2g'(2) = -3 - 4 = -7$
(b) $(5/g)' = (gD(5) - 5g')/g^2 = -5g'/g^2 \implies (5/g)'(2) = -5g'(2)/g(2)^2 = -10/25$
(c) $(6f)'(2) = 6f'(2) = -6$
(d) $(\dfrac{f}{f + g})' = \dfrac{(f + g)f' - f(f' + g')}{(f + g)^2} = \dfrac{gf' - fg'}{(f + g)^2}$
Thus, at $x = 2$ we obtain $\dfrac{(-5)(-1) - (3)(2)}{(3 - 5)^2} = -\dfrac{1}{4}$.

49. $dy/dx = (8x - 1)(x^2 + 4x + 7)(3x^2) + (8x - 1)(x^3 - 5)(2x + 4) + (x^3 - 5)(x^2 + 4x + 7)(8)$

52. First, we find the points of intersection of the flight path and the hillside. $f(x) = x/5 \implies -0.016x^2 + 1.6x = 0.2x \implies 0.016x^2 - 1.4x = x(0.016x - 1.4) = 0$. Thus, $x = 0$ and $x = 1.4/0.016 = 87.5$.
(a) Here we want $f'(0)$. $f'(x) = -0.032x + 1.6$. $f'(0) = 1.6$.
(b) The point of impact is at $x = 87.5$, so here we need $f'(87.5) = -0.032(87.5) + 1.6 = -2.8 + 1.6 = -1.2$.

(c) The distance above the ground on the hill is F(x) = (flight elevation) - (hillside elevation) = $(-0.016x^2 + 1.6x)$ - x/5 = $-0.016x^2 + 1.4x$. The maximum occurs when F'(x) = 0. Now, F'(x) = $-0.032x + 1.4 = 0$ if x = 1.4/0.032 = 43.75. The maximum is F(43.75) = 30.625.

55. The velocity v(t) = s'(t) = 2 + 2t. (a) v(1) = 4, v(4) = 10, v(8) = 18 (ft/sec). (b) s(t) = 50 $\iff$ $6 + 2t + t^2 = 50$ $\iff$ $t^2 + 2t - 44 = 0$. By the quadratic formula, t = $(-2 \pm \sqrt{4 - 4(-44)})/2$ = $(-2 \pm \sqrt{180})/2$ = $(-2 \pm 6\sqrt{5})/2$ = $-1 \pm 3\sqrt{5}$ sec. Since $0 \le t \le 10$, we must choose only the "+" sign. Thus t = $-1 + 3\sqrt{5}$ and v($-1 + 3\sqrt{5}$) = $2 + 2(-1 + 3\sqrt{5})$ = $6\sqrt{5}$.

58. (a) s = $-16t^2 + 16t$ = $-16t(t - 1) = 0$ if t = 0,1. Thus, the hang time is 1 sec. (b) v(t) = s'(t) = $-32t + 16$. v(0) = 16. The maximum height occurs when v(t) = 0 $\Rightarrow$ t = 1/2. The maximum height is s(1/2) = $-16/4 + 16/2 = 4$. (c) s = $-\frac{16}{6}t^2 + 16t$ = $-\frac{16}{6}t(t - 6) = 0$ if t = 0,6 $\Rightarrow$ hang time 6 sec. v(t) = $-\frac{16}{3}t + 16$ $\Rightarrow$ v(0) = 16 and v(t) = 0 at t = 3. Thus, the maximum height is s(3) = $-144/6 + 48 = 24$.

EXERCISES 3.3

1. (a) With r(t) = $3t^{1/3}$, r'(8) is sought. By the example mentioned in the hint, $D_t(t^{1/3}) = t^{-2/3}/3$. Thus, r'(t) = $t^{-2/3}$ and r'(8) = $1/\sqrt[3]{64}$ = 1/4.
   (b) V = $\frac{4}{3}\pi r^3$ = $\frac{4}{3}\pi(3t^{1/3})^3$ = $36\pi t$ $\Rightarrow$ $\frac{dV}{dt}$ = 36π for all t.
   (c) A = $4\pi r^2$ = $36\pi t^{2/3}$. At this point in the text, we do not know how to compute ($D_t(t^{2/3})$ directly (except by the definition). However, we can use the Product Rule: $D_t(t^{2/3})$ = $D_t(t^{1/3}t^{1/3})$ = $t^{1/3}D_t(t^{1/3})$ + $t^{1/3}D_t(t^{1/3})$ = $2t^{1/3}\frac{1}{3}t^{-2/3}$ = $\frac{2}{3}t^{-1/3}$. Thus, $\frac{dA}{dt}$ = $36\pi(\frac{2}{3}t^{-1/3})$, and at t = 8, $\frac{dA}{dt}$ = $24\pi/\sqrt[3]{8}$ = 12π.

4. We seek T'(t) at t = 2, 5, 9. T'(t) = $4 - 3/(t + 1)^2$.
   (a) T'(2) = $4 - 3/3^2$ = $4 - 1/3 = 11/3$
   (b) T'(5) = $4 - 3/6^2$ = $4 - 1/12 = 47/12$
   (c) T'(9) = $4 - 3/10^2$ = 3.97

7. P = $12t^2 - t^4 + 5$ $\Rightarrow$ $\frac{dP}{dt}$ = $24t - 4t^3$ = $4t(6 - t^2)$, both for $t \ge 0$. The growth rate stops when $\frac{dP}{dt}$ = 0 $\Rightarrow$ t = 0, $\sqrt{6}$ (rejecting $-\sqrt{6}$). Thus, the intervals in which dP/dt $\ne$ 0 are $(0,\sqrt{6})$ and $(\sqrt{6},\infty)$. Selecting the test value 1 in the first and 3 in the second, we find that $\frac{dP}{dt}$(1) = 20 > 0, $\frac{dP}{dt}$(3) = -36 < 0. Thus, $\frac{dP}{dt}$ > 0 on $(0,\sqrt{6})$, $\frac{dP}{dt}$ < 0 on $(\sqrt{6},\infty)$.

$$dr \, \tfrac{1}{2\pi} = \frac{(c')(2\pi) - (c)(2\pi)'}{(2\pi)^2} = \frac{2\pi - 2c}{4\pi^2}$$

10. $v(t) = s'(t) = 2t + 3$, $a(t) = v'(t) = 2$, $-2 \le t \le 2$. Since
    $v(t) = 2(t + 3/2)$ is < 0 for $-2 \le t < -3/2$ and is > 0 for
    $-3/2 < t \le 2$, the motion is to the left from $s(-2) = -8$ to
    $s(-3/2) = -33/4$ from $t = -2$ to $t = -3/2$, then to the right to
    $s(2) = 4$ from $t = -3/2$ to $t = 2$.

13. $v(t) = 8t^3 - 12t = 4t(2t^2 - 3)$, $a(t) = 24t^2 - 12$, $-2 \le t \le 2$. On
    the intervals $[-2, -\sqrt{3/2})$ and $(0, \sqrt{3/2})$, $v(t) < 0$ and the motion is
    to the left. On $(-\sqrt{3/2}, 0)$ and $(\sqrt{3/2}, 2]$, $v(t) > 0$ and the motion
    is to the right.

16. $v(t) = 10t \Rightarrow v(1) = 10$, $v(2) = 20$ ft/sec. $v(t) = 28 \Rightarrow 10t =$
    $28 \Rightarrow t = 2.8$.

19. If r = the radius and C = the circumference, then $C = 2\pi r$ or
    $r = C/2\pi$. Thus, $\frac{dr}{dC} = \frac{1}{2\pi}$ is the rate of change of r with respect
    to C, the constant $1/2\pi$.

22. We seek dq/dp from the relation $\frac{1}{f} = \frac{1}{q} + \frac{1}{p}$. We solve for q by
    rewriting this equation as $\frac{1}{q} = \frac{1}{f} - \frac{1}{p}$, and then inverting each side
    to obtain: $q = 1/(\frac{1}{f} - \frac{1}{p})$. By the quotient rule

    $\frac{dq}{dp} = -D_p(\frac{1}{f} - \frac{1}{p})/(\frac{1}{f} - \frac{1}{p})^2 = -(1/p^2)/(\frac{1}{f} - \frac{1}{p})^2$. Replacing the
    denominator by $1/q^2$, we obtain $\frac{dq}{dp} = -\frac{q^2}{p^2}$.

25. (a) $C(100) = 250 + 100(100) + 0.001(1{,}000{,}000) =$
    $250 + 10{,}000 + 1{,}000 = 11{,}250.$
    (b) Average cost $= c(x) = C(x)/x \Rightarrow c(x) = 250/x + 100 + 0.001x^2$,
    and $c(100) = C(100)/100 = 112.5$. Marginal cost $= C'(x) =$
    $100 + 0.003x^2$, $C'(100) = 100 + 0.003(10{,}000) = 130.$

28. $C'(x) = 1 - 10/x^2 \Rightarrow C'(10) = 1 - 10/100 = 0.9$. The cost of
    producing the 11th item is $C(11) - C(10) = (14 + 10/11) -$
    $(13 + 10/10) = 10/11 = 0.9091.$

EXERCISES 3.4

1. (a) $\Delta y = f(x+\Delta x) - f(x) = [2(x+\Delta x)^2 - 4(x+\Delta x) + 5] - [2x^2-4x+5]$
    $= [2x^2+4x\Delta x + 2(\Delta x)^2 - 4x - 4\Delta x + 5] - [2x^2-4x+5]$
    $= 4x\Delta x + 2(\Delta x)^2 - 4\Delta x.$
    (b) $x = 2$, $\Delta x = -0.2 \Rightarrow \Delta y = 4(2)(-0.2) + 2(-0.2)^2 - 4(-0.2) = -1.6 + 0.08$
    $+ 0.8 = -0.72.$

4.   (a) $\Delta y = f(x+\Delta x) - f(x) = \dfrac{1}{2+x+\Delta x} - \dfrac{1}{2+x} = \dfrac{(2+x) - (2+x+\Delta x)}{(2+x+\Delta x)(2+x)} = \dfrac{-\Delta x}{(2+x+\Delta x)(2+x)}$.

     (b) $x = 0,\ \Delta x = -0.03 \implies \Delta y = \dfrac{-(-0.03)}{(2+0-0.03)(2+0)} = \dfrac{0.03}{(1.97)(2)} = \dfrac{0.03}{3.94} \approx 0.0076$.

7.   (a) $\Delta y = \dfrac{1}{x+\Delta x} - \dfrac{1}{x} = \dfrac{x-(x+\Delta x)}{x(x+\Delta x)} = -\dfrac{\Delta x}{x(x+\Delta x)}$.

     (b) $dy = D_x y\ \Delta x = -\dfrac{1}{x^2}\ \Delta x$.

     (c) $dy - \Delta y = -\dfrac{\Delta x}{x^2} + \dfrac{\Delta x}{x(x+\Delta x)} = \dfrac{\Delta x}{x}\left(\dfrac{1}{x+\Delta x} - \dfrac{1}{x}\right) = \dfrac{\Delta x}{x}\left(\dfrac{-\Delta x}{x(x+\Delta x)}\right) = -\dfrac{\Delta x^2}{x^2(x+\Delta x)}$.

10.   (a) $\Delta y = f(x+\Delta x) - f(x) = 8 - 8 = 0$.   (b) $D_x y = D_x 8 = 0 \implies dy = D_x y\ \Delta x = 0(\Delta x) = 0$.   (c) $\Delta y - dy = 0 - 0 = 0$.

13.   Let $y = f(x)$. The change $\Delta y = f(1.03) - f(1)$ is to be estimated. Here $x = 1$, $x+\Delta x = 1.03$, so $\Delta x = .03$.   Thus, $\Delta y \approx dy = f'(x)\Delta x = f'(1)(.03)$. $f'(x) = 20x^4 - 24x^3 + 6x \implies f'(1) = 2$, and $\Delta y \approx 2(.03) = .06$.

16. $dS = F'(t)\Delta t = D_t\left(\dfrac{1}{2-t^2}\right)\Delta t = \dfrac{2t}{(2-t^2)^2}\Delta t$. If t changes from 1 to 1.02, we take $t = 1$, $\Delta t = .02$. Thus $\Delta S \approx dS = F'(1)\Delta t = \dfrac{2}{(2-1^2)^2}(.02) = 2(.02) = .04$.

18.   If A is the area of the square of side x, then $A = x^2$ where $x = 1'$ with error Δx where $|\Delta x| \le \dfrac{1}{16}{}'' = \dfrac{1}{192}{}'$. The actual error $\Delta A \approx dA = 2x\Delta x$ so that $|\Delta A| \approx |dA| = 2|\Delta x| \le \dfrac{2}{192} = \dfrac{1}{96}$ sq. ft. The last figure is the maximum possible error. The average error is $\approx dA/A \approx (1/96)/1^2 = 1/96$. The % error is $\approx \dfrac{1}{96} \cdot 100 \approx 1.04\%$.

19.   If V is the volume of a cube of edge x, then $V = x^3$. Here $x = 10$, $\Delta x = .1$. Thus $\Delta V \approx dV = 3x^2\Delta x = 300(.1) = 30$ in^3. Exactly, $\Delta V = 10.1^3 - 10^3 = 1030.301 - 1000 = 30.301$ in^3.

22.   If r is the radius of the cylinder and hemisphere, the silo volume is the volume of the cylinder + the volume of the hemisphere $= \pi r^2 h + \dfrac{2}{3}\pi r^3$. If C is the circumference of the cylinder then $C = 2\pi r$, or $r = C/2\pi$, and $C = 30 \pm .5$ ft. So, in terms of C, using $h = 50$, the silo volume is $V(C) =$

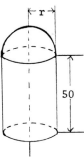

$50\pi\left(\dfrac{C}{2\pi}\right)^2 + \dfrac{2}{3}\pi\left(\dfrac{C}{2\pi}\right)^3 = \dfrac{25}{2\pi}C^2 + \dfrac{1}{12\pi^2}C^3$ and $V(30) =$

$\dfrac{25}{2\pi}(30)^2 + \dfrac{1}{12\pi^2}(30)^3 = \dfrac{900}{2\pi^2}(25\pi + 5) \approx 3808.96$ ft^3.

The error in this calculation is $\Delta V \approx dV = V'(C)\Delta C$. To get the maximum error we can proceed as in #18 above, or, alternately, take $C = 30$, the measured value, and $\Delta C = 0.5$, the maximum error in C. $V'(C) = \dfrac{25C}{\pi} + \dfrac{C^2}{4\pi^2} \implies$ the maximum error in V is $V'(30)(.5)$

$= [\frac{25(30)}{\pi} + \frac{900}{4\pi^2}] (.5) = \frac{150}{4\pi^2} (10\pi + 3) \approx 130.8$. The

average error is $\approx \frac{dV}{V} = \frac{(10\pi + 3)}{60(5\pi + 1)} \approx 0.0343$, and the % error is $\approx 3.43\%$.

25. Here, we are given the % change in F and are to find the Δs that produces it.
Recall: % change $= \frac{\Delta F}{F}(100) \approx \frac{dF}{F}(100) = \frac{F'(s)\Delta s}{F}(100)$. Let $K = gm_1m_2$, a constant. Then $F(s) = Ks^{-2}$, and $F'(s) = -2Ks^{-3}$, so that $\frac{F'(s)}{F} = \frac{-2Ks^{-3}}{Ks^{-2}} = -\frac{2}{s}$.
Using this above, setting s = 20 and setting the result equal to 10 (the de-
sired % change) we obtain: $\frac{-2\Delta s}{20}(100) = 10 \iff -10\Delta s = 10 \iff \Delta s = -1$.

28. Let $f(x) = x^4 - 3x^3 + 4x^2 - 5$ so that $f(2.01)$ is desired. Taking x = 2,
$\Delta x = .01$, we know $\Delta y = f(2.01) - f(2) \approx dy = f'(2)(.01)$ or $N = f(2.01) \approx$
$f'(2)(.01) + f(2)$. Now, $f'(x) = 4x^3 - 9x^2 + 8x$, $f'(2) = 4(8) - 9(4) +$
$8(2) = 12$ and $f(2) = 16 - 3(8) + 4(4) - 5 = 3$. Thus $N \approx 12(.01) + 3 = 3.12$.
The exact value of N works out to be 3.12100501.

31. Let P = pressure difference, and r = radius. Then $P = K/r^4 = Kr^{-4}$, where K
is a constant. A 10% reduction in r means that $\Delta r = dr = -0.1 r$. Then
$dP = -4Kr^{-5}dr = -4Kr^{-5}(-0.1 r) = 0.4Kr^{-4}$. The % change in P is
$\approx \frac{dP}{P}(100) = \frac{0.4Kr^{-4}}{Kr^{-4}}(100) = 40\%$.

EXERCISES 3.5

1. $f'(x) = 3(x^2 - 3x + 8)^2 D_x(x^2 - 3x + 8) = 3(x^2 - 3x + 8)^2(2x - 3)$.

4. $k'(x) = -3(5x^2 - 2x + 1)^{-4}D_x(5x^2 - 2x + 1) = -3(5x^2 - 2x + 1)^{-4}(10x - 2)$.

7. $D_x(8x^3 - 2x^2 + x - 7)^5 = 5(8x^3 - 2x^2 + x - 7)^4 D_x(8x^3 - 2x^2 + x - 7)$
$= 5(8x^3 - 2x^2 + x - 7)^4(24x^2 - 4x + 1)$.

10. $K'(x) = (-1)(3x^2 - 5x + 7)^{-2} D_x(3x^2 - 5x + 7) = -(3x^2 - 5x + 7)^{-2} (6x - 5)$.

13. $N'(x) = (6x - 7)^3 D_x(8x^2 + 9)^2 + (8x^2 + 9)^2 D_x(6x-7)^3$
$= (6x - 7)^3 \cdot 2(8x^2 + 9) D_x(8x^2 + 9) + (8x^2 + 9)^2 \cdot 3(6x - 7)^2 D_x(6x-7)$
$= (6x - 7)^3 \cdot 2(8x^2 + 9)(16x) + (8x^2 + 9)^2 \cdot 3(6x - 7)^2 \cdot 6$
$= (6x - 7)^2(8x^2 + 9)[32x(6x - 7) + 18(8x^2 + 9)]$
$= (6x - 7)^2(8x^2 + 9)(336x^2 - 224x + 162)$.

16. $S'(t) = 3(\frac{3t+4}{6t-7})^2 D_t(\frac{3t+4}{6t-7}) = 3(\frac{3t+4}{6t-7})^2 [\frac{(6t-7)(3)-(3t+4)(6)}{(6t-7)^2}]$
$= 3(\frac{3t+4}{6t-7})^2 \frac{-45}{(6t-7)^2} = \frac{-135(3t+4)^2}{(6t-7)^4}$.

19. $f'(x) = 2(\frac{3x^2-5}{2x^2+7})D_x(\frac{3x^2-5}{2x^2+7}) = 2(\frac{3x^2-5}{2x^2+7})[\frac{(2x^2+7)(6x)-(3x^2-5)(4x)}{(2x^2+7)^2}]$

$= 2(\frac{3x^2-5}{2x^2+7})\frac{62x}{(2x^2+7)^2} = \frac{124x(3x^2-5)}{(2x^2+7)^3}$.

22. $F'(v) = (-3)(v^{-1}-2v^{-2})^{-4}D_v(v^{-1}-2v^{-2}) = -3(v^{-1}-2v^{-2})^{-4}(-v^{-2}+4v^{-3})$.

25. $F'(t) = [2t(2t+1)^2]D_t(2t+3)^3 + (2t+3)^3D_t[2t(2t+1)^2]$

$= [2t(2t+1)^2]3(2t+3)^2 \cdot 2 + (2t+3)^3[2t\cdot2(2t+1)\cdot2 + (2t+1)^2\cdot2]$

(by the product and chain rules)

$= 2(2t+1)(2t+3)^2(24t^2 + 26t + 3)$.

28. $y' = 5(x + 1/x)^4 D_x(x + 1/x) = 5(x + 1/x)^4(1 - 1/x^2)$

(a) $x = 1 \Longrightarrow y' = 5(2)^4(1-1) = 0$. Thus the tangent line is $y = 32$.

(b) The tangent line is horizontal if $y'(x) = 0$. Thus the equation $5(x + 1/x)^4(1 - 1/x^2) = 0$ or, after multiplying by $x^6/5$, $(x^2+1)^4(x^2-1) = 0$ with solution $x = \pm1$. The points are: $(1,32)$, $(-1,-32)$.

31. $dy = 10(x^4 - 3x^2 + 1)^9 D_x(x^4 - 3x^2 + 1)dx = 10(x^4 - 3x^2 + 1)^9(4x^3 - 6x)dx$.

Now, $x = 1$, $x + \Delta x = 1.01 \Longrightarrow \Delta x = dx = .01$ and $\Delta y \approx dy = 10(-1)^9(-2)(.01) = 0.2$.

34. If $v = F(u)$ and $u = G(t)$, then $v = F(G(t))$ and $\frac{dv}{dt} = F'(G(t))G'(t) = \frac{dv}{du} \cdot \frac{du}{dt}$.

Therefore with $v = (u^4 + 2u^2 + 1)^3$, $u = 4t^2$, we get $\frac{dv}{dt} = 3(u^4 + 2u^2 + 1)^2 \cdot (4u^3 + 4u) \cdot 8t$.

37. $\frac{dw}{dt} = \frac{dw}{dx}\frac{dx}{dt} = \frac{300(6400)^2}{(6400 + x)^3}\frac{dx}{dt}$. Setting $x = 1000$ and $\frac{dx}{dt} = 6$ we obtain $\frac{dw}{dt} = -\frac{300(6400)^2}{(7400)^3}(6) \approx -0.1819$.

40. $p(3) = q(r(3)) = q(3) = -2$. $p'(3) = q'(r(3))r'(3) = q'(3)r'(3) = (6)(4) = 24$.

43. We first observe that $D_xf(-x) = f'(-x)D_x(-x) = f'(-x)(-1) = -f'(-x)$.

If f is even then $f(-x) = f(x)$. Differentiating this identity: $-f'(-x) = f'(x)$ or $f'(-x) = -f'(x)$, and f' is odd.

If f is odd, then $f(-x) = -f(x)$. Differentiating this identity: $-f'(-x) = -f'(x)$ or $f'(-x) = f'(x)$, and f' is even.

EXERCISES 3.6

1. The object in #1-8 is to solve the given equation for y in terms of x. Thus $3x - 2y + 4 \stackrel{.}{=} 2x^2 + 3y - 7x \Longrightarrow -2x^2 + 10x + 4 = 5y$ and $y = (-2/5)x^2 + 2x + (4/5)$ is one implicit function defined by the equation. Its domain is $\mathbb{R}$.

4. $-3x^2 + 4y^2 = -12 \iff y^2 = (3x^2 - 12)/4$. So, $y = (1/2)\sqrt{3x^2 - 12}$ is one function defined by the equation. The domain is {x: $3x^2 - 12 \geq 0$} = $(-\infty,-2] \cup [2,\infty)$.

7. $\sqrt{x} + \sqrt{y} = 1 \iff \sqrt{y} = 1 - \sqrt{x} \implies y = 1 - 2\sqrt{x} + x$. The domain is tricky. Obviously $x \geq 0$. But from the 2nd equation $\sqrt{y} = 1 - \sqrt{x} \geq 0$ or $\sqrt{x} \leq 1$, which gives $0 \leq x \leq 1$ as the domain.

10. $4x^3 - 2y^3 = x \implies 12x^2 - 6y^2 y' = 1 \implies y' = (12x^2-1)/6y^2$.

13. $5x^2 - xy - 4y^2 = 0 \implies 10x - (xy'+y) - 8yy' = 0$ (The middle term is $D_x(xy)$) $\implies (-x-8y)y' = y - 10x \implies y' = (10x-y)/(x+8y)$.

16. $x = y + 2y^2 + 3y^3 \implies 1 = (1 + 4y + 9y^2)y' \implies y' = 1/(1+4y+9y^2)$.

19. $(y^2 - 9)^4 = (4x^2 + 3x - 1)^2 \implies 4(y^2 - 9)^3(2yy') = 2(4x^2 + 3x - 1)(8x + 3)$ $\implies y' = (4x^2 + 3x - 1)(8x + 3)/4y(y^2 - 9)^3$.

22. $x^3 + y^3 - 3axy \implies 3x^2 + 3y^2 y' - 3axy' - 3ay \implies (3y^2 - 3ax)y' = 3ay - 3x^2$. Setting $a = 4$, $x = y = 6$ yields $(108 - 72)y' = (72 - 108) \implies y' = -1$.

25. $xy + 16 = 0 \implies xy' + y = 0 \implies y' = -y/x$. At $P(-2,8)$, $y' = -8/(-2) = 4$, and the tangent line is $y - 8 = 4(x + 2)$.

28. $1/x + 3/y = 1 \implies -1/x^2 - (3/y^2)y' = 0 \implies y' = -y^2/3x^2$. At $P(2,6)$, $y' = -36/12 = -3 > y - 6 = -3(x-2)$ is the tangent line equation.

31. (a) To see why the answer is "infinitely many," note that $y = \pm\sqrt[4]{1 - x^4}$, $-1 \leq x \leq 1$. Now, let c be <u>any</u> number in $(-1,1)$ and let $f_c(x) = \sqrt[4]{1 - x^4}$ for $-1 \leq x \leq c$, $f_c(x) = -\sqrt[4]{1 - x^4}$ for $c < x \leq 1$. These infinitely many (discontinuous) functions all satisfy $f_c^4(x) = 1 - x^4$, or $x^4 + y^4 = 1$ where $y = f_c(x)$.

     (c) $y^2 + \sqrt{x} \geq 0 \implies y^2 + \sqrt{x} + 4 \geq 4 \implies y^2 + \sqrt{x} + 4 = 0$ has no solutions.

34. We are given $f(1) = 2$, $f(1.1) = b$, and we wish to estimate b. With $\Delta x = 0.1$, $\Delta y = f(1.1) - f(1) \approx f'(1)(0.1)$ or $b - 2 \approx f'(1)(0.1) \implies b \approx 2 + f'(1)(0.1)$. We obtain $f'(1)$ by implicit differentiation. $x^3 + xy + y^4 = 19 \implies 3x^2 + xy' + y + 4y^3 y' = 0$. Setting $x = 1$, $y = 2$ we obtain $3 + y' + 2 + 32y' = 0 \implies y' = -5/33 = f'(1)$. Thus $b \approx 2 - (5/33)(0.1) \approx 1.98485$.

EXERCISES 3.7

1. $f(x) = \sqrt[3]{x^2} + 4\sqrt{x^3} = x^{2/3} + 4x^{3/2} \implies f'(x) = \frac{2}{3}x^{-1/3} + 4(\frac{3}{2})x^{1/2} = \frac{2}{3\sqrt[3]{x}} + 6\sqrt{x}$.

4. $D_z[(2z^2 - 9z + 8)^{-2/3}] = (-2/3)(2z^2 - 9z + 8)^{-5/3} D_z(2z^2 - 9z + 8) =$

   $(-2/3)(2z^2 - 9z + 8)^{-5/3}(4z - 9).$

7. $D_x(\sqrt{2x}) = \sqrt{2}D_x(x^{1/2}) = (\sqrt{2}/2)x^{-1/2} = 1/\sqrt{2x}.$

10. $f(t) = \sqrt[3]{t^2} - 1/\sqrt{t^3} = t^{2/3} - t^{-3/2} \Rightarrow f'(t) = \frac{2}{3}t^{-1/3} + \frac{3}{2}t^{-5/2} = 2/3\sqrt[3]{t} + 3/2\sqrt{t^5}.$

13. $D_x(\sqrt{4x^2 - 7x + 4}) = D_x(4x^2 - 7x + 4)^{1/2} = \frac{1}{2}(4x^2 - 7x + 4)^{-1/2}D_x(4x^2 - 7x + 4)$

   $= (8x - 7)/2\sqrt{4x^2 - 7x + 4}.$

16. $D_v(1/\sqrt{v^4 + 7v^2})^3 = D_v(v^4 + 7v^2)^{-3/2} = -\frac{3}{2}(v^4 + 7v^2)^{-5/2}(4v^3 + 14v).$

19. $D_s((s^2+9)^{1/4}(4s+5)^4) = (s^2+9)^{1/4}D_s(4s+5)^4 + (4s+5)^4 D_s(s^2+9)^{1/4}$

   $= (s^2+9)^{1/4} 4(4s+5)^3(4) + (4s+5)^4(1/4)(s^2+9)^{-3/4}(2s)$

   $= 16\sqrt[4]{s^2+9}(4s+5)^3 + s(4s+5)^4/2\sqrt[4]{(s^2+9)^3}.$

22. $f'(w) = D_w(w^3(9w + 1)^5)^{1/2} = \frac{1}{2}(w^3(9w + 1)^5)^{-1/2} D_w(w^3(9w + 1)^5) =$

   $\dfrac{w^3 \cdot 5(9w + 1)^4 9 + 3w^2(9w + 1)^5}{2(w^3(9w + 1)^5)^{1/2}} = \dfrac{3w^2(9w + 1)^4[15w + (9w + 1)]}{2w^{3/2}(9w + 1)^{5/2}} =$

   $(3/2)w^{1/2}(9w + 1)^{3/2}(24w + 1).$

25. $y' = D_x(2x^2 + 1)^{1/2} = \frac{1}{2}(2x^2 + 1)^{-1/2}(4x) = 2x/\sqrt{2x^2 + 1}.$ At
   $P(-1,\sqrt{3})$, $y' = -2/\sqrt{3}$, and the tangent line is $y - \sqrt{3} =$
   $(-2/\sqrt{3})(x + 1).$

28. The given line has slope $-1/2$ so that the desired slope is 2.
   $y' = \frac{5}{3}x^{2/3} + \frac{1}{3}x^{-2/3} = 2 \Rightarrow 5x^{4/3} + 1 = 6x^{2/3}$. Let $z = x^{2/3}$ to
   obtain $5z^2 + 1 = 6z \Rightarrow 5z^2 - 6z + 1 = 0 \Rightarrow (5z - 1)(z - 1) = 0$
   $\Rightarrow z = 1/5, 1 \Rightarrow x^2 = 1/5^3, 1 \Rightarrow x = \pm 1/5\sqrt{5}, \pm 1.$

31. First note that $D_x \sqrt{xy} = 1/2(xy)^{-\frac{1}{2}}(xy'+y) = (xy'+y)/2\sqrt{xy}.$
   Then, differentiating $6x + \sqrt{xy} - 3y = 4$, we obtain $6 +$
   $(xy'+y)/2\sqrt{xy} - 3y' = 0 \Rightarrow 12\sqrt{xy} + xy' + y - 6\sqrt{xy} y' = 0 \Rightarrow$
   $(x - 6\sqrt{xy})y' = -(12\sqrt{xy}+y).$

34. Using implicit differentiation, $pv^{1.4} = c \Rightarrow p(1.4v^{0.4}) + \dfrac{dp}{dv}v^{1.4}$
   $= 0 \Rightarrow \dfrac{dp}{dv} = \dfrac{-1.4pv^{0.4}}{v^{1.4}} = -1.4\dfrac{p}{v}.$

37. $f(x) = |1 - x| = |x - 1| = [(x - 1)^2]^{1/2} \Rightarrow f'(x) =$
   $\frac{1}{2}[(x - 1)^2]^{-1/2}2(x - 1) = (x - 1)/|x - 1| = 1$ if $x > 1$, -1 if
   $x < 1$. $f'(1)$ does not exist since the left-hand derivative there
   is -1 whereas the right-hand derivative is $+1$.

40. $f(x) = x/|x| = x/(x^2)^{1/2} \Rightarrow f'(x) = [(x^2)^{1/2} - x\frac{1}{2}(x^2)^{-1/2}(2x)]/x^2$

$= \dfrac{(|x| - x^2/|x|)}{x^2} = \dfrac{(|x|^2 - x^2)}{x^2|x|}$ if $x \neq 0$. Since $|x|^2 = x^2$,

$f'(x) = 0$ if $x \neq 0$. (0 is excluded since it is not even in the domain of f.)

43. $k'(r) = 5(4r + 7)^4 D_r(4r + 7) = 20(4r + 7)^4$, $k''(r) =$
$20(4)(4r + 7)^3 D_r(4r + 7) = 80(4r + 7)^3 \cdot 4$.

46. $g(x) = 3x^8 - 2x^5 \Rightarrow g'(x) = 24x^7 - 10x^4 \Rightarrow g''(x) = 168x^6 - 40x^3$.

49. $D_x y = \dfrac{(3x+1)D_x(2x-3) - (2x-3)D_x(3x+1)}{(3x+1)^2} = \dfrac{(3x+1)(2)-(2x-3)(3)}{(3x+1)^2} = 11(3x+1)^{-2}$.

$D_x^2 y = -2(11)(3x+1)^{-3}D_x(3x+1) = -66(3x+1)^{-3}$.

$D_x^3 y = (-3)(-66)(3x+1)^{-4}D_x(3x+1) = 198(3)(3x+1)^{-4}$.

52. $x^2 3y^2 y' + y^3 2x = 0 \Rightarrow y' = -2y/3x \Rightarrow$

$y'' = \dfrac{-6xy' + 6y}{9x^2} = \dfrac{-6x(-2y/3x) + 6y}{9x^2} = \dfrac{10y}{9x^2}$

EXERCISES 3.8

1. Let x and y be the distances shown. Then $x^2 + y^2 = 20^2$; it is given that $dx/dt = 3$, and dy/dt is desired when $y = 8$ (and thus $x = \sqrt{400-64} = \sqrt{336}$). $x^2 + y^2 = 400 \Rightarrow$
$2x\dfrac{dx}{dt} + 2y\dfrac{dy}{dt} = 0 \Rightarrow \dfrac{dy}{dt} = -\dfrac{x}{y}\dfrac{dx}{dt} = -\dfrac{\sqrt{336}}{8}(3) \approx -6.9$
ft/sec. (Negative since it's falling and y is decreasing.)

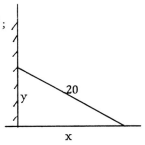

4. Let x be the distance of the first girl east of A, y the distance of the second girl north of A, z the distance between them. Then $z^2 = x^2 + y^2$;
$2z\dfrac{dz}{dt} = 2x\dfrac{dx}{dt} + 2y\dfrac{dy}{dt}$. When $t = 2$ min $= 120$ sec, $x = 1200$ since $\dfrac{dx}{dt} = 10$. The second girl has only been running for 60 seconds, and so $y = 480$ since $\dfrac{dy}{dt} = 8$. At that time $z = \sqrt{1200^2 + 480^2} = 120\sqrt{116}$ and substituting these values:
$2(120)\sqrt{116}\dfrac{dz}{dt} = 2(1200)(10) + 2(480)(8) = 240(132) \Rightarrow \dfrac{dz}{dt} = \dfrac{132}{\sqrt{116}}$
≈ 12.3 ft/sec.

7. Let L = the thickness of the ice so that $\dfrac{dL}{dt} = -\dfrac{1}{4}$ in/hr. Let V = volume of the ice so that $\dfrac{dV}{dt}$ is desired when $L = 2$ in. Now, V = (volume of a hemisphere of radius (120+L)" - volume of a hemisphere of radius 120") $=$
$\dfrac{1}{2}(\dfrac{4}{3})\pi(120+L)^3 - \dfrac{1}{2}(\dfrac{4}{3})\pi(120)^3$. $\dfrac{dV}{dt} = \dfrac{dV}{dL}\dfrac{dL}{dt} = 2\pi(120+L)^2\dfrac{dL}{dt}$, which, when $L = 2$
yields $\dfrac{dV}{dt} = 2\pi(122)^2(-\dfrac{1}{4}) \approx -23{,}380$ in^3/hr.

10. Let y be the height of the base of the balloon and L the length
     of rope let out. Then $20^2 + y^2 = L^2$, we are given that $dL/dt = 5$,
     and we seek dy/dt when $L = 500$, at which instant $y = \sqrt{500^2 - 20^2}$
     $= \sqrt{249,600}$. Now, $20^2 + y^2 = L^2 \Rightarrow 2y \frac{dy}{dt} = 2L \frac{dL}{dt} \Rightarrow \frac{dy}{dt} = \frac{L}{y} \frac{dL}{dt}$
     $= \frac{500}{\sqrt{249,600}}(5) \approx 5.004$ ft/sec.

13. Let V be the volume and h the depth of the water. If A is
     the area of the wetted triangular region at the end of the
     trough as pictured, then $V = 8A$. The region is an equi-
     lateral triangle of side length s and altitude h. Then
     $s^2 = h^2 + s^2/4 \Rightarrow s = 2h/\sqrt{3}$. Also, $A = sh/2 = h^2/\sqrt{3}$ and
     $V = 8h^2/\sqrt{3}$. Thus $\frac{dV}{dt} = \frac{16h}{\sqrt{3}} \frac{dh}{dt}$. Using $\frac{dV}{dt} = 5$, $h = 8'' =$
     $\frac{2'}{3}$, we get $5 = \frac{32}{3\sqrt{3}} \frac{dh}{dt}$ or $\frac{dh}{dt} = \frac{15\sqrt{3}}{32} \approx .8$ ft/min ≈ 9.75 in/min.

16. We know that $A = \pi r^2$, we are given that $\frac{dr}{dt} = 6$ ft/min, and we
     seek $\frac{dA}{dt}$ when $r = 150$. $A = \pi r^2 \Rightarrow \frac{dA}{dt} = 2\pi r \frac{dr}{dt} = 2\pi(150)(6) =$
     1800π ft^2/min.

19. Let C be the circumference and r the radius. Then $C = 2\pi r$. We are given
     that $\frac{dr}{dt} = 0.5$ m/sec, and we seek $\frac{dC}{dt}$ when $r = 4$. $\frac{dC}{dt} = 2\pi \frac{dr}{dt} = 2\pi(.5) =$
     π m/sec for all r.

22. At the given instant, it is given that $p = 40$, $\frac{dp}{dt} = 3$, $v = 60$, and we are to
     find $\frac{dv}{dt}$ then. Using the product rule, $pv^{1.4} = c \Rightarrow 1.4 \, pv^{0.4} \frac{dv}{dt} + v^{1.4} \frac{dp}{dt}$
     $= 0$ so that $\frac{dv}{dt} = -\frac{v^{1.4}}{1.4pv^{0.4}} \frac{dp}{dt} = -\frac{v}{1.4p} \frac{dp}{dt}$. So: $\frac{dv}{dt} = -\frac{60}{1.4(40)}(3) = -\frac{45}{14} \approx$
     -3.2 cm^3/sec., negative since v decreases as p increases (at constant temp.).

25. Let $t = 0$ be the instant at which the first stone is dropped. Then the second
     is dropped 2 seconds later at $t = 2$. If $2 \le t \le 3$, the first stone has fallen
     $16t^2$ ft. However, the second stone has been falling for only $t-2$ seconds,
     and has fallen, therefore, $16(t-2)^2$ ft. The distance, L, between them is
     then $L = 16t^2 - 16(t-2)^2 = 64t - 64$ and $\frac{dL}{dt} = 64$ for all t (i.e. the distance
     between them steadily increases at the rate of 64 ft/sec.).

28. Let t = # of hours past 10 a.m.,

     y = # of miles car is north of P,

     x = # of miles plane is <u>west</u> of P,

     z = altitude of plane in miles, and

     L = distance between car and plane.

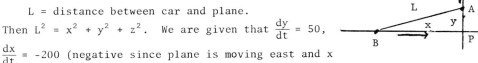

Then $L^2 = x^2 + y^2 + z^2$. We are given that $\frac{dy}{dt} = 50$,

$\frac{dx}{dt} = -200$ (negative since plane is moving east and x

is decreasing), $\frac{dz}{dt} = 0$ (altitude is constant). We have to find $\frac{dL}{dt}$ when $t = \frac{1}{4}$

(10:15 a.m. is $\frac{1}{4}$ hour past 10). At $t = \frac{1}{4}$, $y = 50(\frac{1}{4}) = 12.5$, $x = 100 - 200(\frac{1}{4})$

= 50 (it started 100 miles west of P when t = 0, then moved east for $\frac{1}{4}$ hour),

$z = 5$, and $L = \sqrt{12.5^2 + 50^2 + 5^2} = \sqrt{2681.25} \approx 51.78$. Now, $L^2 = x^2 + y^2 + z^2$

$\Longrightarrow$ $2L \frac{dL}{dt} = 2x \frac{dx}{dt} + 2y \frac{dy}{dt} + 2z \frac{dz}{dt} \Longrightarrow \frac{dL}{dt} = \frac{1}{L} (x \frac{dx}{dt} + y \frac{dy}{dt})$ since $\frac{dz}{dt} = 0$.

Substituting: $\frac{dL}{dt} = \frac{1}{\sqrt{2681.25}} (50(-200) + \frac{25}{2}(50)) = \frac{-9375}{\sqrt{2681.25}} \approx -181$ mph.

31. **Let x be the distance of the plane from the point on the runway**
     which is 300 ft, from the control tower, and let L be the dis-
     tance from the plane to the tower. We are given that dx/dt =
     100 mph, and we seek dL/dt when x = 300 ft. L = $(300^2 + 20^2 + x^2)^{\frac{1}{2}}$
     $\Longrightarrow$ dL/dt = $x(300^2 + 20^2 + x^2)^{-\frac{1}{2}}$ dx/dt. Substituting for x and
     dx/dt, we obtain dL/dt = $300 (180,400)^{-\frac{1}{2}}$ (100) $\approx$ 70.63 mph.

EXERCISES 3.9

1. To find $\sqrt[3]{2}$, we must solve $f(x) = x^3 - 2 = 0$. Since $f(1) = -1$ and $f(2) = 6$, the

    root lies between 1 and 2. With the initial guess of $x_1 = 1$ and using $x_{n+1} = $

    $x_n - \frac{f(x_n)}{f'(x_n)} = x_n - \frac{(x_n^3 - 2)}{3x_n^2}$, we obtain the following results as tabulated,

    after rounding off to 4 places.

n	x_n	$f(x_n)$
1	1.0000	-1.0000
2	1.3333	0.3704
3	1.2639	0.0190
4	1.2599	0.0001
5	1.2599	0.0000

Thus $\sqrt[3]{2}$ = 1.2599 to 4 decimal places.

4. Preliminary analysis: To find the interval in which the largest root of
    $f(x) = 2x^3 - 4x^2 - 3x + 1 = 0$ lies, we begin by evaluating f at convenient x
    values. We get $f(0) = 1$, $f(1) = -4$, $f(2) = -5$, $f(3) = 10$, $f(4) = 53$. From
    this point on, the values of f become increasingly large. Thus, the largest
    root lies between 2 and 3. Our initial guess will be $x_1 = 2$, and the for-
    mula to be used is: $x_{n+1} = x_n - \dfrac{2x_n^3 - 4x_n^2 - 3x_n + 1}{6x_n^2 - 8x_n - 3}$.

n	x_n	$f(x_n)$
1	2.0000	-5.0000
2	3.0000	10.0000
3	2.6296	1.8188
4	2.5254	0.1256
5	2.5171	0.0008
6	2.5170	0.0000
7	2.5170	0.0000

    Thus the largest root is 2.5170 to 4 places.

7. Preliminary analysis: With $f(x) = x^5 + x^2 - 9x - 3$, we have $f(-2) = -13$ and
    $f(-1) = 6$ and, thus, there is a root between -2 and -1. Our initial guess
    will be $x_1 = -2$, and the formula is: $x_{n+1} = x_n - \dfrac{x_n^5 + x_n^2 - 9x_n - 3}{5x_n^4 + 2x_n - 9}$

n	x_n	$f(x_n)$
1	-2.0000	-13.0000
2	-1.8060	-2.6959
3	-1.7395	-0.2462
4	-1.7321	-0.0016
5	-1.7321	-0.0016

    Thus, the desired root is -1.7321 to 4 places. (It is interesting to note
    that if x_1 had been chosen to be -1, the root obtained would have been
    -0.3222.)

10. Preliminary analysis: To find the interval in which the largest zero of
    $f(x) = x^3 - 36x - 84$ lies, we begin by evaluating f at convenient x's. We
    find that $f(0) = -84$, $f(2) = -148$, $f(4) = -164$, $f(6) = -84$, $f(7) = 7$,
    $f(8) = 140$, and, from this point on, the values of f become increasingly
    larger. Thus the largest zero of f lies in $[6,7]$. Our initial guess will
    be $x_1 = 7$ and the formula is $x_{n+1} = x_n - \dfrac{x_n^3 - 36x_n - 84}{3x_n^2 - 36}$.

n	x_n	$f(x_n)$
1	7.00000	-7.00000
2	6.93694	0.08327
3	6.93617	0.00001
4	6.93617	0.00000

Thus, the desired root is 6.93617 to 5 places.

13.   Preliminary analysis: $x = -1$ is one root of $f(x) = x^4 - x - 2 = 0$ obtained by checking the four possible rational roots ± 1, ± 2.   Thus, there is a factor of $(x+1)$, and by long division we find that $x^4 - x - 2 = (x+1)(x^3 - x^2 + x - 2)$.   The cubic factor, $c(x) = x^3 - x^2 + x - 2$, has only one real root.   This can be determined by noting that $c(x) < 0$ if $x < 0$, $c(0) = -2$, $c(1) = -1$, $c(2) = 4$, $c(3) = 19$, and the values of $c(x)$ get increasingly positive from that point on.   Since $c(1) < 0$ and $c(2) > 0$, $c(x)$ has its root between 1 and 2, and the 2nd root of $f(x)$ lies there also.   Our initial guess will be $x_1 = 1$, and the

formula is: $x_{n+1} = x_n - \dfrac{x_n^4 - x_n - 2}{4x_n^3 - 1}$.

n	x_n	$f(x_n)$
1	1.00	-2.00
2	1.67	4.05
3	1.44	0.81
4	1.36	0.07
5	1.35	0.00
6	1.35	0.00

Thus, the roots are -1 and 1.35 to 2 places.

16.   Preliminary analysis: We begin by making a table of values of $f(x) = x^3 + 2x^2 - 8x - 3$ to determine the intervals in which the roots lie.

x	-5	-4	-3	-2	-1	0	1	2	3	4
f(x)	-38	-3	12	13	6	-3	-8	-3	18	61

From the table (and the intermediate value theorem) the roots of f are in the intervals $[-4,-3]$, $[-1,0]$, and $[2,3]$.   Our initial guesses will be -4, -1, and 2, respectively, and the formula is

$x_{n+1} = x_n - \dfrac{x_n^3 + 2x_n^2 - 8x_n - 3}{3x_n^2 + 4x_n - 8}$.

	Root in [-4,-3]			Root in [-1,0]			Root in [2,3]	
n	x_n	$f(x_n)$	n	x_n	$f(x_n)$	n	x_n	$f(x_n)$
1	-4.00	-3.00	1	-1.00	6.00	1	2.00	-3.00
2	-3.88	-0.15	2	-0.33	-0.15	2	2.25	0.52
3	-3.87	0.00	3	-0.35	0.00	3	2.22	0.00
4	-3.87	0.00	4	-0.35	0.00	4	2.22	0.00

Thus, the three roots are -3.87, -0.35, and 2.22 to 2 places.

EXERCISES 3.10 (Review)

1. $f'(x) = \lim_{h \to 0} \frac{1}{h}(\frac{4}{3(x+h)^2+2} - \frac{4}{3x^2+2}) = \lim_{h \to 0} \frac{4}{h} \frac{-(6xh-3h^2)}{(3(x+h)^2+2)(3x^2+2)} = \frac{-24x}{(3x^2+2)^2}$.

4. $D_x(x^4-x^2+1)^{-1} = (-1)(x^4-x^2+1)^{-2}D_x(x^4-x^2+1) = -(4x^3-2x)/(x^4-x^2+1)^2$.

7. $D_z(7z^2-4z+3)^{1/3} = (1/3)(7z^2-4z+3)^{-2/3}D_z(7z^2-4z+3) = (14z-4)/3\sqrt[3]{(7z^2-4z+3)^2}$.

10. $D_x((1/6)(3x^2-1)^4) = (4/6)(3x^2-1)^3(6x) = 4x(3x^2-1)^3$.

13. $D_x(\sqrt[5]{(3x+2)^4}) = D_x(3x+2)^{4/5} = (4/5)(3x+2)^{-1/5}(3)$.

16. Multiply out to get $g(w) = (w^2-4w+3)/(w^2+4w+3)$.   Then

$g'(w) = \frac{(w^2+4w+3)(2w-4) - (w^2-4w+3)(2w+4)}{(w+1)^2(w+3)^2} = (8w^2-24)/(w+1)^2(w+3)^2$.

19. $g'(y) = (7y - 2)^{-2}D_y(2y + 1)^{2/3} + (D_y(7y - 2)^{-2})(2y + 1)^{2/3} =$

$(7y - 2)^{-2} \frac{2}{3}(2y + 1)^{-1/3}(2) + (-2)(7y - 2)^{-3}(7)(2y + 1)^{2/3}$.

22. $H'(t) = 6(t^6 + t^{-6})^5D_t(t^6 + t^{-6}) = 6(t^6 + t^{-6})^5(6t^5 - 6t^{-7})$.

25. $f'(x) = [(3x+2)^{1/2}D_x(2x+3)^{1/3} - (2x+3)^{1/3}D_x(3x+2)^{1/2}]/(3x+2)$

$= \frac{(3x+2)^{1/2}(\frac{1}{3})(2x+3)^{-2/3}(2) - (2x+3)^{1/3}(\frac{1}{2})(3x+2)^{-1/2}(3)}{3x+2}$

$= (3x+2)^{-1/2}(2x+3)^{-2/3}[(2/3)(3x+2)-(3/2)(2x+3)]/(3x+2)$

which may be reduced to $-(x + 19/6)/(3x+2)^{3/2}(2x+3)^{2/3}$.

28. $F'(t) = [(t^2+2)(10t) - (5t^2-7)(2t)]/(t^2+2)^2 = 34t/(t^2+2)^2$.

31. $5x^3 - 2x^2y^2 + 4y^3 - 7 = 0 \implies 15x^2 - 2x^2 \cdot 2yy' - 4xy^2 + 12y^2y' = 0 \implies$

$(15x^2 - 4xy^2) + (12y^2 - 4x^2y)y' = 0$.

34. $y^2 - x^{1/2}y^{1/2} + 3x = 2 \implies 2yy' - (1/2)x^{1/2}y^{-1/2}y' - (1/2)y^{1/2}x^{-1/2} + 3 = 0$

$\implies (2y - (1/2)\sqrt{x/y})y' - ((1/2)\sqrt{y/x} - 3) = 0$

$\implies y' = ((1/2)\sqrt{y/x} - 3)/(2y - (1/2)\sqrt{x/y}).$

37. $x^2y - y^3 = 8 \implies x^2y' + 2xy - 3y^2y' = 0 \implies y' = -2xy/(x^2 - 3y^2).$ At
$P(-3,1)$, $y' = 6/6 = 1$. Thus the tangent line is $y-1 = (1)(x-3).$

40. $y = 2x^3 - x^2 - 3x \implies y' = 6x^2 - 2x - 3.$ (a) $y' = 0 \implies 6x^2 - 2x - 3 = 0$

$\implies x = \dfrac{2 \pm \sqrt{4 - 4(6)(-3)}}{12} = \dfrac{2 \pm \sqrt{76}}{12} = \dfrac{1 \pm \sqrt{19}}{6}.$ (b) $y = 0 \implies 2x^3 - x^2$

$- 3x = 0 \implies x(2x^2 - x - 3) = x(x + 1)(2x - 3) = 0 \implies x = -1, 0, 3/2.$
$y'(-1) = 6 + 2 - 3 = 5$, $y'(0) = -3$ and $y'(3/2) = 15/2.$

43. $x^2 + 4xy - y^2 = 8 \implies 2x + 4xy' + 4y - 2yy' = 0 \implies y' = (2x+4y)/(2y-4x)$
$= (x+2y)/(y-2x) \implies y'' = [(y-2x)(1+2y')-(x+2y)(y'-2)]/(y-2x)^2$
$= 5(y - xy')/(y - 2x)^2.$ Substituting for y' from the 2nd line and simplifying yields the given answer.

46. $y' = [(x^2+1)(5) - 5x(2x)]/(x^2+1)^2 = 5(1-x^2)/(x^2+1)^2.$ Thus $dy =$
$[5(1-x^2)/(x^2+1)^2]\Delta x.$ If x changes from 2 to 1.98, we select $x = 2$, $\Delta x =$
$-.02$ and $\Delta y \approx dy = (5(1-4)/5^2)(-.02) = (-3/5)(-.02) = (.6)(.02) = 0.012.$
Exactly, $\Delta y = f(1.98)-f(2) = \dfrac{9.9}{4.9204} - \dfrac{10}{5} \approx 0.0120315.$

49. Let $h(x) = g(f(x)).$ If x changes from -1 to -1.01, then select $x = -1$,
$\Delta x = -.01$ and $\Delta y \approx dy = h'(x)\Delta x = g'(f(x))f'(x)\Delta x = g'(f(-1))f'(-1)\Delta x.$ Now,
$f'(x) = 6x^2 + 2x - 1$, $f'(-1) = 6-2-1 = 3$, $g'(x) = 5x^4 + 12x^2 + 2$, $f(-1) =$
$-2+1+1+1 = 1$, $g'(f(-1)) = g'(1) = 5+12+2 = 19.$ Thus $\Delta y \approx (19)(3)(-.01) =$
$-0.57.$

52. $C(x) = 1000 + 2x + .005x^2 \implies C(2000) = 25,000.$
$c(x) = C(x)/x = 1000/x + 2 + .005x \implies c(2000) = 12.5.$
$C'(x) = 2 + .01x \implies C'(2000) = 22.$
$c'(x) = -1000/x^2 + .005 \implies c'(2000) = .00475.$

55. We are given that $\dfrac{\Delta T}{T}(100) = 0.5$ and we seek $\dfrac{\Delta R}{R}(100) \approx \dfrac{dR}{R}(100)$
$= \dfrac{4kT^3\Delta T}{kT^4}(100) = 4\dfrac{\Delta T}{T}(100) = 4(.5) = 2\%.$

58. $y^2 = 2x^3 \implies 2y\dfrac{dy}{dt} = 6x^2\dfrac{dx}{dt}.$ At $(2,4)$, $\dfrac{dy}{dt} = x = 2$ and thus $8(2) = 6(2)^2\dfrac{dx}{dt}$
$\implies \dfrac{dx}{dt} = \dfrac{16}{24} = \dfrac{2}{3}.$

61. $pv = c \implies p + \dfrac{dp}{dv}v = 0$ (differentiating implicitly with respect
to v). Thus $\dfrac{dp}{dv} = -\dfrac{p}{v}.$

63   Since $f(-1) = 4$ and $f(0) = -5$, there is a root between -1 and 0.   Our
initial guess will be $x_1 = 0$.

n	x_n	$f(x_n)$
1	0.0000	-5.0000
2	-0.7143	-0.4373
3	-0.7586	0.0294
4	-0.7560	0.0001
5	-0.7560	0.0000

CHAPTER 4

EXTREMA AND ANTIDERIVATIVES

EXERCISES 4.1

1. $f'(x) = -12x - 6x^2 = -6x(x+2)$. Since $f'(x)$ exists for all x, the only critical numbers are solutions of $f'(x) = 0$, or $-6x(x+2) = 0 \Rightarrow x = 0$, $x = -2$, both of which are in the given interval $[-3,1]$. Now we calculate the values of $f(x)$ at $x = 0$ and $x = -2$, the critical numbers, and at $x = -3$ and $x = 1$, the end points of the interval. The absolute maximum of f on $[-3,1]$ is the largest of these values, and the absolute minimum of f is the smallest. $f(x) = 5 - 6x^2 - 2x^3 \Rightarrow f(0) = 5$, $f(-2) = 5 - 6(4) - 2(-8) = -3$, $f(-3) = 5 - 6(9) - 2(-27) = 5$, $f(1) = 5 - 6 - 2 = -3$. Thus the maximum is 5 attained at $x = 0$ and -3; the minimum is -3 attained at $x = 1$ and -2.

4. $f'(x) = 4x^3 - 10x = 2x(2x^2-5)$, which exists for all x. Thus the only critical numbers are solutions of $f'(x) = 0$, or $2x(2x^2 - 5) = 0 \Rightarrow x = 0$ and $x = \pm\sqrt{5/2}$. Of these, $-\sqrt{5/2}$ is not in the given interval, $[0,2]$, but 0 and $\sqrt{5/2} \approx 1.6$ are. Again, we calculate $f(x)$ at the critical numbers in the interval and at the end points. ($x = 0$ is both, but we consider it only an end point since, strictly speaking, $f'(0)$ does not exist--only the right-hand derivative does exist.) $f(0) = 4$, $f(2) = 16 - 5(4) + 4 = 0$, $f(\sqrt{5/2}) = (\sqrt{5/2})^4 - 5(\sqrt{5/2})^2 + 4 = 25/4 - 25/2 + 4 = -9/4$. Thus the maximum is $f(0) = 4$, and the minimum is $f(\sqrt{5/2}) = -9/4$.

6. For $f(x) = |x|$, $f'(0)$ does not exist by Example 3, Sec. 3.1, but $f'(x) = 1$ if $x > 0$, $f'(x) = -1$ if $x < 0$. Thus $x = 0$ is the only critical point, and the graph has no tangent line at $(0,0)$. Since $f(0) = 0$, f has a local minimum at $x = 0$ because $f(x) = |x| \geq 0 = f(0)$ for all x in any interval (a,b) containing 0.

10. $g'(x) = 2$, exists for all x and is never 0. Thus, no critical numbers.

13. $F'(w) = 4w^3 - 32 = 4(w^3-8) = 0$ only if $w = 2$, the only critical number.

16. $M'(x) = (2x-1)/3(x^2-x-2)^{2/3} = (2x-1)/3\sqrt[3]{(x-2)(x+1)^2}$. Thus $M'(x) = 0$ if $x = 1/2$, and $M'(x)$ fails to exist at $x = -1$ and $x = 2$.

19. $G'(x) = (-2x^2 + 6x - 18)/(x^2-9)^2$ fails to exist at $x = \pm 3$, but neither of these are in the domain of G. Setting $G'(x) = 0$, we obtain no real solutions since $b^2 - 4ac = 36 - 144 < 0$. Thus, no critical numbers.

22. $g'(x) = 3x^2 - 6/x^2$ fails to exist at $x = 0$, but 0 is not in the domain of g and is not a critical number. $g'(x) = 0 \Rightarrow 3x^4 - 6 = 0 \Rightarrow x^4 = 2 \Rightarrow x = \pm \sqrt[4]{2}$, the only critical numbers.

25. $f'(x) = (x + 5)^4 \cdot 3(2x - 3)^2 \cdot 2 + 4(x + 5)^3(2x - 3)^3 = (x + 5)^3(2x - 3)^2[6(x + 5) + 4(2x - 3)] = (x + 5)^3(2x - 3)^2(14x + 18) = 0$ if $x = -5$, $3/2$, and $-18/14 = -9/7$.

28. If $f(x) = k$ for all x in (a,b) and if $c \in$ (a,b), then $f(x) = f(c)$ for all x there, and so $f(x) \geq f(c)$ and $f(x) \leq f(c)$ are both true.

34. If $f(x)$ is a polynomial of degree n, then $f'(x)$ has degree n-1 and can have at most n-1 distinct zeros. Since $f'(x)$, as a polynomial, exists everywhere, these are the only possible critical points.

EXERCISES 4.2

1. $f'(0)$ doesn't exist by Example 3, Section 3.1.

4. If $[a] = n$ and $b-a \geq 1$, then $b \geq a+1$ and $[b] \geq n+1$. Thus $f(b) - f(a) = [b] - [a] \geq 1$. Trying to solve $f(b) - f(a) = f'(c)(b-a)$ we get $f'(c) = \frac{f(b)-f(a)}{b-a} \geq \frac{1}{b-a}$. This equation has no solutions since $f'(x) = 0$ if x is not an integer and $f'(x)$ does not exist if x is an integer. (See #22, Section 3.1.) There is no contradiction since $b-a \geq 1 \Longrightarrow$ [a,b] contains at least one integer at which f is neither continuous nor differentiable.

7. As a polynomial function, f is continuous and differentiable everywhere. Rolle's Theorem applies since $f(3) = f(-3) = 118$. $f'(x) = 4x^3 + 8x = 4x(x^2+2) = 0$ only at $x = c = 0$, the desired solution.

10. As a polynomial function, f is continuous and differentiable everywhere. With $a = 1$, $b = 3$ we obtain: $f(3) - f(1) = f'(c)(3-1) \Longleftrightarrow 37 - 3 = (10c-3) \cdot 2 \Longleftrightarrow 17 = 10c - 3 \Longleftrightarrow c = 2$.

13. $f(x) = x^{2/3}$ is continuous on [-8,8], but $f'(0)$ does not exist.

$(f'(0) = \lim_{h \to 0} \frac{f(h) - f(0)}{h} = \lim_{h \to 0} \frac{h^{2/3} - 0}{h} = \lim_{h \to 0} \frac{1}{h^{1/3}}$ which does not exist.)

Thus f does not satisfy the hypotheses of the Mean Value Theorem.

16. $f(x) = 1 - 3x^{1/3}$ is continuous on [-8,-1] and differentiable on (-8,-1). ($f'(0)$ does not exist, but $0 \notin (-8,-1)$.) $f(-1) - f(-8) = f'(c)(-1-(-8)) \Longleftrightarrow 4-7 = (-c^{-2/3})(7) \Longleftrightarrow c^{-2/3} = 3/7 \Longleftrightarrow c^{2/3} = 7/3 \Longleftrightarrow c^2 = (7/3)^3 \Longleftrightarrow c = \pm \sqrt{(7/3)^3}$. The negative root must be chosen for c to be in (-8,-1). Thus $c = -\sqrt{(7/3)^3} \approx -3.6$.

20. $f'(x)$ is a linear polynomial (i.e. of degree one). Thus $f(b) - f(a) = f'(c)(b-a)$ is a linear equation in the unknown c.

22. Hint: If $f(x_1) = f(x_2) = 0$, then $f'(z_1) = 0$ for some z_1 between x_1 and x_2 by Rolle's Theorem. Now, $f'(x)$ is a quadratic polynomial with at most 2 zeros. What would happen if $f(x) = 0$ at more than 3 points?

25. Hint: Let f(x) be the number of miles travelled from A to B t hours after the start of the trip. To make the trip in one hour means f(1) - f(0) = 50. Now, recall that, since f is a distance, f' is the velocity, or speedometer reading.

28. Let [a,b] denote the time interval. Then $Q(b) - Q(a) = 10 - 2 = 8$, and $b - a = 15$. Using $I = Q'(t)$, we solve $Q(b) - Q(a) = Q'(c)(b - a)$ to obtain $8 = 15I(c) \Rightarrow I(c) = 8/15 =$ average current during this interval. Since $8/15 > 1/2$, the current must have exceeded 1/2 in order for the average to be greater than 1/2.

EXERCISES 4.3

4. (Graph)

1. $f'(x) = -7-8x = 0 \Rightarrow x = -7/8$ is the only critical point, and the intervals to be considered are $(-\infty,-7/8)$ and $(-7/8,\infty)$. Choosing $k = -1$ in the 1st, we have $f'(k) = f'(-1) = -7+8 = 1 > 0$; hence $f'(x) > 0$ on $(-\infty,-7/8)$ and f is increasing on $(-\infty,-7/8]$. Choosing $k = 0$ in the 2nd interval, $f'(k) = f'(0) = -7 < 0$; hence $f'(x) < 0$ on $(-7/8,\infty)$ and f is increasing on $[-7/8,\infty)$. Thus $f(-7/8)$ is a local maximum.

4. $f'(x) = 3x^2 - 2x - 40 = (3x+10)(x-4) = 0$ if $x = -10/3$ and 4. Tabulating our work:

Interval	k	f'(k)	f'(x)	f(x)
$(-\infty,-10/3)$	-4	16	+	increasing on $(-\infty,-10/3]$
$(-10/3,4)$	0	-40	-	decreasing on $[-10/3,4]$
$(4,\infty)$	5	25	+	increasing on $[4,\infty)$

Thus $f(-10/3) = 2516/27$ is a local maximum, and $f(4) = -104$ is a local minimum.

7. $f'(x) = \frac{4}{3}(x^{1/3} + x^{-2/3}) = 4(x+1)/3x^{2/3}$. Thus critical numbers, $x = -1$ and $x = 0$.

Interval	k	f'(k)	f'(x)	f(x)
$(-\infty,-1)$	-8	-7/3	-	decreasing on $(-\infty,1]$
$(-1,0)$	-1/8	14/3	+	increasing on $[-1,0]$
$(0,\infty)$	1	8/3	+	increasing on $[0,\infty)$

Thus a local minimum $f(-1) = -3$. (No extremum at $x = 0$. Only a vertical tangent.)

10. $f'(x) = x(-2x)/2\sqrt{4-x^2} + \sqrt{4-x^2} = 2(2-x^2)/\sqrt{4-x^2} = 0$ at $x = \pm\sqrt{2}$.

Interval	k	f'(k)	f'(x)	f(x)
$(-2,-\sqrt{2})$	$-\sqrt{3}$	-2	-	decreasing on $[-2,-\sqrt{2}]$
$(-\sqrt{2},\sqrt{2})$	0	2	+	increasing on $[-\sqrt{2},\sqrt{2}]$
$(\sqrt{2},2)$	$\sqrt{3}$	-2	-	decreasing on $[\sqrt{2},2]$

Thus $f(\sqrt{2}) = 2$ is a local maximum, and $f(-\sqrt{2}) = -2$ is a local minimum.

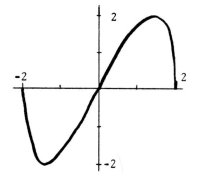

13. $f'(x) = 3x^2 - 3/x^2 = 3(x^4-1)/x^2 = 3(x^2+1)(x^2-1)/x^2 = 0$ at $x = \pm 1$. ($x = 0$ is not a critical number even though $f'(0)$ doesn't exist since 0 is not in the domain of f. $f(x)$ becomes arbitrarily large and positive as $x \to 0^+$, large and negative as $x \to 0^-$.) Tabulating as before:

Interval	k	f'(k)	f'(x)	f(x)
$(-\infty,-1)$	-2	45/4	+	increasing on $(-\infty,-1]$
$(-1,0)$	-1/2	-45/4	-	decreasing on $[-1,0)$
$(0,1)$	1/2	-45/4	-	decreasing on $(0,1]$
$(1,\infty)$	2	45/4	+	increasing on $[1,\infty)$

Thus $f(-1) = -4$ is a local maximum, and $f(1) = 4$ is a local minimum.

16. $f'(x) = 4(x^2-10x)^3 D_x(x^2-10x) = 4(x^2-10x)^3(2x-10)$
$= 8[x(x-10)]^3(x-5) = 8x^3(x-10)^3(x-5) = 0$ at $x = 0,5,10.$

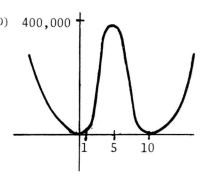

Interval	k	f'(k)	f'(x)	f(x)
$(-\infty,0)$	-1	$-48 \cdot 11^3$	-	decreasing on $(-\infty,0]$
$(0,5)$	1	$32 \cdot 9^3$	+	increasing on $[0,5]$
$(5,10)$	6	$8 \cdot 6^3 \cdot (-4)^3$	-	decreasing on $[5,10]$
$(10,\infty)$	11	$48 \cdot 11^3$	+	increasing on $[10,\infty)$

Thus $f(0) = 0$ and $f(10) = 0$ are local minima, and $f(5) = 25^4$ is a local maximum.

19. $f'(x) = (x-2)^3 \cdot 4(x+1)^3 + (x+1)^4 \cdot 3(x-2)^2 = (x-2)^2(x+1)^3(7x-5) = 0$ at $x = 2$, -1, and 5/7.

Interval	k	f'(k)	f'(x)	f(x)
$(-\infty,-1)$	-2	$64 \cdot 19$	+	increasing on $(-\infty,-1]$
$(-1,5/7)$	0	-20	-	decreasing on $[-1,5/7]$
$(5/7,2)$	1	16	+	increasing on $[5/7,2]$
$(2,\infty)$	3	$64 \cdot 16$	+	increasing on $[2,\infty)$

Thus local maximum $f(-1) = 0$ and a local minimum $f(5/7) = (-9/7)^3(12/7)^4 \approx$ -18.4. (No extremum at $x = 2$, only a horizontal tangent.)

22. $f'(x) = [(x-1)(2x) - (x^2+3)]/(x-1)^2 = (x-3)(x+1)/(x-1)^2 = 0$ at $x = -1$ and 3, the only critical numbers.

Interval	k	f'(k)	f'(x)	f(x)
$(-\infty,-1)$	-2	5/9	+	increasing on $(-\infty,-1]$
$(-1,1)$	0	-3	-	decreasing on $[-1,1)$
$(1,3)$	2	-3	-	decreasing on $(1,3]$
$(3,\infty)$	4	5/9	+	increasing on $[3,\infty)$

Thus $f(-1) = -2$ is a local maximum, and $f(3) = 6$ is a local minimum.

23. Since f has only one local maximum, $f(-7/8) = 8\frac{1}{16}$ this will be the absolute maximum on any interval containing -7/8. The absolute minimum will be at an end point. (a) On $[-1,1]$, $f(-1) = 8$, $f(1) = -6$. Thus, absolute maximum $f(-7/8)$ and absolute minimum $f(1)$. (b) On $[-4,2]$, $f(-4) = -31$, $f(2) = -25$. Thus, absolute maximum $f(-7/8)$ and absolute minimum $f(-4)$. (c) On $[0,5]$, f is decreasing. Thus, absolute maximum $f(0)$ and absolute minimum $f(5)$.

26. f here has a local maximum $f(-10/3)$ and a local minimum $f(4)$.
(a) On $[-1,1]$ f is decreasing. Thus, absolute maximum $f(-1) = 46$ and absolute minimum $f(1) = -32$.
(b) $[-4,2]$ contains $x = -10/3$, and the local maximum of f there will be the absolute maximum on this interval. Since $f(-4) = 88$ and $f(2) = -68$, the absolute minimum is $f(2)$.

(c) [0,5] contains x = 4, and the local minimum of f there will be the absolute minimum on this interval. Since f(0) = 8 and f(5) = -92, the absolute
maximum if f(0).

28.

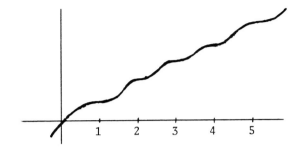

31. Hint. $P = I^2 R = V^2 R/(R + r)^2 \Rightarrow dP/dR = V^2(r - R)/(R + r)^3 = 0$
     if R = r.

34. There will be 3 fencing strips of length x and 4 of length y used
     to construct the partitioned cages. Since 1000 ft of fencing are
     to be used, we obtain 3x + 4y = 1000, or y = (1000 - 3x)/4. Next
     A = xy = x(1000 - 3x)/4 = $(1000x - 3x^2)/4$. Since x $\geq$ 0 and y $\geq$ 0,
     we must have 1000 - 3x > 0 or x $\leq$ 1000/3. Thus, we seek the
     maximum of A for 0 $\leq$ x $\leq$ 1000/3. (Note that A = 0 if x = 0 or
     1000/3.)   A' = (1000 - 6x)/4 =0 if x = 1000/6 = $166\frac{2}{3}$ ft. Since
     A > 0 for this x, the maximum of A occurs at this x. The corres-
     ponding y = (1000 - 500)/4 = 125 ft.

EXERCISES 4.4

1. $f'(x) = 3x^2 - 4x + 1 = (3x-1)(x-1)$, $f''(x) = 6x-4 = 6(x - 2/3)$. The critical
    numbers are x = 1/3, 1 and f''(1/3) = -2. $\therefore$ local maximum f(1/3) = 31/27;
    f''(1) = 2 $\therefore$ local minimum f(1) = 1. Since f''(x) > 0 if x > 2/3 and f''(x)
    < 0 if x < 2/3, the graph is CU on (-∞,2/3), CD on (2/3,∞), and has a point
    of inflection at x = 2/3.

4. $f'(x) = 16x - 8x^3 = 8x(2-x^2)$, $f''(x) = 16-24x^2 =$
    $8(2-3x^2)$. The critical numbers are x = 0, $+\sqrt{2}$ and
    f''(0) = 16 $\therefore$ local minimum f(0) = 0; $f''(+\sqrt{2}) =$
    -32, $\therefore$ local maxima $f(+\sqrt{2}) = 8$. Next, f''(x) > 0
    $\Leftrightarrow 2 - 3x^2 > 0 \Leftrightarrow x^2 < 2/3 \Leftrightarrow |x| < \sqrt{2/3}$.
    Thus the graph is CU on $(-\sqrt{2/3},\sqrt{2/3})$, CD on
    $(-\infty,-\sqrt{2/3})$, and $(\sqrt{2/3},\infty)$ with points of inflection
    at x = $+\sqrt{2/3}$.

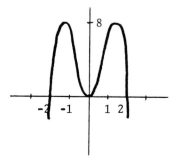

7. $f(x) = (x^2-1)^2 = x^4 - 2x^2 + 1 \implies f'(x) = 4x^3 - 4x = 4x(x^2-1) \implies f''(x) = 12x^2 - 4 = 4(3x^2 - 1)$. From $f'(x) = 0$, the critical numbers are $x = 0, \pm 1$. $f''(0) = -4$. $\therefore$ local maximum $f(0) = 1$. $f''(\pm 1) = 8$. $\therefore$ local minima $f(\pm 1) = 0$. Next, $f''(x) < 0 \iff 3x^2 - 1 < 0 \iff x^2 < 1/3 \iff |x| < 1/\sqrt{3}$. Thus the graph is CU on $(-\infty, -1/\sqrt{3})$ and $(1/\sqrt{3}, \infty)$, CD on $(-1\sqrt{3}, 1/\sqrt{3})$ with points of inflection at $x = \pm 1/\sqrt{3}$.

10. $f(x) = (x+4)/\sqrt{x} = x^{1/2} + 4x^{-1/2} \implies f'(x) = \frac{1}{2}x^{-1/2} - 2x^{-3/2} = (x-4)/2x^{3/2} \implies$ $f''(x) = (x^{3/2} - \frac{3}{2}(x-4)x^{1/2})/2x^3 = \sqrt{x}(12-x)/4x^3$. The only critical number is $x = 4$ (not $x = 0$ since 0 is not in the domain of f), and $f''(4) = 1/8$. $\therefore$ local minimum $f(4) = 4$. Noting that $f''(x) > 0$ if $0 < x < 12$, and $f''(x) < 0$ if $x > 12$, the graph is CU on $(0,12)$, CD on $(12,\infty)$, with a point of inflection at $x = 12$.

10. (Graph)   16. (Graph)

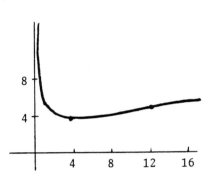

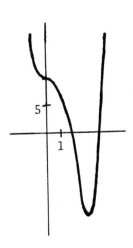

13. $f'(x) = \dfrac{(x^2+1) - 2x^2}{(x^2+1)^2} = \dfrac{1-x^2}{(x^2+1)^2}$. $f''(x) = \dfrac{(x^2+1)^2(-2x) - (1-x^2)2(x^2+1)2x}{(x^2+1)^4}$

$= \dfrac{2x^3-6x}{(x^2+1)^3} = \dfrac{2x(x^2-3)}{(x^2+1)^3}$. From $f'(x) = 0$, $x = \pm 1$ are the only critical numbers. $f''(1) = -4/8$. $\therefore$ a local maximum $f(1) = 1/2$. $f''(-1) = 4/8$. $\therefore$ a local minimum $f(-1) = -1/2$. Now, $f''(x) = 0$ if $x = 0, \pm\sqrt{3}$. We tabulate our work as follows:

Interval	k	f''(k)	f''(x)	Concavity
$(-\infty, -\sqrt{3})$	-2	-4/125	-	CD
$(-\sqrt{3}, 0)$	-1	1/2	+	CU
$(0, \sqrt{3})$	1	-1/2	-	CD
$(\sqrt{3}, \infty)$	2	4/125	+	CU

Thus the graph has PI's at $x = -\sqrt{3}, 0,$ and $\sqrt{3}$.

16. $f'(x) = 4x^3 - 12x^2 = 4x^2(x-3) = 0$ at $x = 0,3$. $f''(x) = 12x^2 - 24x = 12x(x-2)$.
     $f''(3) = 36 \Longrightarrow f(3) = -17$ is a local minimum. $f''(0) = 0$ means that the 2nd
     derivative test is not applicable. Trying the 1st derivative test, if $x < 0$
     then $f'(x) < 0$, and if $0 < x < 3$ then $f'(x) < 0$ also. Thus f is decreasing
     on $(-\infty,3)$ and has no local extremum at $x = 0$, only a horizontal tangent.
     Now, $f''(x) = 0$ at $x \doteq 0,2$. We tabulate:

Interval	k	f''(k)	f''(x)	Concavity
$(-\infty,0)$	-1	36	+	CU
$(0,2)$	1	-12	-	CD
$(2,\infty)$	3	36	+	CU

     Thus, PI's at $x = 0$ and $x = \overset{\bullet}{2}$. (The graph is next to that of #10 above.)

22.

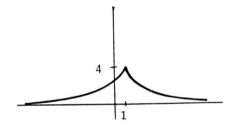

28. HINT. $f''(x)$ is a linear polynomial.
     Thus $f''(x) = 0$ has only one solu-
     tion. Show $f''(x)$ changes sign there.

32. $p(x) = 80 - \sqrt{x-1}$, $C(x) = 75x + 2\sqrt{x-1}$.
     (a) $p'(x) = -1/2\sqrt{x-1}$.   (b) $R(x) = xp(x) = 80x - x\sqrt{x-1}$.
     (c) $P(x) = R(x) - C(x) = 5x - (x+2)\sqrt{x-1}$.
     (d) $P'(x) = 5 - \dfrac{(x+2)}{2\sqrt{x-1}} - \sqrt{x-1} = \dfrac{10\sqrt{x-1} - (x+2) - 2(x-1)}{2\sqrt{x-1}} = (10\sqrt{x-1}-3x)/2\sqrt{x-1}$.
     (e) $P'(x) = 0$ when $10\sqrt{x-1} = 3x \iff 100(x-1) = 9x^2 \iff 9x^2 - 100x + 100 = 0$
     $\iff (9x-10)(x-10) = 0 \Longrightarrow x = 10/9$ and $x = 10$. We calculate $P(1) = 5$,
     $P(10/9) = 122/27 \approx 4.52$, $P(10) = 14$ and observe that for $x > 10$, $P'(x) < 0$
     and $P(x)$ decreases, eventually becoming negative. Thus the maximum profit
     is 14 when 10 items are produced.
     (f) $C'(x) = 75 + 1/\sqrt{x-1} \Longrightarrow C'(10) = 75 + 1/3$.

35. Here, $C(x) = 500 + .02x + .001x^2$ and $p(x) = 8$. Thus $R(x) = 8x$ and $P(x) =$
     $R(x) - C(x) = 7.98x - 500 - .001x^2$. $P'(x) = 7.98 - .002x = 0$ when $x =$
     $7.98/.002 = 3990$. $P''(x) < 0 \Longrightarrow$ this produces a maximum of $P(x)$. The
     maximum profit is $P(3990) = 7.98(3990) - 500 - .001(3990)^2 = 31,840.20 -$
     $500 - 15,920.10 = 15,420.10$ dollars.

EXERCISES 4.5

1. Let x be the length of the sides of the base and y the height.
    The volume is $x^2y = 4$ so that $y = 4/x^2$. The surface area, S, is

to be minimized. S = 4(Area of one side) + (Area of base) =
$4xy + x^2 = 16/x + x^2 \Rightarrow S' = -16/x^2 + 2x = 0 \Rightarrow -16 + 2x^3 = 0 \Rightarrow$
x = 2 and y = 4/4 = 1. S is minimized since $S'' = 32/x^2 + 2 > 0$
at x = 2.

4. Let r be the base radius, h the height and V the volume of the
cylinder. Then $V = \pi r^2 h = 1 \Rightarrow h = 1/\pi r^2$. In #3, we minimized
the surface area S = (curved area) + (base area) = $2\pi rh + \pi r^2$.
In this exercise, however, the base is cut out of a square, and
the area, M, of metal used is to be minimized. The smallest
square we can use has side length equal to the base diameter, 2r.
Thus, M = (curved area) + (area of square from which base is cut)
$= 2\pi rh + (2r)^2 = 2\pi r(1/\pi r^2) + 4r^2 = 2/r + 4r^2 \Rightarrow M' = -2/r^2 + 8r$
$= 0$ if $8r = 2/r^2 \Rightarrow 8r^3 = 2 \Rightarrow r = \sqrt[3]{1/4}$. $M'' = 4/r^2 > 0 \Rightarrow$ min.

7. At a time t hours after 1:00 p.m., B is 10t
miles west of its location at 1:00 p.m. and
A is 30 - 15t miles south of B's 1:00 p.m.
location. If f(t) is the square of the
distance between A and B at this time, then
$\sqrt{f(t)}$ (or f(t)) is to be minimized.
$f(t) = (10t)^2 + (30 - 15t)^2 = 325t^2 - 900t +$
$900 \Rightarrow f'(t) = 650t - 900 = 0$ when t = 18/13
hours, or t = 1 hour 23 1/13 minutes past 1:00 p.m.; thus,
minimum f(t) at 2:23:05 p.m. $f''(t) = 650 \Rightarrow$ this is a minimum.

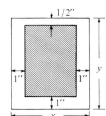

10. With x and y the page dimensions as shown,
xy = 90, or y = 90/x. If the printed area is
A, then A = (x - 2)(y - 3/2) = (x - 2)(90/x - 3/2)
$= 93 - 180/x - 3x/2 \Rightarrow A' = 180/x^2 - 3/2 = 0$
$\Rightarrow 3x^2 = 360 \Rightarrow x = \sqrt{120} \approx 11$ inches and
$y = 90/\sqrt{120} \approx 8.2$ inches.

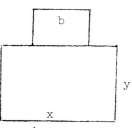

13. Let x,y be the field dimensions as shown. If
b is the barn length, then x - b ft. of
fencing will be used on the side next to the
barn. Thus, 2y + x + (x - b) = 500; or
y = 250 + b/2 - x. A = xy = (250 + b/2)x
$- x^2 \Rightarrow A' = 250 - b/2 - 2x = 0$ if x =
125 - b/4. (A'' < 0 $\Rightarrow$ max.) The corresponding y is
(250 + b/2) - x = (250 + b/2) - (125 + b/4) = 125 + b/4, the
same as x. Thus, the optimal shape is square.

16. In #15, the income per room per day was I(x) = 80 if $1 \leq x \leq 30$
or 80 - (x - 30) = 110 - x if $31 \leq x \leq 60$. Thus, the revenue
function R(x) is R(x) = xI(x) = 80x if $1 \leq x \leq 30$, or $110x - x^2$
if $31 \leq x \leq 60$. In this exercise, however, this revenue function
must be decreased to reflect the cleaning cost of \$6 per room
occupied per day. Since x is the number of rooms rented, this
cost is 6x and, thus, R(x) = 74x if $1 \leq x \leq 30$, $104x - x^2$ if
$31 \leq x \leq 60$. R' = 74 if $1 \leq x \leq 30$, 104 - 2x if $31 \leq x \leq 60$,
R' = 0 if x = 52. (In #15, R was maximized if x = 55.)

19. Placing the circle as shown (with equation $x^2 + y^2 = a^2$)
and the rectangle in the upper half, if (x,y) is the
point shown, then the area, A, is A = 2xy.

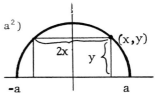

METHOD 1 The Explicit Method. From the equation of the circle, y =
$(a^2-x^2)^{1/2}$ and A = $2x(a^2-x^2)^{1/2}$, $0 \leq x \leq a$. A' = $2(a^2-2x^2)/\sqrt{a^2-x^2}$ = 0
when x = $a/\sqrt{2}$ and thus y = $a/\sqrt{2}$ also. This produces a maximum for A since
A = 0 when x = 0 or x = a and A > 0 otherwise. The dimensions of the rectangle of maximum area are x = $a/\sqrt{2}$ by 2y = $\sqrt{2}a$.

METHOD 2 The Implicit Method. As above, A = 2xy and A' = 2xy' + 2y. We obtain y' by differentiating the equation of the circle implicity: 2x + 2yy'
= 0 or y' = -x/y. Substituting into A' we get A' = $-2x^2/y + 2y = 2(y^2-x^2)/y$
= 0 when x = y. Substituting this into the equation of the circle we get
$2x^2 = a^2$ or x = y = $a/\sqrt{2}$ as in Method 1.

Solving for $\frac{1}{2}h$ gives us r of sphere!
so

Figuring for
± height of cyl!
so it's $\frac{h^2}{4}$,

not $\frac{h^2}{2}$!

22. Let r and h be the base radius and height of the
cylinder. Then $r^2 = a^2 - h^2/4$. If V is the
volume of the cylinder, then V = $\pi r^2 h$ =
$\pi(a^2h - h^3/4)$, for $0 \leq h \leq a$. V' = $\pi(a^2 - 3h^2/4)$
= 0 if $3h^2 = 4a^2$ or h = $2a/\sqrt{3}$. The corresponding
r is $\sqrt{2/3}a$. This value of h produces a maximum
for V on physical grounds or by an analysis as
in #7 above.

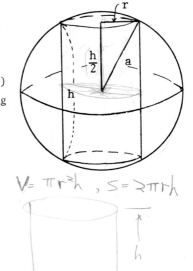

$\frac{h^2}{4} = \left(\frac{h}{2}\right)^2$!

$V = \pi r^2 h$, $S = 2\pi r h$

$V = \frac{4}{3}\pi r^3$
$S = 4\pi r^2$

25. If S is the strength, w the width, d the
     depth, then S = kwd² where k is a propor-
     tionality constant. With the circular
     cross-section placed as shown, the equation
     x² + y² = a², and (x,y) is the point shown,
     then w = 2x, d = 2y, y² = a² - x² ⟹
     S = k(2x)4y² = 8k(a²x - x³) ⟹ S' = 8k(a² - 3x²)

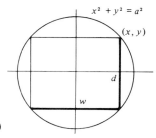

     = 0 if x = a/√3 ⟹ y = (√(2/3))a and w = 2a/√3, d = 2√2a/√3.

28. The volume is given by $V = (1/3)\pi r^2 h$. Hence $36\pi = (1/3)\pi r^2 h$
     $\Rightarrow h = 108/r^2$. The surface area S of the cone is given by
     $S = \pi r \sqrt{r^2 + h^2} = \pi r \sqrt{r^2 + (108/r^2)^2} = \pi r^{-1}(r^6 + 108^2)^{1/2}$.
     $S'(r) = \pi r^{-1}(1/2)(r^6 + 108^2)^{-1/2}(6r^5) + (r^6 + 108^2)^{1/2}(-\pi r^{-2})$
     $= [4\pi r^6 - 2\pi(108)^2]/2r^2(r^6 + 108)^{1/2}$. $S'(r) = 0 \Leftrightarrow 4\pi r^6$
     $= 2\pi(108)^2 \Leftrightarrow r^6 = (1/2)(108)^2 \Leftrightarrow r = (2)^{-1/6}(2)^{2/3}(27)^{1/3}$
     $= 3\sqrt{2}$. Since $S'(r) < 0$ if $r < 3\sqrt{2}$ and $S'(r) > 0$ if $r > 3\sqrt{2}$,
     $r = 3\sqrt{2}$ is the x-coordinate of a relative maximum point by the
     first derivative test. Then $h = 108/r^2 \Rightarrow h = 6$.

31. Let x be the width of the base of the rec-
     tangle and y be the height of the rectangle.
     The sides of the triangle are each length x
     and hence the height of the triangle is
     (√3/2)x. Area of the triangle is (√3/4)x²
     and area of the rectangle is xy. But
     3x + 2y = 12 ⟹ y = (1/2)(12 - 3x). So
     total area A(x) = (√3/4)x² + (x)(1/2)(12 - 3x)
     = ((√3 - 6)/4)x² + 6x. A'(x) = ((√3 - 6)/2)x + 6 ⟹ A'(x) = 0
     ⟺ x = 12/(6 - √3). A"(x) = (√3 - 6)/2 < 0 ⟹ max.

34. Let x and y be the dimensions of the rectangle with
     the rotation about an edge of length y. Then 2x +
     2y = p and the volume, V, of the cylinder is V =
     $\pi x^2 y = \pi x^2 (\frac{p}{2} - x) = \pi(\frac{px^2}{2} - x^3)$. $V' = \pi(px - 3x^2)$

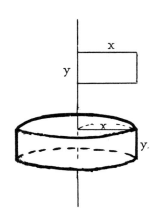

     = 0 when x = 0 or x = p/3. The maximum of V occurs
     when x = p/3 (V = 0 when x = 0) and the correspond-
     ing y is p/6. Thus, for maximum volume, the base,
     x, should be twice the height y.

37. Let x be the side of the square base, g the girth (g = 4x),
    and ℓ the length in inches. Postal regulations require
    ℓ + g ≤ 108. Clearly, to maximize the volume, V, we must
    take ℓ + g = 108, or ℓ + 4x = 108 ⟹ ℓ = 108 - 4x. Then V
    = $x^2\ell$ = $108x^2 - 4x^3$, 0 < x < 27. dV/dx = $216x - 12x^2$ =
    12x(18-x) = 0 when x = 18. (V"(18) = -216 ⟹ V is maximized.) The cor-
    responding length is ℓ = 108 - 4(18) = 36.

40. First note that 8 wells producing 1600 barrels daily means that the average
    production is 1600/8 = 200 bbl/well/day. Let x be the number of additional
    wells. Then the total number of wells is 8+x and the average daily produc-
    tion per well is 200-10x. Then the total production, P, is (# of wells) ·
    (average production per well), or P = (8+x)(200-10x) = $1600 + 120x - 10x^2$.
    Clearly x ≥ 0 and x ≤ 20 for at x = 20, the average yield per well is 0. To
    maximize P we compute P' = 120 - 20x = 0 at x = 6. P" < 0 ⟹ 6 additional
    wells maximizes P.

43. (a) If x is the average spacing between cars, then each car
    occupies a 12 + x ft. block of space on the bridge. Since the
    bridge is one mile in length, there are [5280/(12 + x)] such
    blocks. x ≥ d ⟹ [5280/(12 + x)] ≤ [5280/(12 + d)] = N.

    (b) The N cars, occupying one mile of space, will take
    (1 mile/ν mil/hr) = 1/ν hrs to drive off the bridge. Thus in
    one hour, F = Nν = ν(5280/(12 + d)).

    (c) F = $5280v/(12 + 0.025v^2)$ ⟹ F' = $5280(12 - 0.025v^2)/(12 + 0.025v^2)^2$
    = 0 if v^2 = 12/0.025 = 480 ⟹ ν ≈ 21.9 mi/hr.

EXERCISES 4.6

1. $\lim\limits_{x\to\infty} \dfrac{5x^2 - 3x + 1}{2x^2 + 4x - 7} = \lim\limits_{x\to\infty} \dfrac{5 - 3/x + 1/x^2}{2 + 4/x - 7/x^2} = \dfrac{5 - 0 + 0}{2 + 0 - 0} = \dfrac{5}{2}.$

4. $\lim\limits_{x\to-\infty} \dfrac{(3x+4)(x-1)}{(2x+7)(x+2)} = \lim\limits_{x\to-\infty} \dfrac{3x^2 + x - 4}{2x^2 + 11x + 14} = \lim\limits_{x\to-\infty} \dfrac{3 + (1/x)-(4/x^2)}{2 + (11/x)+(14/x^2)} = \dfrac{3}{2}.$

7. $\lim\limits_{x\to\infty} \dfrac{\sqrt{4x+1}}{10-3x} = \lim\limits_{x\to\infty} \dfrac{\sqrt{(4/x) + (1/x^2)}}{10/x-3} = \dfrac{0}{-3} = 0.$ (Since x→∞, we may assume x > 0.

   Thus x = $\sqrt{x^2}$ when dividing numerator and denominator by x.)

10. $\lim\limits_{x\to4^+} \dfrac{5}{4-x}$ = -∞ since, as x → 4^+, x > 4, 4-x < 0, whereas the numerator,

    5, is > 0. $\lim\limits_{x\to4^-} \dfrac{5}{4-x}$ = ∞ since, now, x → 4^- ⟹ x < 4, 4-x > 0. Thus, x = 4 is a vertical asymptote. $\lim\limits_{x\to+\infty}\dfrac{5}{4-x} = 0 \implies$

y = 0 is a horizontal asymptote. For the graph

we also need: $f'(x) = 5/(4-x)^2 > 0 \implies$ f is

increasing on $(-\infty,4)$ and $(4,\infty)$. $f''(x) =$

$10/(4-x)^3$ is > 0 if $x < 4$ and < 0 if $x > 4$.

Thus the graph is CU on $(-\infty,4)$, CD on $(4,\infty)$ with

no PI's.

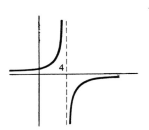

13. If x is nearly -8, then the numerator, 3x, is nearly -24 < 0 whereas the de-

nominator is always positive. Thus $\lim\limits_{x\to-8^+}\dfrac{3x}{(x+8)^2} = \lim\limits_{x\to-8^-}\dfrac{3x}{(x+8)^2} = -\infty$. Thus,

x = -8 is a vertical asymptote. y = 0 is a horizontal asymptote since

$\lim\limits_{x\to+\infty}\dfrac{3x}{(x+8)^2} = 0$. For the details of the graph, we need f'(x) and f''(x)

analyzed. $f'(x) = (24-3x)/(x+8)^3$, after simplification, so that $x = 24/3 =$

8 is the only critical number. As in Sec. 4.3, we find that f is decreasing

on $(-\infty,-8)$ and on $[8,\infty)$, increasing on $(-8,8]$ so that $f(8) = 3/32$ is a local

maximum. f''(x) simplifies to $6(x-16)/(x+8)^4 < 0$ if $x < 16$ and > 0 if $x > 16$.

Hence, the graph is CD on $(-\infty,-8)$ and $(-8,16)$, and is CU on $(16,\infty)$ with a PI

at x = 16.

16. $f(x) = 4x/(x^2 - 4x + 3) = 4x/(x-1)(x-3)$.

For x near 1 or 3, 4x > 0, and the sign of

f(x) is that of (x-1)(x-3). As $x \to 1^-$,

x-1 < 0, x-3 < 0 $\implies$ f(x) > 0 and the limit

is ∞. As $x \to 1^+$ or $x \to 3^-$, x-1 > 0, x-3 <

0 $\implies$ f(x) < 0, and both limits are -∞.

As $x \to 3^+$, x-1 > 0, x-3 > 0 $\implies$ f(x) > 0

and the limit is ∞. Thus x = 1 and x = 3

are vertical asymptotes. y = 0 is a hori-

zontal asymptote since $\lim\limits_{x\to+\infty}$ f(x) = 0.

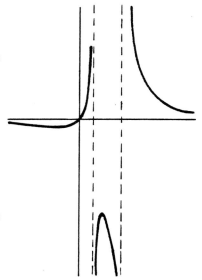

Also $f'(x) = 4(3-x^2)/(x^2-4x+3)^2 = 0$ at

$x = +\sqrt{3}$, and $f''(x) = 8(x^3-9x+12)/(x^2-4x+3)^3$

$\implies f(\sqrt{3}) \approx -7.5$ is a local maximum,

$f(-\sqrt{3}) \approx -0.53$ is a local minimum and

there is a PI between x = -4 and x = -3.

19. The denominator, x^2-4, is zero at $x = +2$ which are the vertical asymptotes

since $\lim\limits_{x\to-2^+}$ f(x) = $\lim\limits_{x\to2^-}$ f(x) = -∞ and $\lim\limits_{x\to-2^-}$ f(x) = $\lim\limits_{x\to2^+}$ f(x) = ∞.

$\lim\limits_{x\to+\infty}\dfrac{1}{x^2-4} = 0 \implies$ y = 0 is a horizontal asymptote.

22. Since the denominator, x^2+1, is never 0, f(x) is defined and continuous for all x. Thus there are no vertical asymptotes. y = 0 is a horozontal asymptote since $\lim\limits_{x \to +\infty} \dfrac{3x}{x^2+1}$

$= \lim\limits_{x \to \infty} \dfrac{3/x}{1+1/x^2} = \dfrac{0}{1} = 0.$

25. f(x) = (x+2)(x+1)/(x+3)(x-1) $\to \pm\infty$ as x $\to$ -3 and as x $\to$ 1, and these are vertical asymptotes. y = 1 is a horizontal asymptote since f(x) = $(1+3/x+2/x^2)/(1+2/x-3/x^2) \to 1$ as x $\to \pm\infty$.

27. HINT. f(x) = (x+4)/(x²-16) = 1/(x-4) if x ≠ -4. Thus x = -4 is not a vertical asymptote since a finite limit exists as x $\to$ -4.

28. $f(x) = \dfrac{\sqrt[3]{(4-x)(4+x)}}{4-x} = \sqrt[3]{\dfrac{(4+x)}{(4-x)^2}} \to \infty$ as x $\to$ 4 so that x = 4 is a vertical asymptote. Dividing numerator and denominator of the original f(x) formula by x $f(x) = \dfrac{\sqrt[3]{16/x^2 - 1/x}}{4/x - 1} \to \dfrac{0}{-1}$ as x $\to \pm\infty$. Thus y = 0 is a horizontal asymptote.

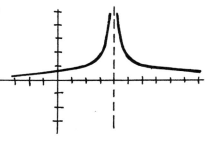

31. Since f(x) = $(8-x^3)/2x^2 \to \infty$ as x $\to$ 0, x = 0 is a vertical asymptote. Next, we may write f(x) = $\dfrac{8}{2x^2} - \dfrac{x^3}{2x^2} = -\dfrac{1}{2}x + \dfrac{4}{x^2}$. Thus y = $-\dfrac{1}{2}x$ is an oblique asymptote since $4/x^2 \to 0$ as x $\to \pm\infty$.

34. The denominator of f(x) is $2x^3 - 8x = 2x(x^2-4) = 0$ at x = 0, x = -2 and x = 2, which are the three vertical asymptotes since f(x) = $(1-x^4)/(2x^3 - 8x) \to \pm\infty$ as x approaches these values from left or right. Next, by long division, we obtain f(x) = $-\dfrac{1}{2}x + \dfrac{1-4x^2}{2x^3-8x}$. Thus y = $-\dfrac{1}{2}x$ is an oblique asymptote since $(1-4x^2)/(2x^3-8x) \to 0$ as x $\to \pm\infty$.

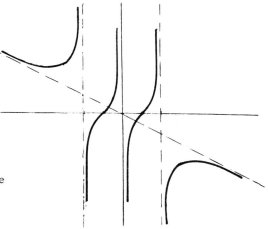

37. f'(x) = $-9x/\sqrt{16 - 9x^2} \Rightarrow |f'(x)| \Rightarrow \infty$ as x $\to -\dfrac{4^+}{3}$ and as x $\to \dfrac{4^-}{3}$. Thus, there are vertical tangents at (±4/3, 3).

40. $a = \lim\limits_{S \to \infty} R$. Thus, when S is large, $R \approx a$, nearly constant.

EXERCISES 4.7

NOTE: In Exercises 1-22, F will denote the most general antiderivative of the given function. (4.31), etc. are the formulas used to find F. Remember: The answer F can always be checked by differentiating, F' should be f.

1. $f(x) = 9x^2 - 4x + 3 \implies F(x) = \dfrac{9x^3}{3} - \dfrac{4x^2}{2} + 3x + C = 3x^3 - 2x^2 + 3x + C.$

4. $f(x) = 10x^4 - 6x^3 + 5 \implies F(x) = \dfrac{10x^5}{5} - \dfrac{6x^4}{4} + 5x + C = 2x^5 - \dfrac{3}{2}x^4 + 5x + C.$

7. $f(x) = 3\sqrt{x} + 1/\sqrt{x} = 3x^{1/2} + x^{-1/2} \implies F(x) = \dfrac{3x^{3/2}}{(3/2)} + \dfrac{x^{1/2}}{(1/2)} + C = 2x^{3/2} + 2x^{1/2} + C.$

10. $f(x) = 3x^5 - \sqrt[3]{x^5} = 3x^5 - x^{5/3} \implies F(x) = \dfrac{3x^6}{6} - \dfrac{x^{8/3}}{(8/3)} + C = \dfrac{x^6}{2} - \dfrac{3}{8}x^{8/3} + C.$

13. $f(x) = \dfrac{(3x+4)^2}{x^4} = \dfrac{9x^2 + 24x + 16}{x^4} = 9x^{-2} + 24x^{-3} + 16x^{-4} \implies F(x) = \dfrac{9x^{-1}}{-1}$

$+ \dfrac{24x^{-2}}{-2} + \dfrac{16x^{-3}}{-3} + C = -\dfrac{9}{x} - \dfrac{12}{x^2} - \dfrac{16}{3x^3} + C = (-1/x^3)(9x^2 + 12x + 16/3) + C.$

16. $f(x) = (2x-5)(3x+1) = 6x^2 - 13x - 5 \implies F(x) = \dfrac{6}{3}x^3 - \dfrac{13}{2}x^2 - 5x + C = 2x^3 -$

$\dfrac{13}{2}x^2 - 5x + C.$

19. $f(x) = \sqrt[5]{32x^4} = 2x^{4/5} \implies F(x) = \dfrac{2x^{9/5}}{9/5} + C = \dfrac{10}{9}x^{9/5} + C.$

22. $f(x) = \dfrac{x^3 + 3x^2 - 9x - 2}{x - 2} = \dfrac{(x-2)(x^2 + 5x + 1)}{x - 2} = x^2 + 5x + 1 \text{ (if } x \neq 2) \implies$

$F(x) = \dfrac{x^3}{3} + \dfrac{5}{2}x^2 + x + C \text{ (if } x \neq 2).$

25. $f''(x) = 4x-1 \implies f'(x) = 2x^2 - x + C.$ $f'(2) = -2 \implies 8 - 2 + C = -2 \implies C$

$= -8.$ Thus, $f'(x) = 2x^2 - x - 8 \implies f(x) = \dfrac{2}{3}x^3 - \dfrac{x^2}{2} - 8x + D.$ $f(1) = 3$

$\implies \dfrac{2}{3} - \dfrac{1}{2} - 8 + D = 3 \implies D = 11 - 1/6 = 65/6.$

28. $f''(t) = t^{2/3} \implies f'(t) = \dfrac{3}{5}t^{5/3} + C.$ $f'(1) = 2 \implies \dfrac{3}{5} + C = 2$, or $C = \dfrac{7}{5}.$

Thus $f'(t) = \dfrac{3}{5}t^{5/3} + \dfrac{7}{5} \implies f(t) = \dfrac{3}{5} \cdot \dfrac{3}{8}t^{8/3} + \dfrac{7}{5}t + D = \dfrac{9}{40}t^{8/3} + \dfrac{7}{5}t + D.$

$f(1) = 3 \implies \dfrac{9}{40} + \dfrac{7}{5} + D = 3 \implies D = \dfrac{120 - 9 - 56}{40} = \dfrac{55}{40} = \dfrac{11}{8}.$

31. $a(t) = v'(t) = -32 \implies v(t) = -32t + C.$ $v(0) = 80 \implies C = 80$ and $v(t) = s'(t) = -32t + 80 \implies s(t) = -16t^2 + 80t + D.$ $s(0) = 240 \implies D = 240.$

34. We represent the motion as in Example 6--origin at ground level, positive direction upward. Then $v(0) = 0$, $s(0) = 1000$. $a(t) = -32 \Rightarrow v(t) = -32t + C$. But $C = 0$ from $v(0) = 0$. So, $v(t) = -32t$ and $s(t) = -16t^2 + 1000$, where s is the distance above ground. Thus the distance <u>fallen</u> in t seconds is $16t^2$, and $v(3) = -96$. It strikes the ground when $s(t)$ becomes 0, i.e. when $16t^2 = 1000$ or $t = \sqrt{250}/2 \approx 7.9$ sec. (Alternate method: place origin at the 1000 ft. level. Then $s(0) = 0$, $s(t) = -16t^2$, and it hits the ground when $s(t) = -1000$.)

37. The same setup as in #34 is needed so that $s(t)$ is the distance above ground. The initial conditions are $v(0) = v_o$, $s(0) = s_o$. $a(t) = -g \Rightarrow v(t) = -gt + C$. $v(0) = v_o \Rightarrow C = v_o \Rightarrow v(t) = -gt + v_o \Rightarrow s(t) = -\frac{1}{2}gt^2 + v_o t + D$. $s(0) = s_o \Rightarrow D = s_o$.

40. Let $a(t) = k$, unknown. The initial and final conditions are $v(0) = 60$ mph = 88 ft/sec and $v(9) = 0$. $a(t) = k \Rightarrow v(t) = kt + C$. $v(0) = 88 \Rightarrow C = 88 \Rightarrow v(t) = kt + 88$. $v(9) = 0 \Rightarrow 9k + 88 = 0 \Rightarrow k = -88/9$ ft/sec^2 = $-60/9$ mph/sec.

43. $\frac{dV}{dt} = 3\sqrt{t} + \frac{t}{4} = 3t^{1/2} + \frac{t}{4} \Rightarrow V = \frac{3t^{3/2}}{(3/2)} + \frac{t^2}{8} + C = 2\sqrt{t^3} + \frac{t^2}{8} + C$. Setting $t = 4$ and $V = 20$: $20 = 2\sqrt{64} + \frac{16}{8} + C = 16 + 2 + C \Rightarrow C = 2$.

46. $dA/dt = 2.74 - 0.11t - 0.01t^2 \Rightarrow A(t) = 2.74t - 0.055t^2 - \frac{0.01}{3}t^3 + C$. The number of gallons used between 1980 ($t = 0$) and 1984 ($t = 4$) is $A(4) - A(0) = 2.74(4) - 0.055(16) - 0.01(64)/3 + C - C \approx 9.87$ billion gallons.

47. $C'(x) = 20 - .015x \Rightarrow C(x) = 20x - .0075x^2 + D$. $C(1) = 25 = 20 - .0075 + D \Rightarrow D = 5.0075$. $C(50) = 20(50) - .0075(2500) + 5.0075 = 1000 - 18.75 + 5.0075 \approx 986.26$.

48. $C'(x) = 2x^{-1/3} \Rightarrow C(x) = 3x^{2/3} + C$. $C(8) = 20 \Rightarrow 12 + C = 20 \Rightarrow C = 8 \Rightarrow C(x) = 3x^{2/3} + 8$.

EXERCISES 4.8 (Review)

1. $f(4) - f(0) = f'(c)(4 - 0) \Rightarrow 85 - 1 = (3c^2 + 2c + 1)4 \Rightarrow 3c^2 + 2c + 1 = 21 \Rightarrow 3c^2 + 2c - 20 = 0 \Rightarrow c = (-2 + \sqrt{4 + 240})/6 = (-1 + \sqrt{61})/3 \approx 2.3$. (The second solution was not between 0 and 4, and was disregarded.)

4.

$$f'(x) = \frac{-2x}{(1+x^2)^2} \quad \begin{cases} > 0 \text{ if } x < 0 \\ \\ < 0 \text{ if } x > 0 \end{cases}$$

Thus f is increasing on $(-\infty, 0]$, decreasing on $[0, \infty)$, and has a local maximum $f(0) = 1$.

7.   (Continuation of #4)

$$f''(x) = \frac{2(3x^2-1)}{(1+x^2)^3} \quad \begin{cases} > 0 \text{ if } |x| > 1/\sqrt{3} \\ \\ < 0 \text{ if } |x| < 1 \sqrt{3} \end{cases}$$

Thus the graph is CU on $(-\infty, -1/\sqrt{3})$ and $(1/\sqrt{3}, \infty)$, CD on $(-1/\sqrt{3}, 1/\sqrt{3})$ with points of inflection at $x = \pm 1/\sqrt{3}$. $f''(0) = -2$ confirms that there is a

10. First note that $x^3 - 6x = x(x^2 - 6)$ is ≥ 0 if $-\sqrt{6} \leq x \leq 0$ or $\sqrt{6} \leq x$, and is ≤ 0 if $x \leq -\sqrt{6}$ or $0 \leq x \leq \sqrt{6}$.   Next, $f(x) = |x^3 - 6x| = ((x^2 + 6x)^2)^{1/2} \Longrightarrow$
$f'(x) = \frac{1}{2}((x^3 - 6x)^2)^{-1/2}D_x[(x^3 - 6x)^2] = \frac{(x^3 - 6x)(3x^2 - 6)}{\sqrt{(x^3 - 6x)^2}} =$
$\frac{3(x^3 - 6x)(x^2 - 2)}{|x^3 - 6x|}$.   The critical numbers of f are 0, $\pm\sqrt{6}$ at which f' fails to exist (at each point the left and right derivatives differ) and $x = \pm\sqrt{2}$ at which f' is 0.   From the sign analysis at the start of the problem, $f'(x) = 3(x^2 - 2)$, $f''(x) = 6x$ if $-\sqrt{6} < x < 0$ or $\sqrt{6} < x$ where $x^3 - 6 > 0$.   Moreover, $f'(x) = -3(x^2 - 2)$, $f''(x) = -6x$ if $x < -\sqrt{6}$ or $0 < x < \sqrt{6}$, where $x^3 - 6x < 0$.   The analysis is then tabulated.

Interval	$(-\infty, -\sqrt{6})$	$(-\sqrt{6}, -\sqrt{2})$	$(-\sqrt{2}, 0)$	$(0, \sqrt{2})$	$(\sqrt{2}, \sqrt{6})$	$(\sqrt{6}, \infty)$
k	-3	-2	-1	1	2	3
f'(k)	-21	6	-3	3	-6	21
f'(x)	-	+	-	+	-	+
Variation of f	Dec'g	Inc'g	Dec'g	Inc'g	Dec'g	Inc'g

Thus f has local minima at $x = 0$, $\pm\sqrt{6}$ and local maxima at $x = \pm\sqrt{2}$.   From the f''(x) formula above, the graph is CD on $(-\sqrt{6}, 0)$ and $(0, \sqrt{6})$ and is CU on $(-\infty, -\sqrt{6})$ and $(\sqrt{6}, \infty)$.

16. Let x be the reduction in monthly charge C so that $C = 20 - x$. Then the number N of subscribers is $N = 5000 + 500x = 500(10 + x)$.   Then the monthly revenue is $R = NC = 500(10 + x)(20 - x) = 500(200 + 10x - x^2) \Longrightarrow R' = 500(10 - 2x) = 0$ when $x = 5$, $C = 15$, $N = 7500$, and $R = 112,500$. $R'' < 0 \Longrightarrow$ max.

19. $\lim\limits_{x\to-\infty} \dfrac{(2x-5)(3x+1)}{(x+7)(4x-9)} = \lim\limits_{x\to-\infty} \dfrac{(2-5/x)(3+1/x)}{(1+7/x)(4-9/x)} = \dfrac{(2)(3)}{(1)(4)} = \dfrac{3}{2}.$

22. $\lim\limits_{x\to-3} \sqrt[3]{\dfrac{x+3}{x^3+27}} = \lim\limits_{x\to-3} \sqrt[3]{\dfrac{x+3}{(x+3)(x^2-3x+9)}} = \lim\limits_{x\to-3} \sqrt[3]{\dfrac{1}{x^2-3x+9}} = \sqrt[3]{\dfrac{1}{9+9+9}} = \sqrt[3]{\dfrac{1}{27}} = \dfrac{1}{3}.$

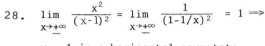

25. As $x \to 0^+$, $\sqrt{x} \to 0$ and is positive. Thus $1/\sqrt{x} \to \infty$ and $\sqrt{x} - 1/\sqrt{x} \to -\infty$.

28. $\lim\limits_{x\to+\infty} \dfrac{x^2}{(x-1)^2} = \lim\limits_{x\to+\infty} \dfrac{1}{(1-1/x)^2} = 1 \implies$

y = 1 is a horizontal asymptote.

$\lim\limits_{x\to1} \dfrac{x^2}{(x-1)^2} = \infty \implies$ x = 1 is a vertical

asymptote. For the graph, $f'(x) =$
$-2x/(x-1)^3 = 0$ only at x = 0. $f''(x) =$
$2(2x+1)/(x-1)^4$. $f''(0) = 2 > 0 \implies$
f(0) = 0 is a local minimum. There
is a PI at x = -1/2.

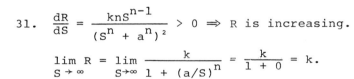

31. $\dfrac{dR}{dS} = \dfrac{knS^{n-1}}{(S^n + a^n)^2} > 0 \implies$ R is increasing.

$\lim\limits_{S\to\infty} R = \lim\limits_{S\to\infty} \dfrac{k}{1 + (a/S)^n} = \dfrac{k}{1+0} = k.$

34. $f(x) = 3x^5 + 2x^3 - x \implies F(x) = \dfrac{3}{6}x^6 + \dfrac{2}{4}x^4 - \dfrac{1}{2}x^2 + C = (x^6 + x^4 - x^2)/2 + C.$

37. $f(x) = (2x+1)^3 = 8x^3 + 12x^2 + 6x + 1 \implies F(x) = \dfrac{8}{4}x^4 + \dfrac{12}{3}x^3 + \dfrac{6}{2}x^2 + x + C$
$= 2x^4 + 4x^3 + 3x^2 + x + C.$

40. With the origin at ground level we have initial conditions
v(0) = -30, s(0) = 900. Thus, a(t) = -32 $\implies$ v(t) = -32t - 30
$\implies$ s(t) = -16t^2 - 30t + 900. v(5) = -160 - 30 = -190. The
ground is struck when s(t) = 0. Using the quadratic formula,
rejecting the negative root, t = $[30 - \sqrt{900 + 64(900)}]/(-32)$
$\overset{\sim}{=}$ $30(-1 + \sqrt{65})/32 \approx 6.6$ sec.

CHAPTER 5

THE DEFINITE INTEGRAL

<u>EXERCISES 5.1</u>

1. $\displaystyle\sum_{k=1}^{5} (3k-10) = (3\cdot1-10) + (3\cdot2-10) + (3\cdot3-10) + (3\cdot4-10) + (3\cdot5-10) =$

$-7-4-1+2+5 = -5.$

4. Observe that $[1+(-1)^n] = 2$ if n is even and $= 0$ if n is odd. Thus $\displaystyle\sum_{n=1}^{10} [1+(-1)^n]$

$= 0+2+0+2+0+2+0+2+0+2 = 10.$

7. $\displaystyle\sum_{i=1}^{8} 2^i = 2^1 + 2^2 + 2^3 + 2^4 + 2^5 + 2^6 + 2^7 + 2^8 = 2+4+8+16+32+64+128+256 = 510.$

10. $\displaystyle\sum_{k=1}^{1000} 2 = 2+2+\ldots+2, 1000$ times yielding $2(1000) = 2000.$

13. $\displaystyle\sum_{k=1}^{n} (k^2+3k+5) = \sum_{k=1}^{n} k^2 + 3 \sum_{k=1}^{n} k + \sum_{k=1}^{n} 5 = \frac{n(n+1)(2n+1)}{6} + \frac{3n(n+1)}{2} + 5n$

$= \frac{n}{6}[(n+1)(2n+1) + 9(n+1) + 30] = n(n^2+6n+20)/3.$

16. $\displaystyle\sum_{k=1}^{n} (k^3+2k^2-k+4) = \sum_{k=1}^{n} k^3 + 2 \sum_{k=1}^{n} k^2 - \sum_{k=1}^{n} k + \sum_{k=1}^{n} 4 = (\frac{n(n+1)}{2})^2 +$

$\frac{2n(n+1)(2n+1)}{6} - \frac{n(n+1)}{2} + 4n = n[3n(n+1)^2 + 4(n+1)(2n+1) - 6(n+1) + 48]/12 =$

$n(3n^3 + 14n^2 + 9n + 46)/12.$

19. $a = 0$, $b = 5 \Longrightarrow \Delta x = (b-a)/n = 5/n$, and the subdividing points are $x_0 = 0$,

$x_1 = a + \Delta x = 5/n$, $x_2 = a + 2\Delta x = 10/n$, $\ldots$, $x_k = a + k\Delta x = 5k/n$, $\ldots$, $x_n = 5$.

Since $f(x) = x^2$ is increasing on $[0,5]$, the minimum value of f on $[x_{k-1}, x_k]$

occurs at the left end point, x_{k-1}, and the maximum value of f occurs at the

right end point, x_k. Thus $u_k = x_{k-1} = 5(k-1)/n$ and $v_k = x_k = 5k/n$.

(a) Here we need $f(u_k) = f(x_{k-1}) = f(\frac{5(k-1)}{n}) =$

$[\frac{5(k-1)}{n}]^2 = \frac{25(k-1)^2}{n^2}$. Then

$\displaystyle\sum_{k=1}^{n} f(u_k)\Delta x = \sum_{k=1}^{n} \frac{25(k-1)^2}{n^2} \cdot \frac{5}{n} = \frac{125}{n^3} \sum_{k=1}^{n} (k-1)^2.$

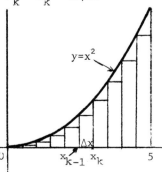

To evaluate the sum we write $\displaystyle\sum_{k=1}^{n} (k-1)^2 =$

$$\sum_{k=1}^{n} (k^2 - 2k + 1) = \sum_{k=1}^{n} k^2 - 2\sum_{k=1}^{n} k + \sum_{k=1}^{n} 1 =$$

$[\dfrac{n(n+1)(2n+1)}{6} - 2(\dfrac{n(n+1)}{2}) + n]$. We used (5.5ii) to evaluate the first sum,

(5.5i) for the second sum, and (5.2) (with c = 1) for the third. Substituting

the value for the sum into the previous expression, we obtain $\displaystyle\sum_{k=1}^{n} f(u_k)\Delta x =$

$\dfrac{125}{n^3}[\dfrac{n(n+1)(2n+1)}{6} - n(n+1) + n] = \dfrac{125(n+1)(2n+1)}{6n^2} - \dfrac{125(n+1)}{n^2} + \dfrac{125}{n^2} =$

$\dfrac{125}{6}(1 + \dfrac{1}{n})(2 + \dfrac{1}{n}) - 125(\dfrac{1}{n} + \dfrac{1}{n^2}) + \dfrac{125}{n^2}$. As $\Delta x \to 0$, $\dfrac{5}{n} \to 0$, $n \to \infty$. Also $\dfrac{1}{n} \to 0$

and $\dfrac{1}{n^2} \to 0$ so that $(1 + \dfrac{1}{n}) \to 1$ and $(2 + \dfrac{1}{n}) \to 2$. Moreover, in the limit as

$n \to \infty$, the sums, $\displaystyle\sum_{k=1}^{n} f(u_k)\Delta x$, approach the area. Thus, $\displaystyle\sum_{k=1}^{n} f(u_k)\Delta x =$

$\dfrac{125}{6}(1 + \dfrac{1}{n})(2 + \dfrac{1}{n}) - 125(\dfrac{1}{n} + \dfrac{1}{n^2}) + \dfrac{125}{n^2} \to \dfrac{125}{6}(1)(2) - 125(0+0) - 125(0) = \dfrac{125}{3} = A.$

(b) Here we need $f(v_k) = f(x_k) = f(5k/n) = (5k/n)^2$

$= 25k^2/n^2$. Then $\displaystyle\sum_{k=1}^{n} f(v_k)\Delta x = \sum_{k=1}^{n} \dfrac{25k^2}{n^2} \cdot \dfrac{5}{n} =$

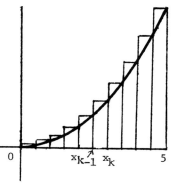

$\dfrac{125}{n^3} \displaystyle\sum_{k=1}^{n} k^2 = \dfrac{125}{n^3} \cdot \dfrac{n(n+1)(2n+1)}{6} = \dfrac{125}{6}(1 + \dfrac{1}{n}) \cdot$

$(2 + \dfrac{1}{n})$, where we used (5.5ii) to evaluate the sum.

As above, these sums approach the area as $n \to \infty$

and we obtain $A = \dfrac{125}{6}(1)(2) = \dfrac{125}{3}$.

22. $a = -2$, $b = 6 \implies \Delta x = (6-(-2))/n = 8/n$, and the subdividing points are $x_o = a$
      $= -2$, $x_1 = -2 + 8/n$, ..., $x_k = a + k\Delta x = -2 + 8k/n$, ..., $x_n = 6$. Since f is
      constant, any point in $[x_{k-1}, x_k]$ can be used for u_k or v_k. $\displaystyle\sum_{k=1}^{n} f(u_k)\Delta x =$

$$\sum_{k=1}^{n} f(v_k)\Delta x = \sum_{k=1}^{n} 7 \cdot \frac{8}{n} = \frac{56}{n} \sum_{k=1}^{n} 1 = \frac{56}{n}(n) = 56 \text{ for all } \Delta x \text{ and } n. \quad A = \lim_{n\to\infty} 56$$

$= 56$ as expected since the region is a rectangle with base $6-(-2) = 8$ and height 7.

25. Here $a = 1$, $b = 2$, $\Delta x = 1/n$, and the subdividing points are $x_o = 1$, $x_1 = 1 + 1/n$, ..., $x_k = 1 + k/n$, ..., $x_n = 2$. Since $f(x) = x^3 + 1$ is increasing, the minimum value of f on $[x_{k-1}, x]$ occurs at the left end point, x_{k-1}, and the maximum value of f occurs at the right end point, x_k. Thus $u_k = x_{k-1} = 1 + (k-1)/n$ and $v_k = x_k = 1 + k/n$.

(a) Here we need $f(u_k) = f(x_{k-1}) = f(1 + \frac{k-1}{n}) = (1 + \frac{k-1}{n})^3 + 1$. Recalling

that $(c+d)^3 = c^3 + 3c^2d + 3cd^2 + d^3$ we obtain $f(u_k) =$

$$[1 + \frac{3(k-1)}{n} + \frac{3(k-1)^2}{n^2} + \frac{(k-1)^3}{n^3}] + 1. \text{ Then } \sum_{k=1}^{n} f(u_k)\Delta x = \sum_{k=1}^{n} [2 + \frac{3(k-1)}{n}$$

$$+ \frac{3(k-1)^2}{n^2} + \frac{(k-1)^3}{n^3}] \cdot \frac{1}{n} = \frac{1}{n} \sum_{k=1}^{n} 2 + \frac{1}{n} \sum_{k=1}^{n} [\frac{3(k-1)}{n} + \frac{3(k-1)^2}{n^2} + \frac{(k-1)^3}{n^3}].$$

We must now evaluate these sums. We could evaluate the second sum by expanding all the $(k-1)$ expressions as in No. 19(a) above; however, there is an easier

way. Observe that $\sum_{k=1}^{n} (k-1)^m = 0^m + 1^m + 2^m + \ldots + (n-1)^m = \sum_{j=1}^{n-1} j^m$ for any

positive number m. Thus, (replacing the first sum by 2n using (5.2)),

$$\sum_{k=1}^{n} f(u_k)\Delta x = (\frac{1}{n})2n + \frac{1}{n} \sum_{k=1}^{n} [\frac{3(k-1)}{n} + \frac{3(k-1)^2}{n^2} + \frac{(k-1)^3}{n^3}] = 2 + \frac{1}{n} \sum_{j=1}^{n-1} (\frac{3j}{n}$$

$$+ \frac{3j^2}{n^2} + \frac{j^3}{n^3}) = 2 + \frac{3}{n^2} \sum_{j=1}^{n-1} j + \frac{3}{n^3} \sum_{j=1}^{n-1} j^2 + \frac{1}{n^4} \sum_{j=1}^{n-1} j^3. \text{ Now we use}$$

(5.5(i)-(iii)) with n replaced by n-1 (since the upper limit of summation is

now n-1) to obtain $\sum_{k=1}^{n} f(u_k)\Delta x = 2 + \frac{3}{n^2}\frac{(n-1)n}{2} + \frac{3}{n^3}\frac{(n-1)n(2n-1)}{6} +$

$\frac{1}{n^4}(\frac{(n-1)n}{2})^2 = 2 + \frac{3}{2}\frac{n-1}{n} + \frac{3}{6}\frac{n-1}{n}\frac{2n-1}{n} + \frac{1}{4}(\frac{n-1}{n})^2$. As $\Delta x \to 0$, $\frac{1}{n} \to 0$, and $n \to \infty$,

also $\frac{n-1}{n} = 1 - \frac{1}{n} \to 1$ and $\frac{2n-1}{n} = 2 - \frac{1}{n} \to 2$ and the sums $\sum_{k=1}^{n} f(u_k)\Delta x$ approach

the area, A, as n approaches ∞. Thus, letting n approach ∞ in the above ex-

pression for $\sum_{k=1}^{n} f(u_k)\Delta x$, we obtain $A = \lim_{n\to\infty} \sum_{k=1}^{n} f(u_k)\Delta x = \lim_{n\to\infty} [2 + \frac{3}{2}\frac{n-1}{n} +$

$$\frac{1}{2}\frac{n-1}{n}\frac{2n-1}{n} + \frac{1}{4}(\frac{n-1}{n})^2] = 2 + \frac{3}{2}(1) + \frac{1}{2}(1)(2) + \frac{1}{4}(1)^2 = 2 + \frac{3}{2} + 1 + \frac{1}{4} =$$

$$\frac{8+6+4+1}{4} = \frac{19}{4}.$$

(b) $f(v_k) = f(x_k) = x_k^3 + 1 = (1 + \frac{k}{n})^3 + 1 = 2 + \frac{3}{n}k + \frac{3}{n^2}k^2 + \frac{k^3}{n^3}$. Then

recalling $\Delta x = 1/n$, $\sum_{k=1}^{n} f(v_k)\Delta x = \sum_{k=1}^{n} (2 + \frac{3}{n}k + \frac{3}{n^2}k^2 + \frac{k^3}{n^3}) \cdot \frac{1}{n} =$

$$\frac{1}{n}\sum_{k=1}^{n} 2 + \frac{3}{n^2}\sum_{k=1}^{n} k + \frac{3}{n^3}\sum_{k=1}^{n} k^2 + \frac{1}{n^4}\sum_{k=1}^{n} k^3 = \frac{1}{n}(2n) + \frac{3}{n^2}\cdot\frac{n(n+1)}{2} + \frac{3}{n^3}\cdot$$

$$\frac{n(n+1)(2n+1)}{6} + \frac{1}{n^4}(\frac{n(n+1)}{2})^2 = 2 + \frac{3}{2}\frac{n+1}{n} + \frac{1}{2}\frac{(n+1)(2n+1)}{n^2} + \frac{1}{4}\frac{(n+1)^2}{n^2}.$$ As

$\Delta x \to 0$, $n \to \infty$, and, as above, $A = 2 + \frac{3}{2}(1) + \frac{1}{2}(1)(2) + \frac{1}{4}(1)^2 =$

$$\frac{8+6+4+1}{4} = \frac{19}{4}.$$

EXERCISES 5.2

1. $x_o = 0$, $x_1 = 1.1$, $x_2 = 2.6$, $x_3 = 3.7$, $x_4 = 4.1$ and $x_5 = 5$. Thus $\Delta x_1 = x_1-x_0$ = 1.1, $\Delta x_2 = x_2-x_1 = 1.5$, $\Delta x_3 = x_3-x_2 = 1.1$, $\Delta x_4 = x_4-x_3 = 0.4$ and $\Delta x_5 = x_5-x_4 = 0.9$. The largest of the Δx_k's is $\Delta x_2 = 1.5 \Longrightarrow \|P\| = 1.5$.

4. $x_o = 1$, $x_1 = 1.6$, $x_2 = 2$, $x_3 = 3.5$, $x_4 = 4$. $\Delta x_1 = x_1-x_o = 0.6$, $\Delta x_2 = x_2-x_1$ =0.4, $\Delta x_3 = x_3-x_2 = 1.5$, $\Delta x_4 = x_4-x_3 = 0.5$. $\|P\| = 1.5$ since the largest Δx_k is $\Delta x_3 = 1.5$.

7. Since this is a regular partition, all Δx_k's are equal to $(b-a)/n = (6-0)/6$ = 1. With w_k as the midpoint of $[x_{k-1},x_k] = [k-1,k]$, we have $R_p =$

$$\sum_{k=1}^{6} f(w_k)\Delta x_k = f(\frac{1}{2}) + f(\frac{3}{2}) + f(\frac{5}{2}) + f(\frac{7}{2}) + f(\frac{9}{2}) + f(\frac{11}{2}) = (8 - \frac{1}{2}(\frac{1}{4})) +$$

$$(8 - \frac{1}{2}(\frac{9}{4})) + (8 - \frac{1}{2}(\frac{25}{4})) + (8 - \frac{1}{2}(\frac{49}{4})) + (8 - \frac{1}{2}(\frac{81}{4})) + (8 - \frac{1}{2}(\frac{121}{4})) =$$

$$48 - \frac{1}{2}\frac{(1+9+25+49+81+121)}{4} = 48 - \frac{286}{8} = \frac{49}{4}.$$

10. Here $\Delta x_1 = \Delta x_2 = \Delta x_3 = \Delta x_4 = 2$, $\Delta x_5 = 7$, and $R_p = f(1)\cdot 2 + f(4)\cdot 2 + f(5)\cdot 2 +$ $f(9)\cdot 2 + f(9)\cdot 7 = 1\cdot 2 + \sqrt{4}\cdot 2 + \sqrt{5}\cdot 2 + \sqrt{9}\cdot 2 + \sqrt{9}\cdot 7 = 2 + 4 + 2\sqrt{5} + 6 + 21 =$ $33 + 2\sqrt{5} \approx 37.47.$

13. The expression involving w_k in the sum is $f(w_k)$.   Here,

$f(w_k) = 2\pi w_k(1+w_k{}^3)$ so that $f(x) = 2\pi x(1+x^3)$. The interval being partitioned

is $[0,4]$.   Thus,   the limit is $\displaystyle\int_0^4 2\pi x(1+x^3)\,dx$.

16. Since $a = b$, the value of the integral is 0.

19. $f(x) = 2x+6$ is continuous and ≥ 0 on $[-3,2]$.   Thus, $\displaystyle\int_{-3}^2 2x+6\ dx$ is

the area under the graph of f from -3 to 2. The graph of $f(x) = 2x+6$ is a
line from $(-3,0)$ to $(2,10)$, and the region is a triangle with base $2-(-3) =$
5 and height 10. Thus the integral $=$ area $= (1/2)5(10) = 25$.

22. $f(x) = \sqrt{a^2-x^2}$ is continuous and ≥ 0 on $[-a,a]$.   Thus, the integral is the
area under the graph of f from $-a$ to a. The graph of f is the upper half of
the circle of radius a centered at the origin. $(y = \sqrt{a^2-x^2} \implies y \geq 0$ and
$y^2 = a^2 - x^2$, or $x^2 + y^2 = a^2$.) The region is, thus, a semi-circle of radius
a and area $(1/2)\pi a^2$, which is the value of the integral.

EXERCISES 5.3

1. $\displaystyle\int_{-2}^4 5\ dx = 5(4-(-2)) = 5(6) = 30$.

4. $\displaystyle\int_4^{-3} dx = -\displaystyle\int_{-3}^4 dx = -1(4-(-3)) = -7$

7. By (5.15) $\displaystyle\int_1^4 (3x^2+5)\,dx = \displaystyle\int_1^4 3x^2\,dx + \displaystyle\int_1^4 5\ dx$. By (5.14) and (5.13) we get:

$3\displaystyle\int_1^4 x^2\,dx + 5(4-1) = 3(21) + 5(3) = 63 + 15 = 78$.

10. Using (5.15), (5.14) and (5.13), $\displaystyle\int_1^4 (3x+2)^2\ dx = \displaystyle\int_1^4 (9x^2+12x+4)\,dx = \displaystyle\int_1^4 9x^2\,dx$

$+ \displaystyle\int_1^4 12x\ dx + \displaystyle\int_1^4 4\ dx = 9\displaystyle\int_1^4 x^2\,dx + 12\displaystyle\int_1^4 x\ dx + 4(4-1) = 9(21) + 12(15/2) + 12$

$= 291$.

13. Rewriting the sum as $\displaystyle\int_{-3}^5 f(x)\,dx + \displaystyle\int_5^1 f(x)\,dx$, we use (5.17) to combine these

to get $\displaystyle\int_{-3}^1 f(x)\,dx$.

16. $\int_{-2}^{6} f(x)dx - \int_{-2}^{2} f(x)dx = \left[\int_{-2}^{2} f(x)dx + \int_{2}^{6} f(x)dx\right] - \int_{-2}^{2} f(x)dx = \int_{2}^{6} f(x)dx,$

where (5.17) was used to write the first integral as the sum in the brackets.

19. (a) We seek $z \in (-1,8)$ such that $\int_{-1}^{8} 3\sqrt{x+1}\,dx = 3\sqrt{z+1}(8-(-1))$, or $54 = 27\sqrt{z+1} \Rightarrow \sqrt{z+1} = 2 \Rightarrow z+1 = 4 \Rightarrow z = 3$.

(b) The average value of f is $(1/9)\int_{-1}^{8} 3\sqrt{x+1}\,dx = 54/9 = 6$, or equivalently, $f(z) = f(3) = 3\sqrt{3+1} = 6$.

22. (a) We seek $z \in (1,4)$ such that $\int_{1}^{4}(2 + 3\sqrt{x})dx = (2+3\sqrt{z})(4-1)$, or $20 = 3(2 + 3\sqrt{z}) \Rightarrow \sqrt{z} = (20/3 - 2)/3 = 14/9 \Rightarrow z = 196/81 \approx 2.42$.

(b) The average value of f is (integral value)/(length of interval) = 20/3 or, equivalently, $f(z) = 2 + 3\sqrt{196/81} = 20/3$.

EXERCISES 5.4

1. $\int_{1}^{4}(x^2-4x-3)dx = \frac{x^3}{3} - 2x^2 - 3x \Big]_{1}^{4} = [\frac{64}{3} - 32 - 12] - [\frac{1}{3} - 2 - 3] = -18.$

4. $\int_{0}^{2}(w^4-2w^3)dw = \frac{w^5}{5} - \frac{w^4}{2} \Big]_{0}^{2} = [\frac{32}{5} - \frac{16}{2}] - [0-0] = -8/5.$

7. $\int_{1}^{2}\frac{5}{8x^6}dx = \frac{5}{8}\int_{1}^{2} x^{-6}dx = \frac{5}{8}[\frac{x^{-5}}{-5}]_{1}^{2} = -\frac{1}{8}[2^{-5} - 1^{-5}] = -\frac{1}{8}[\frac{1}{32} - 1] = \frac{31}{256}.$

10. $\int_{-1}^{-2}\frac{2s-7}{s^3}ds = \int_{-1}^{-2}(2s^{-2} - 7s^{-3})ds = -2s^{-1} + \frac{7}{2}s^{-2}]_{-1}^{-2} = -\frac{2}{s} + \frac{7}{2s^2}]_{-1}^{-2} =$

$[\frac{-2}{-2} + \frac{7}{2(4)}] - [\frac{-2}{-1} + \frac{7}{2}] = 1 + \frac{7}{8} - 2 - \frac{7}{2} = -\frac{29}{8}.$

13. $\int_{0}^{1}(2x-3)(5x+1)dx = \int_{0}^{1}(10x^2-13x-3)dx = \frac{10}{3}x^3 - \frac{13}{2}x^2 - 3x]_{0}^{1} = \frac{10}{3} - \frac{13}{2} - 3$

$= -\frac{37}{6}.$

16. The value is 0 since the upper and lower limits are the same.

17. HINT: $(x^2-1)/(x-1) = x+1$ if $x \neq 1$.

19. See #16.

22. $\int_{-2}^{-1}(r - \frac{1}{r})^2 dr = \int_{-2}^{-1}(r^2 - 2r(\frac{1}{r}) + \frac{1}{r^2})dr = \int_{-2}^{-1}(r^2 - 2 + r^{-2})dr =$

$\frac{r^3}{3} - 2r - r^{-1}]_{-2}^{-1} = [\frac{(-1)^3}{3} - 2(-1) - \frac{1}{(-1)}] - [\frac{(-2)^3}{3} - 2(-2) - \frac{1}{(-2)}] =$

$$[- \frac{1}{3} + 2 + 1] - [- \frac{8}{3} + 4 + \frac{1}{2}] = \frac{7}{3} - \frac{3}{2} = \frac{5}{6}.$$

25. $\int_0^4 \sqrt{3t} \, (\sqrt{t} + \sqrt{3}) dt = \sqrt{3} \int_0^4 \sqrt{t}(\sqrt{t} + \sqrt{3}) dt = \sqrt{3} \int_0^4 (t + \sqrt{3} \, t^{1/2}) dt =$

$$\sqrt{3}[\frac{t^2}{2} + \frac{\sqrt{3}}{3/2} t^{3/2}]_0^4 = \sqrt{3}\{[\frac{16}{2} + \frac{2\sqrt{3}}{3} \cdot 4^{3/2}] - [0-0]\} = \sqrt{3}(8 + \frac{2}{\sqrt{3}}(8)) = 8(\sqrt{3} + 2).$$

28. $D_x \int_0^x (5t+3)^2 dt = D_x \int_0^x (25t^2 + 30t + 9) dt = D_x [\frac{25}{3}t^3 + 15t^2 + 9t]_0^x =$

$D_x [\frac{25}{3}x^3 + 15x^2 + 9x] = 25x^2 + 30x + 9 = (5x+3)^2.$

31. Since $f(x) = x^2 + 1 > 0$, the area $= \int_{-1}^2 (x^2+1) dx = \frac{x^3}{3} + x]_{-1}^2 = (\frac{8}{3} + 2) -$

$(- \frac{1}{3} - 1) = 6.$

37. $\int_0^4 (\sqrt{x}+1) dx = \int_0^4 (x^{1/2}+1) dx = \frac{2}{3}x^{3/2} + x]_0^4 = \frac{2}{3}(8) + 4 = \frac{28}{3}.$ Thus we seek

$z \in (0,4)$ such that $(\sqrt{z}+1)4 = \frac{28}{3}$ or $\sqrt{z} + 1 = \frac{7}{3} \implies \sqrt{z} = \frac{4}{3} \implies z = \frac{16}{9}.$

40. $\int_1^9 3x^{-2} dx = -3x^{-1}]_1^9 = -3(\frac{1}{9} - 1) = -3(- \frac{8}{9}) = \frac{8}{3}.$ $\frac{3}{z^2}(9-1) = \frac{8}{3} \iff 8z^2 = 72 \iff$

$z^2 = 9 \implies z = 3.$ ($z = -3$ is not in $(1,9)$.

41. $b-a = 2 - (-1) = 3$, and the average value is $\frac{1}{3} \int_{-1}^2 (x^2 + 3x - 1) dx =$

$\frac{1}{3}[\frac{x^3}{3} + \frac{3x^2}{2} - x]_{-1}^2 = \frac{1}{3}(\frac{20}{3} - \frac{13}{6}) = \frac{1}{3} \cdot \frac{27}{6} = \frac{3}{2}.$

43. Since $v(t)$ is an antiderivative of $a(t)$, the average value of

$a(t)$ on $[t_1, t_2]$ is $\frac{1}{t_2 - t_1} \int_{t_1}^{t_2} a(t) dt = \frac{v(t_2) - v(t_1)}{t_2 - t_1}.$

46. Since $R(0) = 0$, $\frac{1}{n} \int_0^n R'(x) dx = \frac{R(n) - R(0)}{n} = \frac{R(n)}{n}.$

49. Let $u = g(x)$. Then by the Chain Rule, $D_x \int_a^u f(t) dt =$

$D_u \int_a^u f(t) dt \, D_x u = f(u)g'(x) = f(g(x))g'(x).$

EXERCISES 5.5

$$D_x \int_a^u f(t) \, dt = D_u \int_a^u f(t) \, dt$$

1. Let $u = 3x+1$ so that $du = 3 \, dx$ or $dx = (1/3)du$. $\int (3x+1)^4 dx = (1/3)\int u^4 \, du =$

$u^5/15 + C = \frac{(3x+1)^5}{15} + C.$

4. Using the first method of Example 3, let $u = 9 - z^2$, so $du = -2z\,dz$. Then

$$\int \sqrt{9-z^2}\; z\;dz = (-1/2)\int (9-z^2)^{1/2}(-2z\,dz) = -\frac{1}{2}\int u^{1/2}du = -\frac{1}{2}(\frac{2}{3})\,u^{3/2} + C$$

$$= -\frac{1}{3}(9-z^2)^{3/2} + C.$$

7. Let $u = 1-2s^2$, $du = -4s\,ds$, $s\,ds = -\frac{1}{4}du$. $\int \frac{s}{\sqrt[3]{1-2s^2}}\,ds = -\frac{1}{4}\int u^{-1/3}du =$

$$-\frac{1}{4}(\frac{3}{2})u^{2/3} + C = -\frac{3}{8}(1-2s^2)^{2/3} + C.$$

10. Let $v = 1 + 1/u$, $dv = -1/u^2\,du$. Then $\int (1 + \frac{1}{u})^{-3}(\frac{1}{u^2})du = -\int v^{-3}\,dv = \frac{v^{-2}}{2} + C$

$$= \frac{1}{2}(1 + \frac{1}{u})^{-2} + C.$$

13. Let $u = t^2-1$, $du = 2t\,dt$. When $t = 1$ or -1, $u = 0$. Thus $\int_{-1}^{1}(t^2-1)^3 t\,dt = \frac{1}{2}\int_0^0 u^3\,du = 0.$

16. Let $u = x^2+9$. Then $du = 2x\,dx$, $x\,dx = (1/2)du$. $x = 0 \Rightarrow u = 9$. $x = 4 \Rightarrow u = 25$. $\int_0^4 \frac{x}{\sqrt{x^2+9}}\,dx = \int_0^4 (x^2+9)^{-1/2}x\,dx = \int_9^{25} u^{-1/2}(1/2)du = u^{1/2}\big]_9^{25} = \sqrt{25} - \sqrt{9} = 5 - 3 = 2.$

19. Let $u = 8x+5$, $du = 8\,dx$ so that $dx = \frac{1}{8}du$. Then $\int 5(8x+5)^{1/2}dx = \frac{5}{8}\int u^{1/2}\,du$

$$= \frac{5}{8}\frac{u^{3/2}}{(3/2)} + C = \frac{5}{8}\cdot\frac{2}{3}(8x+5)^{3/2} + C.$$

22. Let $u = 3-x^4$, $du = -4x^3dx$ so that $x^3dx = -\frac{1}{4}du$. Then $\int (3-x^4)^3\,x^3dx = -\frac{1}{4}\int u^3du = -\frac{1}{4}\frac{u^4}{4} + C = -\frac{1}{16}(3-x^4)^4 + C.$

25. Let I denote the given integral.
   (a) Let $u = \sqrt{x} + 3$. Then $du = (1/2\sqrt{x})dx$ and $dx/\sqrt{x} = 2\,du$. Substituting we obtain: $I = 2\int u^2du = \frac{2}{3}u^3 + C_a = \frac{2}{3}(x^{1/2} + 3)^3 + C_a = \frac{2}{3}(x^{3/2} + 9x + 27x^{1/2}+27) + C_a = \frac{2}{3}x^{3/2} + 6x + 18x^{1/2} + 18 + C_a.$
   (b) $I = \int \frac{x + 6\sqrt{x} + 9}{\sqrt{x}}\,dx = \int (\sqrt{x} + 6 + \frac{9}{\sqrt{x}})dx = \int (x^{1/2} + 6 + 9x^{-1/2})dx = \frac{2}{3}x^{3/2} + 6x + 18x^{1/2} + C_b.$ Comparing answers we see that $C_b = C_a + 18.$

28. Let $u = 3x+2$, $du = 3\,dx$ so that $dx = (1/3)du$. Then $D_x\int (3x+2)^7dx = D_x\int \frac{u^7}{3}\,du = D_x(\frac{u^8}{24} + C) = D_x(\frac{(3x+2)^8}{24} + C) = \frac{8(3x+2)^7(3)}{24} = (3x+2)^7.$

31. An antiderivative of $D_x\sqrt{x^2+16}$ is $\sqrt{x^2+16}$ itself. Thus $\displaystyle\int_0^3 D_x\sqrt{x^2+16}\ dx =$

$\sqrt{x^2+16}\Big]_0^3 = \sqrt{9+16} - \sqrt{16} = 5 - 4 = 1.$

34. Since $f(x)$ is ≥ 0 and continuous on $[1,2]$, the desired area is $A = \displaystyle\int_1^2 f(x)\,dx$

$= \displaystyle\int_1^2 (x^2+1)^{-2}x\ dx.$ Let $u = x^2+1$, $du = 2x\ dx$, $x\ dx = (1/2)du.$ $x = 1 \Longrightarrow$

$u = 2$, $x = 2 \Longrightarrow$ $u = 5.$ Thus $A = \dfrac{1}{2}\displaystyle\int_2^5 u^{-2}\ du = -\dfrac{1}{2}u^{-1}\Big]_2^5 = -\dfrac{1}{2}\left(\dfrac{1}{5} - \dfrac{1}{2}\right) =$

$-\dfrac{1}{2}\left(-\dfrac{3}{10}\right) = \dfrac{3}{20}.$

37. Let $u = x+4$, $du = dx.$ $x = 0 \Longrightarrow u = 4.$ $x = 5 \Longrightarrow u = 9.$ $\displaystyle\int_0^5 \sqrt{x+4}\ dx =$

$\displaystyle\int_4^9 u^{1/2}du = \dfrac{2}{3}u^{3/2}\Big]_4^9 = \dfrac{2}{3}(9^{3/2} - 4^{3/2}) = \dfrac{2}{3}(3^3 - 2^3) = \dfrac{38}{3}.$ Thus we seek

$z \in (0,5)$ (NOT $(4,9)$) such that $\sqrt{z+4}\,(5-0) = \dfrac{38}{3} \Longrightarrow \sqrt{z+4} = \dfrac{38}{15} \Longrightarrow z+4 =$

$\left(\dfrac{38}{15}\right)^2 \Longrightarrow z = \left(\dfrac{38}{15}\right)^2 - 4 \approx 2.418.$

41, 42. Verify by differentiating the function on the right to obtain the integrand on the left side.

EXERCISES 5.6

NOTE: T will denote the trapezoidal approximation and S the Simpson Rule approximation. In the tables, m and r denote the coefficients in T and S, respectively.

1. Here $f(x) = 1/x$, $b = 4$, $a = 1$, $n = 6$, $(b-a)/n = .5$. Our work is arranged in the following table.

k	x_k	$f(x_k)$	m	$mf(x_k)$	r	$rf(x_k)$
0	1.0	1.0000	1	1.0000	1	1.0000
1	1.5	0.6667	2	1.3334	4	2.6668
2	2.0	0.5000	2	1.0000	2	1.0000
3	2.5	0.4000	2	0.8000	4	1.6000
4	3.0	0.3333	2	0.6666	2	0.6666
5	3.5	0.2857	2	0.5714	4	1.1428
6	4.0	0.2500	1	0.2500	1	0.2500
				5.6214		8.3262

(a) $(b-a)/2n = 1/4 \Longrightarrow$ T $= 5.6214/4 = 1.40535 \approx 1.41.$

(b) $(b-a)/3n = 1/6 \Longrightarrow$ S $= 8.3262/6 = 1.3877 \approx 1.39.$

4. Tabulating as above, we have $f(x) = \sqrt{1+x^3}$, $(b-a)/n = .25$.

k	x_k	$f(x_k)$	m	$mf(x_k)$	r	$rf(x_k)$
0	2.00	3.0000	1	3.0000	1	3.0000
1	2.25	3.5200	2	7.0400	4	14.0800
2	2.50	4.0774	2	8.1548	2	8.1548
3	2.75	4.6687	2	9.3374	4	18.6748
4	3.00	5.2915	1	5.2915	1	5.2915
				32.8237		49.2011

Since $(b-a)/2n = 1/8$, $T = 32.8237/8 = 4.1030 \approx 4.10$.

Since $(b-a)/3n = 1/12$, $S = 49.2011/12 = 4.1001 \approx 4.10$.

7. We obtain with $f(x) = \sqrt[3]{x^2 + 8}$, $(b-a)/n = (3/2)/6 = 1/4$:

i	x_i	$f(x_i)$	m	$mf(x_i)$	r	$rf(x_i)$
0	1.0000	2.0801	1	2.0801	1	2.0801
1	1.2500	2.1225	2	4.2451	4	8.4902
2	1.5000	2.1722	2	4.3445	2	4.3445
3	1.7500	2.2282	2	4.4564	4	8.9127
4	2.0000	2.2894	2	4.5789	2	4.5789
5	2.2500	2.3551	2	4.7102	4	9.4204
6	2.5000	2.4244	1	2.4244	1	2.4224
				26.8395		40.2511

(a) Since $(b-a)/2n = 1/8$, $T = (1/8)(26.8395) = 3.3549 \approx 3.35$.

(b) Since $(b-a)/3n = 1/12$, $S = (1/12)(40.2511) = 3.3543 \approx 3.35$.

10. From #2, $f(x) = 1/(1+x)$, $b = 3$, $a = 0$, $n = 8$, $(b-a)/n = 3/8$. Thus, the partitioning points are: $x_0 = 0$, $x_1 = 3/8$, $x_2 = 6/8 = 3/4$, $x_3 = 9/8$, $x_4 = 3/2$, $x_5 = 15/8$, $x_6 = 9/4$, $x_7 = 21/8$ and $x_8 = 3$. The midpoints are: $\overline{x}_1 = 3/16$, $\overline{x}_2 = 9/16$, $\overline{x}_3 = 15/16$, $\overline{x}_4 = 21/16$, $\overline{x}_5 = 27/16$, $\overline{x}_6 = 33/16$, $\overline{x}_7 = 39/16$, and $\overline{x}_8 = 45/16$. By the Midpoint Rule, $\int_0^3 1/(x+1)dx \approx \frac{3}{8}[\frac{1}{1+3/16} + \frac{1}{1+9/16} + \frac{1}{1+15/16} +$

$\frac{1}{1+21/16} + \frac{1}{1+27/16} + \frac{1}{1+33/16} + \frac{1}{1+39/16} + \frac{1}{1+45/16}] =$

$\frac{3}{8}[\frac{16}{19} + \frac{16}{25} + \frac{16}{31} + \frac{16}{37} + \frac{16}{43} + \frac{16}{49} + \frac{16}{55} + \frac{16}{61}] \approx 1.38$

13. For notational convenience, let $A = \int_1^{2.7} \frac{1}{x}\,dx$ and $B = \int_1^{2.8} \frac{1}{x}\,dx$. The trape-

zoidal approximations to A and B are 0.9940 and 1.0304, respectively. (Details omitted.) Since $f'(x) = -x^{-2}$ and $f''(x) = 2x^{-3}$, it follows that $|f''(x)| \leq 2$ if $x \geq 1$. Thus, the error in this approximation to A, with M = 2 and $(b-a)^3/n^2$ written as $(b-a)((b-a)/n)^2 = 1.7(.1)^2$, is $\leq \frac{2}{12}(1.7)(.1)^2 = \frac{.017}{6} \leq$ .003. Similarly, the error in the approximation to B is $\leq \frac{2}{12}(1.8)(.1)^2 = \frac{.018}{6} = .003$. Thus, $0.9910 \leq A \leq 0.9970$, $1.0274 \leq B \leq 1.0334$, and A < 1 < B as desired.

16. From the given data, b = 4, a = 2, n = 10, (b-a)/n = 0.2. We tabulate our work as before.

k	x_k	$f(x_k)$	m	$mf(x_k)$	r	$rf(x_k)$
0	2.0	12.1	1	12.1	1	12.1
1	2.2	11.4	2	22.8	4	45.6
2	2.4	9.7	2	19.4	2	19.4
3	2.6	8.4	2	16.8	4	33.6
4	2.8	6.3	2	12.6	2	12.6
5	3.0	6.2	2	12.4	4	24.8
6	3.2	5.8	2	11.6	2	11.6
7	3.4	5.4	2	10.8	4	21.6
8	3.6	5.1	2	10.2	2	10.2
9	3.8	5.9	2	11.8	4	23.6
10	4.0	5.6	1	5.6	1	5.6
				146.1		220.7

Since (b-a)/2n = .1, T = (.1)(146.1) $\approx$ 14.61.

Since (b-a)/3n = .2/3, S = (.2/3)(220.7) $\approx$ 14.71.

19. From the given data, b = D, a = 0, (b - a)/n = 0.2D. Thus, n = 5, and Simpson's Rule cannot be used since n is odd.
$$\overline{v}_x = \frac{1}{D}\int_0^D v(x)dx \approx \frac{1}{D}\frac{0.2D}{2}[v(0) + 2v(0.2D) + 2v(0.4D) + 2v(0.6D) + 2v(0.8D) + v(D)] = 0.1[0.28 + 2(0.23) + 2(0.19) + 2(0.17) + 2(0.13) + 0.02] = 0.174.$$

22. $f(x) = 1/(1 + x)$, b = 3, a = 0 $\Rightarrow$ $0 \leq x \leq 3$ and b - a = 3.
(a) $f''(x) = 2/(1 + x)^3$. $x \geq 0 \Rightarrow |f''(x)| \leq 2$. Thus, E_T, the error in the trapezoidal approximation, satisfies, with M = 2 and b - a = 3: $E_T \leq \frac{M(b - a)^3}{12n^2} = \frac{2(27)}{12n^2} = \frac{9}{2n^2}$. We seek n such that $E_T \leq 0.001$. This will be true if $9/2n^2 \leq 0.001 \Rightarrow$

$2n^2/9 \geq 1000 \Rightarrow n^2 \geq 4500 \Rightarrow n \geq \sqrt{4500} \approx 67.1 \Rightarrow n = 68$ is the smallest such integer.

(b) $f^{(4)}(x) = 24/(1 + x)^4 \leq 24$ for $x \geq 0$. Thus E_S, the error in the Simpson approximation satisfies with $M = 24$, $b - a = 3$:

$$E_S \leq \frac{M(b - a)^5}{180n^4} = \frac{24(243)}{180n^4} = \frac{162}{5n^4} \leq 0.001 \text{ if } 5n^4/162 \geq 1000 \Rightarrow$$

$\sqrt[4]{162000/5} \approx 13.4 \Rightarrow n = 14$.

EXERCISES 5.7 (Review)

1. $\sum\limits_{k=1}^{5} (k^2+3) = (1^2+3) + (2^2+3) + (3^2+3) + (4^2+3) + (5^2+3) = 4+7+12+19+28 = 70$.

4. Since $[-2,3]$ is to be partitioned into 5 equal subintervals, each is of length 1 and $x_0 = -2$, $x_1 = -1$, $x_2 = 0$, $x_3 = 1$, $x_4 = 2$, $x_5 = 3$. Since w_k is the midpoint of $[x_{k-1}, x_k]$, $w_1 = -3/2$, $w_2 = -1/2$, $w_3 = 1/2$, $w_4 = 3/2$, $w_5 = 5/2$. Since $\Delta x_k = 1$ for $k = 1, \ldots, 5$, we obtain

$$R_p = f(-3/2) + f(-1/2) + f(1/2) + f(3/2) + f(5/2)$$
$$= (1 - \frac{9}{4}) + (1 - \frac{1}{4}) + (1 - \frac{1}{4}) + (1 - \frac{9}{4}) + (1 - \frac{25}{4})$$
$$= 5 - (9+1+1+9+25)/4 = 5 - 45/4 = -25/4.$$

7. $\int_0^1 \sqrt[3]{8x^7} \, dx = \int_0^1 2x^{7/3} \, dx = 2(\frac{3}{10})[x^{10/3}]_0^1 = \frac{6}{10} = \frac{3}{5}$.

10. $\int (x^2+4)^2 dx = \int (x^4 + 8x^2 + 16) dx = \frac{x^5}{5} + \frac{8}{3}x^3 + 16x + C$.

13. Let $t = w^2 + 2w$, $dt = 2(w+1)dw$. When $w = 1$, $t = 3$ and when $w = 2$, $t = 8$. Thus $\int_1^2 \frac{w+1}{\sqrt{w^2+2w}} \, dw = \frac{1}{2} \int_3^8 \frac{1}{\sqrt{t}} \, dt = \frac{1}{2} \int_3^8 t^{-1/2} dt = t^{1/2}]_3^8 = \sqrt{8} - \sqrt{3} \approx 2.828 - 1.732 \approx 1.10$.

16. Noting that $(x^2-x-6)/(x+2) = x-3$ if $x \neq -2$, $\int_1^2 \frac{x^2-x-6}{x+2} \, dx = \int_1^2 (x-3) dx = \frac{x^2}{2} - 3x]_1^2 = -3/2$.

19. Let $u = x^3+1$, $du = 3x^2 dx$. When $x = 0$, $u = 1$, and when $x = 2$, $u = 9$. Then
$$\int_0^2 x^2\sqrt{x^3+1} \, dx = \frac{1}{3} \int_1^9 u^{1/2} \, du = \frac{2}{9}u^{3/2}]_1^9 = \frac{2}{9}[9^{3/2} - 1] = \frac{52}{9}.$$

22. Let $u = 1 - v^{-1}$, $du = v^{-2}dv$ and $\int \frac{\sqrt[4]{1-v^{-1}}}{v^2} \, dv = \int \sqrt[4]{u} \, du = \frac{4}{5}u^{5/4} + C = \frac{4}{5}(1 - v^{-1})^{5/4} + C$.

25. Since an antiderivative of $D_x f(x)$ is $f(x)$ itself, $\int_{-1}^{2} D_x \sqrt{\dfrac{x+1}{x+2}}\, dx =$

$\sqrt{\dfrac{x+1}{x+2}}\,\Big]_{-1}^{2} = \sqrt{\dfrac{3}{4}} - \sqrt{\dfrac{0}{2}} = \dfrac{\sqrt{3}}{2}$.

28. $\displaystyle\int_{0}^{7} D_x \left(\dfrac{x^2}{\sqrt{3x+4}}\right) dx = \dfrac{x^2}{\sqrt{3x+4}}\,\Big]_{0}^{7} = \dfrac{49}{\sqrt{25}} - 0 = \dfrac{49}{5}$.

31. (a) Tabulating our work as before with $f(x) = \sqrt{1+x^4}$ and $(b-a)/n = 10/5 = 2$, we have:

k	x_k	$f(x_k)$	m	$mf(x_k)$
0	0	1.0000	1	1.0000
1	2	4.1231	2	8.2462
2	4	16.0312	2	32.0624
3	6	36.0139	2	72.0278
4	8	64.0078	2	128.0156
5	10	100.0050	1	100.0050
				341.3570

Since $(b-a)/2n = 1$, T = 341.36.

(b) Now, $(b-a)/n = 10/8 = 1.25$.

k	x_k	$f(x_k)$	r	$rf(x_k)$
0	0.00	1.0000	1	1.0000
1	1.25	1.8551	4	7.4204
2	2.50	6.3295	2	12.6590
3	3.75	14.0980	4	56.3920
4	5.00	25.0200	2	50.0400
5	6.25	39.0753	4	156.3012
6	7.50	56.2589	2	112.5178
7	8.75	76.5690	4	306.2760
8	10.00	100.0050	1	100.0050
				802.6114

Since $(b-a)/3n = 10/24$, S = 8026.114/24 $\approx$ 334.42.

CHAPTER 6

APPLICATIONS OF THE DEFINITE INTEGRAL

NOTE: In addition to presenting the solutions of #1, 4, 7, etc., I have included the integral forms of the answers to the remaining odd-numbered problems. In that way, you can determine if the reason for not getting the answer in the back of the text is due to incorrect set-up (i.e. the wrong integral) or to an arithmetic or algebraic mistake in evaluating the correct integral.

EXERCISES 6.1

1. The upper boundary of the region is $y = 1/x^2$; the lower is $y = -x^2$. (See sketches in the text for odd-numbered problems.) Thus $A = \int_1^2 [\frac{1}{x^2} - (-x^2)]\,dx$

$= \int_1^2 (x^{-2} + x^2)\,dx = -x^{-1} + x^3/3 \ ]_1^2 = 17/6.$

3. $A = \int_{-1}^2 ((4+y) - (-y^2))\,dy$.

4. Because of the shape and boundaries of the region, the easiest method is to use y as independent variable.

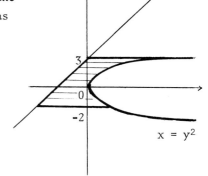

$A = \int_{-2}^3 [y^2 - (y-2)]\,dy$

$= \frac{y^3}{3} - \frac{y^2}{2} + 2y \ ]_{-2}^3$

$= \frac{21}{2} + \frac{26}{3} = \frac{115}{6}$.

5. $A = \int_{-2}^2 (5 - (x^2+1))\,dx$.

7. For the points of intersection, we solve $x^2 = 4x \iff x^2 - 4x = x(x-4) = 0 \iff$ $x = 0,4$. Since $4x \geq x^2$ on $[0,4]$, $A = \int_0^4 (4x - x^2)\,dx = 2x^2 - \frac{x^3}{3} \ ]_0^4 = 32 - \frac{64}{3} = \frac{32}{3}.$

9. $A = \int_{-2}^1 ((1 - x^2) - (x-1))\,dx$.

10. The curves intersect when $3 - x^2 = -x+3 \iff x = 0,1.$

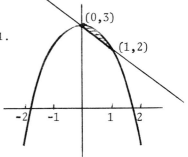

$A = \int_0^1 [(3-x^2) - (-x+3)]\,dx = \int_0^1 (-x^2+x)\,dx =$

$-\frac{x^3}{3} + \frac{x^2}{2} \ ]_0^1 = \frac{1}{6}$.

11. $A = \int_{-\sqrt{3}}^{\sqrt{3}} ((2 - y^2) - (y^2 - 4))\,dy.$

13. The lines $y = x$ and $y = 3x$ intersect at $(0,0)$; $y = x$ and $x+y = 4$ intersect at $(2,2)$; $y = 3x$ and $x+y = 4$ intersect at $(1,3)$. The region must be subdivided no matter which variable is chosen as independent. For $0 \le x \le 1$, the upper boundary is $y = 3x$; the lower is $y = x$. For $1 \le x \le 2$, the upper boundary is $x+y = 4$, the lower is $y = x$.

$$A = \int_0^1 (3x-x)\,dx + \int_1^2 [(-x+4)-x]\,dx = \int_0^1 2x\,dx + \int_1^2 (4-2x)\,dx = x^2]_0^1 + 4x-x^2]_1^2$$

$$= 1 + [(8-4)-(4-1)] = 1 + 4 - 3 = 2.$$

15. $A = \int_{-1}^{0} (x^3 - x)\,dx - \int_0^1 (x^3 - x)\,dx.$

16. Writing $y = x(x^2-x-6) = x(x-3)(x+2)$, we see that $y \ge 0$ if $-2 \le x \le 0$ and $y \le 0$ if $0 \le x \le 3$. Thus, using (6.1) on $[0,3]$, with $f(x) = 0$ and $g(x) = x^3 - x^2 - 6x$

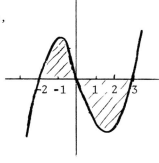

$$A = \int_{-2}^{0} (x^3-x^2-6x)\,dx - \int_0^3 (x^3-x^2-6x)\,dx$$

$$= [\frac{x^4}{4} - \frac{x^3}{3} - 3x^2]_{-2}^{0} + [-\frac{x^4}{4} + \frac{x^3}{3} + 3x^2]_0^3$$

$$= \frac{64}{12} + \frac{63}{4} = \frac{253}{12}.$$

(Note that if you simply integrated y from -2 to 3 without the initial analysis, the "area" would have worked out to be $-125/12$, negative!)

17. $A = -\int_{-2}^{0} (4y - y^3)\,dy + \int_0^2 (4y - y^3)\,dy.$

19. $y = x\sqrt{4-x^2}$ is ≤ 0 if $-2 \le x \le 0$ and is ≥ 0 if $0 \le x \le 2$. Thus

$$A = \int_{-2}^{0} -x\sqrt{4-x^2}\,dx + \int_0^2 x\sqrt{4-x^2}\,dx = \frac{1}{3}(4-x^2)^{3/2}]_{-2}^{0} - \frac{1}{3}(4-x^2)^{3/2}]_0^2 = \frac{2}{3} \cdot 4^{3/2}$$

$= \frac{16}{3}$. (By symmetry, the area of the right half could have been computed and then doubled to get the answer.)

22. $x = \sqrt{1-y^2} \implies x \ge 0$ and $x^2+y^2 = 1 \implies R$ is semicircular with radius $1 \implies A = \pi/2$. If P is a partition of $[-1,1]$, $A = \lim_{\|P\| \to 0} \sum_{k=1}^{n} \sqrt{1-w_k^2}\,\Delta y_k$ where $w_k \in [y_{k-1}, y_k]$.

25. Δx_k in the sum tells us that x is the independent variable, and $f(w_k) =$
$4w_k + 1 \Longrightarrow f(x) = 4x+1$. R is the trapezoidal region under the line y =
$4x+1$, from x = 0 to x = 1. $A = \int_0^1 (4x+1)dx = 2x^2+x]_0^1 = 3$.

28. Δy_k in the sum $\Longrightarrow$ y is the independent variable, and $f(w_k) = \sqrt{3w_k+1} \Longrightarrow f(y)$
$= \sqrt{3y+1}$. R is the region to the right of the y axis and to the left of the
graph of $x = \sqrt{3y+1}$ from y = 0 to y = 1. $A = \int_0^1 \sqrt{3y+1} \, dy$. Let u = 3y+1, du
$= 3 \, dy$. $y = 0 \Longrightarrow u = 1$, $y = 1 \Longrightarrow u = 4$, and $A = \frac{1}{3} \int_1^4 u^{1/2}du = \frac{1}{3} \cdot \frac{2}{3}u^{3/2}]_1^4$
$= \frac{2}{9}(4^{3/2} - 1^{3/2}) = \frac{2}{9}(8-1) = \frac{14}{9}$.

31. $\Delta y_k \Longrightarrow$ y is the independent variable and $f(y) = (5+\sqrt{y})/\sqrt{y} = 5y^{-1/2} + 1$,
$1 \le y \le 4$. $A = \int_1^4 (5y^{-1/2}+1)dy = 10y^{1/2} + y]_1^4 = 24 - 11 = 13$.

33. $A = \int_0^1 [(6-3x^2)-3x]dx + \int_1^2 [3x-(6-3x^2)]dx$.

34. The graphs intersect when $x^2-4 = x+2 \Longleftrightarrow x^2-x-6 = (x+2)(x-3) = 0 \Longrightarrow x =$
$-2,3$. Only x = 3 is in [1,4]. When it is not obvious from the functions f
and g, we can test values (as in Section 4.3) to see which function is larger
on the interval in question. Thus, selecting k = 1 in [1,3] we find that
$f(1) = -3 < g(1) = 3$. Thus $f(x) \le g(x)$ on [1,3]. Similarly, $f(4) = 12 >$
$g(4) = 6 \Longrightarrow f(x) \ge g(x)$ on [3,4]. (Recall f(3) = g(3) so that our interest
is focused on the two intervals [1,3] and [3,4].). Thus $A = \int_1^3 (g(x)-f(x))dx$
$+ \int_3^4 (f(x)-g(x))dx = \int_1^3 (x-x^2+6)dx + \int_3^4 (-x+x^2-6)dx = \frac{x^2}{2} - \frac{x^3}{3} + 6x]_1^3 +$
$[- \frac{x^2}{2} + \frac{x^3}{3} - 6x]_3^4 = \frac{22}{3} + \frac{17}{6} = \frac{61}{6}$.

37. In the table below, $f(x_k)$ denotes the ordinate value of the upper boundary
at x_k, $g(x_k)$ is that of the lower, and $h(x_k)$ is the difference, $h(x_k) =$
$f(x_k) - g(x_k)$ as the author has measured them. (Your answers may differ.)

k	x_k	$f(x_k)$	$g(x_k)$	$h(x_k)$	m	$mh(x_k)$	r	$rh(x_k)$
0	0	0	0	0	1	0	1	0
1	1	1.5	1.0	0.5	2	1.0	4	2.0
2	2	2.0	1.25	0.75	2	1.5	2	1.5
3	3	2.5	1.5	1.0	2	2.0	4	4.0
4	4	3.0	2.0	1.0	2	2.0	2	2.0
5	5	3.5	2.5	1.0	2	2.0	4	4.0
6	6	5.0	5.0	0	1	0	1	0
						8.5		13.5

(a) $(b-a)/2n = 6/12 = 1/2 \implies T = 4.25$.

(b) $(b-a)/3n = 6/18 = 1/3 \implies S = 4.5$.

40. $(b - a)/3n = 80/24 = 10/3$. Using the functional values as your author has measured them, $\int_0^{80} p(x)dx \approx (10/3)[0 + 4(10) + 2(27) + 4(30) + 2(20) + 4(15) + 2(10) + 4(5) + 0] = (10/3)(354) = 1180$.

EXERCISES 6.2

NOTE: When using the disc or washer method to compute the volume of a solid of revolution, you must remember that the partitioning must be done so that a typical rectangle is _perpendicular_ to the line around which the revolution is taking place (the axis of revolution). Thus, if revolution takes place around a vertical line, a typical rectangle must be horizontal so that a y-interval must be partitioned and the bounding curves must be graphs of functions of y. Conversely, if we re-volve about a horizontal line, the rectangle must be vertical so that an x-inter-val must be partitioned and the bounding curves expressed as graphs of functions of x.

1. By (6.3), $V = \pi \int_1^3 (1/x)^2 dx = \pi \int_1^3 x^{-2} dx = \pi[-\frac{1}{x}]_1^3 = \frac{2\pi}{3}$.

3. $V = \pi \int_0^2 y \, dy$.

4. Writing the equation as $x = 1/y$, $1 \leq y \leq 3$, we have by

   (6.4), $V = \pi \int_1^3 (1/y)^2 \, dy = 2\pi/3$.

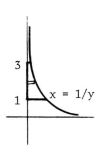

5. $V = \pi \int_0^4 (x^2 - 4x)^2 \, dx$.

7. The curves intersect when $2y = y^2$ or $y = 0,2$. For $0 \le y \le 2$, the right boundary is $x = 2y$; the left is $x = y^2$. Thus a typical rectangle, corresponding to the subinterval $[y_{k-1}, y_k]$ in a partition of $0 \le y \le 2$, sweeps out a "washer" of volume $[\pi(2w_k)^2 - \pi(w_k^2)^2]\Delta y_k$. Summing and passing to the limit as $\|P\| \to 0$, we get

$$V = \pi \int_0^2 [(2y)^2 - (y^2)^2]\,dy = \pi[\tfrac{4}{3}y^3 - \tfrac{1}{5}y^5]\Big|_0^2 = \frac{64\pi}{15}.$$

9. $V = \pi \displaystyle\int_{-\sqrt{2}}^{\sqrt{2}} ((4 - x^2)^2 - x^4)\,dx.$

10. The curves intersect when $x^{1/3} = -x^2 \iff$ $x = -x^6 \iff x(x^5+1) = 0$ yielding $x = -1, 0$. Formula (6.7) cannot be used here since the functions are both negative for $-1 \le x \le 0$. The typical rectangle shown sweeps out a washer with outer radius $-w_k^{1/3}$ and inner radius w_k^2. (As distances, these radii must be ≥ 0.) The washer's volume is

$\pi[(-w_k^{1/3})^2 - (w_k^2)^2]\Delta x_k$. Summing and passing to the limit we obtain

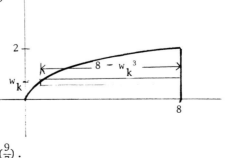

$y = x^{1/3}$

$y = x^2$

$(-1,-1)$

$$V = \int_{-1}^0 (x^{2/3} - x^4)\,dx = \pi[\tfrac{3}{5}x^{5/3} - \tfrac{1}{5}x^5]\Big|_{-1}^0 = \frac{2\pi}{5}.$$

11. $V = \pi \displaystyle\int_{-1}^2 ((y + 2)^2 - y^4)\,dy.$

13. $x = y^3$, $x = 8 \implies y = 2$. Partition the interval $0 \le y \le 2$. The typical rectangle shown sweeps out a disc (when revolved around $x = 8$) of radius $8 - w_k^3$, thickness Δy_k and volume

$\pi(8-w_k^3)^2\Delta y_k$. Thus $V \approx \displaystyle\sum_{k=1}^n \pi(8-w_k^3)^2\Delta y_k$. Passing to the limit as $\|P\| \to 0$, $V =$

$$\pi\int_0^2 (8-y^3)^2\,dy = \pi\int_0^2 (64-16y^3+y^6)\,dy =$$

$$\pi[64y - 4y^4 + \tfrac{1}{7}y^7]\Big|_0^2 = \pi[128 - 64 + \frac{128}{7}] = 64\pi(\tfrac{9}{7}).$$

15. $V = \pi \int_0^8 [3^2 - (3 - x^{1/3})^2]dx.$

16.   Since $x = 2$ is vertical, we must partition the in-
      terval $0 \leq y \leq 1$ and express the boundary with x
      as a function of y. Thus, the left boundary is
      $x = -\sqrt[4]{y}$, and the right boundary is $x = \sqrt[4]{y}$. The
      rectangle shown sweeps out a washer of outer
      radius $2-(-\sqrt[4]{w_k}) = 2+\sqrt[4]{w_k}$, inner radius $2-\sqrt[4]{w_k}$, and

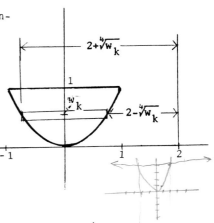

      volume $\pi[(2+\sqrt[4]{w_k})^2 - (2-\sqrt[4]{w_k})^2]\Delta y_k = 8\pi\sqrt[4]{w_k}\Delta y_k.$
      Summing and passing to the limit we get
      $V = 8\pi \int_0^1 y^{1/4}dy = 8\pi \cdot \frac{4}{5} y^{5/4}]_0^1 = \frac{32\pi}{5}.$

17.   (a) $V = \pi \int_{-2}^2 (4-x^2)^2 dx.$ (b) $V = \pi \int_{-2}^2 [(5-x^2)^2 - 1]dx.$ (c) $\pi \int_0^4 [(2+\sqrt{y})^2 -$
      $(2-\sqrt{y})^2]dy.$

19.   The curves intersect when $x^3 = 4x \Longrightarrow x(x^2-4)$
      $= 0 \Longrightarrow x = -2,0,2.$ If $-2 \leq w_k \leq 0$, the
      rectangle sweeps out a washer of outer radius
      $8 - 4w_k$, inner radius $8 - w_k^3$ (both positive
      since $w_k < 0$) and volume $\pi[(8 - 4w_k)^2 -$
      $(8 - w_k^3)^2]\Delta x_k.$ If $0 \leq w_k \leq 2$, the outer
      and inner radii are reversed from the 1st
      case. Thus, the washer volume is
      $\pi[(8 - w_k^3)^2 - (8 - 4w_k)^2]\Delta x_k.$ Summing and
      passing to the limit yields the text answer.

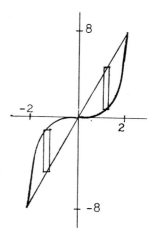

22.   Writing $x-y = 1$ as $y = x-1$, the curves intersect when $1-x^2 = x-1 \Longrightarrow x^2+x-2$
      $= (x+2)(x-1) = 0 \Longrightarrow x = -2,1.$ On the interval $-2 \leq x \leq 1$, $x-1 \leq 1-x^2 < 3.$
      Thus, when revolving a rectangle about $y = 3$, the outer radius is $3 - (w_k-1)$
      $= 4-w_k.$ The inner radius is $3 - (1-w_k^2) = 2 + w_k^2$, and its volume is
      $\pi[(4-w_k)^2 - (2+w_k^2)^2]\Delta x_k.$ As before, this yields $V = \pi \int_{-2}^1 [(4-x)^2 -$
      $(2+x^2)^2]dx.$

25.   Revolve the triangle with vertices $(0,0)$, $(r,0)$ and $(0,h)$ about the y-axis
      using $x = (-r/h)(y-h)$ to obtain $V = \pi \int_0^h \frac{r^2(y-h)^2}{h^2} dy.$

28. Revolve the region between the circle $x = \sqrt{r^2 - y^2}$ and the y-axis, for r-h

    $\le y \le r$, about the y-axis to obtain $V = \pi \int_{r-h}^{r} (r^2-y^2)dy = (\pi h^2/3)(3r-h)$.

31. (b $-$ a)/2n = 6/12 = 1/2. $v = \pi \int_0^6 f^2(x)dx \approx \frac{\pi}{2}[1(2)^2 + 2(1)^2 + 2(2)^2 + 2(4)^2 + 2(2)^2 + 2(2)^2 + 1(1)^2] = \frac{\pi}{2}(63) = 31.5\pi$.

<u>EXERCISES 6.3</u>

NOTE: When using the shell method to compute the volume of a solid of revolution, the partitioning must be done so that a typical rectangle's altitude is <u>parallel</u> to the axis of revolution. Thus, revolution about a vertical line means that an x-interval must be partitioned, the altitude of the rectangle is vertical, and the bounding curves must be expressed as functions of x. Similarly, revolution about a horizontal line means a y-interval must be partitioned and the bounding curves expressed as functions of y.

1. $V = 2\pi \int_0^4 x\sqrt{x}\ dx = 2\pi \int_0^4 x^{3/2}dx = 2\pi(\frac{2}{5})x^{5/2}]_0^4 = \frac{4\pi}{5}(2^5) = \frac{128\pi}{5}$.

3. $V = 2\pi \int_0^2 x(\sqrt{8x} - x^2)dx$.

4. $x^2-5x = x(x-5) \le 0$ on [0,5] since $x \ge 0$, $x-5 \le 0$ there. So, with rotation about the y-axis (vertical), we partition this x interval. The shell swept out by the rectangle shown has altitude $0 - (w_k^2 - 5w_k) = 5w_k - w_k^2$, average radius w_k and thickness Δx_k. Its volume is $2\pi w_k(5w_k-w_k^2)\Delta x_k$. Summing, and passing to the limit as $\|P\| \to 0$ we obtain

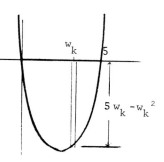

    $V = 2\pi \int_0^5 x(5x-x^2)dx = 2\pi \int_0^5 (5x^2-x^3)dx = 2\pi[\frac{5}{3}x^3 - \frac{x^4}{4}]_0^5 = 2\pi[\frac{625}{3} - \frac{625}{4}] =$

    $\frac{1250\pi}{12} = \frac{625\pi}{6}$.

5. $V = 2\pi \int_4^7 x[(\frac{x}{2} - \frac{3}{2}) - (2x - 12)]dx$.

7. To go around the x-axis and use the shell method, we partition the y-interval, $0 \le y \le 4$. Solving $x^2 = 4y$ for x, we find that the left boundary is $x = -2\sqrt{y}$, and the right boundary is $x = 2\sqrt{y}$. The typical rectangle is horizontal with altitude = (right boundary) - (left boundary) = $2\sqrt{w_k} - (-2\sqrt{w_k}) =$

$4\sqrt{w_k}$. The shell has average radius w_k, thickness Δy_k, and volume

$2\pi w_k(4\sqrt{w_k})\Delta y_k = 8\pi w_k^{3/2}\Delta y_k$. Summing and passing to the limit, we obtain

$$V = 8\pi \int_0^4 y^{3/2}dy = \frac{16\pi}{5} y^{5/2}\Big]_0^4 = \frac{16\pi}{5} \quad (32).$$

9. $V = 2\pi \int_0^6 y(y/2)\,dy.$

10. $2y = x$ intersects $x = 1$ at $y = 1/2$. The right
boundary is $x = 2y$; the left is $x = 1$. Thus,
the altitude is $2w_k - 1$, radius w_k, thickness
Δy_k. The shell volume is $2\pi w_k(2w_k-1)\Delta y_k$. Thus,
as above, $V = 2\pi \int_{1/2}^4 y(2y-1)\,dy = 2\pi[\frac{2}{3}y^3 - \frac{y^2}{2}]_{1/2}^4$

$= 2\pi(\frac{104}{3} + \frac{1}{24}) = \frac{833\pi}{12}$.

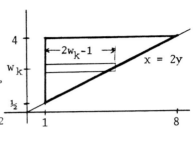

11. $V = 2\pi \int_1^4 (5-y)2y^{-1/2}dy.$

13. (a) Partitioning $[0,2]$, the typical rectangle
shown sweeps out a cylindrical shell of average
radius $3-w_k$, height w_k^2+1, thickness Δx_k, and
volume $2\pi(3-w_k)(w_k^2+1)\Delta x_k$. Summing and passing
to the limit, $V = 2\pi \int_0^2 (3-x)(x^2+1)\,dx = 16\pi$.

(b) Now, the average radius is w_k+1 and, as
above, $V = 2\pi \int_0^2 (x+1)(x^2+1)\,dx = 64\pi/3$.

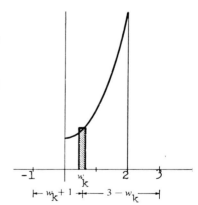

16. (a) Partitioning the x interval $[0,4]$,
the shell swept out by the rectangle has
average radius $4-w_k$, height $\sqrt{w_k}$ and
volume $2\pi(4-w_k)\sqrt{w_k}\,\Delta x_k$. As before,

$V = 2\pi \int_0^4 (4-x)\sqrt{x}\,dx = 256\pi/15.$

(b) Now the average radius is $6-w_k$ and

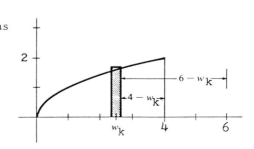

$$V = 2\pi \int_0^4 (6-x)\sqrt{x}\, dx = 192\pi/5.$$ (c) To revolve about $y = 2$ using the shell

method, the equation of the graph must be written as $x = y^2$, and the y

interval $[0,2]$ must be partitioned. Now the typical rectangle is horizon-

tal and sweeps out a shell of average radius $2-w_k$, height $4-w_k^2$, and volume

$2(2-w_k)(4-w_k^2)\Delta y_k$, so that $V = 2\pi \int_0^2 (2-y)(4-y^2)\, dy = 40\pi/3.$

19. Writing the equations as $y = 3-x$ and $y = 3-x^2$, the curves intersect when

$3-x = 3-x^2 \implies x^2-x = x(x-1) = 0 \implies x = 0,1.$ On the interval $[0,1]$, $3-x^2 \geq$

$3-x$. (Use test value $k = 1/2$ to verify this.) Thus when the interval $0 \leq x$

≤ 1 is partitioned, the typical rectangle, when revolved about $x = 2$, sweeps

out a shell of average radius $2-w_k$, height $(3-w_k^2) - (3-w_k) = w_k - w_k^2$,

thickness Δx_k and volume $2\pi(2-w_k)(w_k-w_k^2)\Delta x_k$. Summing and passage to the

limit yields the answer in the text.

22. Because the region is symmetric about the y-axis,
we will consider only the 1st quadrant region,
doubling the result for the entire volume. To use
the shell method when revolving around a horizontal
line ($y = -1$ here), y must be the independent
variable so that the typical rectangle is horizontal. Expression the equations as $x = y^{3/2}$ and
$x = y^{1/2}$, the curves intersect when $y^{3/2} = y^{1/2}$,

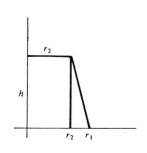

$y \geq 0 \implies y^3 = y$, $y \geq 0 \implies y = 0,1$. Note that

$y^{1/2} \geq y^{3/2}$ on $0 \leq y \leq 1$. Partitioning $0 \leq y \leq 1$, the rectangle sweeps out

a shell of radius $1+w_k$, height $= w_k^{1/2} - w_k^{3/2}$, thickness Δy_k, and volume

$2\pi(1+w_k)(w_k^{1/2} - w_k^{3/2})\Delta y_k$. As usual, and inserting the factor of 2 as de-

scribed in the first line, $V = (2)2\pi \int_0^1 (1+y)(y^{1/2}-y^{3/2})\, dy.$

25. Revolve the trapezoid with vertices $(0,0)$, $(0,h)$,
(r_2,h) and $(r_1,0)$ about the y-axis. The upper

boundary is $y = h$ for $0 \leq x \leq r_2$ and $y =$

$(h/(r_2 - r_1))(x-r_1)$ for $r_2 \leq x \leq r_1$. Thus

$$V = 2\pi \int_0^{r_2} xh\, dx + 2\pi \int_{r_2}^{r_1} x(h/(r_2-r_1))(x-r_1))dx.$$

28. The Δx_k tells us that $0 \le x \le 1$ is being partitioned. The term

$2\pi(w_k{}^5 + w_k{}^{3/2})\Delta x_k = 2\pi w_k[w_k{}^4 - (-w_k{}^{1/2})]\Delta x_k$ is the volume of a cylin-

drical shell with average radius w_k, height $w_k{}^4 - (-w_k{}^{1/2})$ and thickness Δx_k.

Thus the limit is the volume of the solid obtained by revolving the region

bounded by the graphs of $y = x^4$, $y = -\sqrt{x}$ and $x = 1$ about the y-axis.

$V = 2\pi \int_0^1 (x^5 + x^{3/2})dx = 17\pi/15$. (Note that other answers are possible.

For example, writing the term in the sum as $2\pi w_k(w_k{}^4 + \sqrt{w_k})\Delta x_k$, we can cor-

rectly interpret the solid as having been obtained by revolving the region

under the graph of $f(x) = x^4 + \sqrt{x}$, $0 \le x \le 1$, about the y axis.)

EXERCISES 6.4

1. If (x,y) is a point on the circle for which $y \ge 0$ then, because of the sym-
    metry of the circle, the side of the cross-sectional square has length $2y$,
    and the area of the square is $A = 4y^2$. Since $x^2 + y^2 = a^2$, $A = 4(a^2 - x^2)$

    and $V = \int_{-a}^a 4(a^2 - x^2)dx = 4[a^2x - \frac{x^3}{3}]_{-a}^a = \frac{16a^3}{3}$.

3. $V = \int_{-2}^2 [(4-x^2)/2]^2 dx$.

4. If (x,y) is on the curve $y = x^2$, the cross-sectional
    square has side length $(4-y) = 4-x^2$ and area
    $A = (4-x^2)^2$. Thus

    $V = \int_{-2}^2 (4-x^2)^2 dx = \int_{-2}^2 (16 - 8x^2 + x^4)dx$

    $= [16x - \frac{8}{3}x^3 + \frac{1}{5}x^5]_{-2}^2 = 2[32 - \frac{64}{3} + \frac{32}{5}] = \frac{512}{15}$.

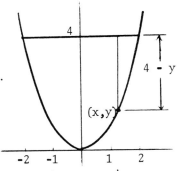

5. $V = \int_0^h (2a^2x^2/h^2)dx$.

7. If (x,y) is on $y^2 = 4x$, the base (diameter) of the
    cross-sectional semi-circle has length $4-x =$
    $4 - y^2/4$. The radius is $(4-y^2/4)/2$ and the area of
    the semi-circle is $A(y) = \frac{\pi}{8}(4 - \frac{y^2}{4})^2$ and

    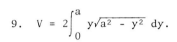

$$V = \frac{\pi}{8} \int_{-4}^{4} (4 - \frac{y^2}{4})^2 dy = \frac{\pi}{8} \int_{-4}^{4} (16 - 2y^2 + \frac{y^4}{16}) dy$$

$$= \frac{\pi}{8}[16y - \frac{2}{3}y^3 + \frac{y^5}{80}]_{-4}^{4} = \frac{128\pi}{15}.$$

9. $V = 2\int_{0}^{a} y\sqrt{a^2 - y^2} \, dy.$

10. Position the x-axis with the origin at
     the point of intersection of the cylin-
     drical axes and perpendicular to the
     plane they lie in. The sketch depicts
     only 1/8 of the entire solid, the nearest
     1/4 of the top half.

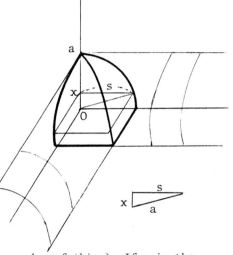

     Restricting our attention to this por-
     tion, we see that for a number $x \epsilon [0,a]$,
     the plane through x perpendicular to the
     x-axis cuts the solid in a square. (The
     facts that the edges of the cut are on
     the cylindrical surface parallel to the
     axis of the cylinder, and that the cylin-
     ders are perpendicular should convince the reader of this.) If s is the
     length of a side, then $s = \sqrt{a^2-x^2}$ (see insert in sketch). The area of the
     square is $A(x) = s^2 = a^2-x^2$ and the volume is $\int_{0}^{a} (a^2-x^2)dx = \frac{2}{3}a^3$. The
     entire solid therefore has volume $8(\frac{2}{3})a^3 = \frac{16a^3}{3}$.

11. $V = b\int_{-a}^{a} \sqrt{a^2-x^2} \, dx.$

13. We place a coordinate line along the 4 cm side with the origin at the vertex
     shown. (See sketches at end of solution.) Consider a point on this line x
     units from the vertex; a plane through this point, perpendicular to the coor-
     dinate line, intersects the solid in a triangular cross-section. Let b be
     the base and h the height of this triangle. We can express b and h in terms
     of x using similar triangles. From the middle sketch below, $b/x = 2/4$, or

b = x/2. From the last sketch, h/x = 3/4, or h = (3/4)x. The area of this

cross-sectional triangle is A(x) = (1/2)bh = 1/2 · x/2 · 3x/4 = $\frac{3}{16}$ x². Thus

$$V = \int_0^4 (3/16)x^2 \, dx = \frac{1}{16}x^3\Big]_0^4 = \frac{64}{16} = 4.$$

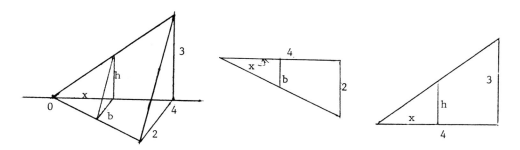

15. $V = \frac{\pi}{8} \int_0^a (a-y)^2 \, dy.$

16. Since a regular hexagon of side y is the union of
6 equilateral triangles of side y, the area of the
hexagon is 6 times the area of such a triangle, or

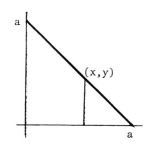

$6(\frac{\sqrt{3}}{4} y^2) = \frac{3\sqrt{3}}{2} y^2$. Positioning the base triangle

as shown, if (x,y) is on the hypotenuse, then

y = -x + a. Thus, the cross-sectional area func-

tion is $A(x) = \frac{3\sqrt{3}}{2} y^2 = \frac{3\sqrt{3}}{2}(-x+a)^2$. Thus,

$$V = \frac{3\sqrt{3}}{2} \int_0^a (-x+a)^2 \, dx = \frac{\sqrt{3}}{2} a^3.$$

EXERCISES 6.5

1. $8x^2 = 27y^3 \implies y = (2/3)x^{2/3} \implies y' = (4/9)x^{-1/3} = f'(x)$. Then $\sqrt{1 + f'(x)^2}$

$= \sqrt{1 + (16/81)x^{-2/3}} = \sqrt{x^{-2/3}(x^{2/3} + 16/81)} = x^{-1/3}\sqrt{x^{2/3} + 16/81}$, and $L_1^8 =$

$\int_1^8 x^{-1/3}\sqrt{x^{2/3} + 16/81} \, dx$. Let $u = x^{2/3} + 16/81$ so that $du = (2/3)x^{-1/3} \, dx$

or $x^{-1/3} \, dx = (3/2)du$. $x = 1 \implies u = 1 + 16/81 = A$. $x = 8 \implies u = 8^{2/3} +$

$16/81 = 4 + 16/81 = B$. Thus $L_1^8 = (3/2)\int_A^B u^{1/2} \, du = u^{3/2}\Big]_A^B = B^{3/2} - A^{3/2} =$

$(4 + 16/81)^{3/2} - (1 + 16/81)^{3/2} \approx 7.29.$

3. $L_1^4 = \int_1^4 \sqrt{1 + (9/4)x}\ dx.$

4. Since $y = 6x^{2/3} + 1$, the calculations are similar to those in #1. The only thing to watch out for is that $\sqrt{x^{2/3}} = -x^{1/3}$ in this problem since $-8 \le x \le -1 < 0$ here. Then $L_{-8}^{-1} = \int_{-8}^{-1} -x^{-1/3}\sqrt{x^{2/3} + 16}\ dx = 20^{3/2} - 17^{3/2}$ with the substitution $u = x^{2/3} + 16$. (See #10 also.)

5. $L_{2/3}^{8/3} = \int_{2/3}^{8/3} \sqrt{1 + (243/32)y}\ dy.$

7. $y' = \dfrac{x^2}{4} - \dfrac{1}{x^2} = f'(x)$. Then $1 + f'(x)^2 = 1 + (\dfrac{x^4}{16} - \dfrac{1}{2} + \dfrac{1}{x^4}) = \dfrac{x^4}{16} + \dfrac{1}{2} + \dfrac{1}{x^4} = (\dfrac{x^2}{4} + \dfrac{1}{x^2})^2$. Thus, $L_1^2 = \int_1^2 (\dfrac{x^2}{4} + x^{-2})dx = \dfrac{x^3}{12} - \dfrac{1}{x}\Big]_1^2 = \dfrac{1}{6} - (-\dfrac{11}{12}) = \dfrac{13}{12}$.

9. $L_1^2 = \int_1^2 (\dfrac{y^4}{6} + \dfrac{3}{2}y^{-4})dy$ (by (6.19)).

10. $x = g(y) = \dfrac{y^4}{16} + \dfrac{1}{2y^2} \implies g'(y) = \dfrac{y^3}{4} - \dfrac{1}{y^3}$. Then $1 + g'(y)^2 = 1 + (\dfrac{y^6}{16} - \dfrac{1}{2} + \dfrac{1}{y^6})$ $= \dfrac{y^6}{16} + \dfrac{1}{2} + \dfrac{1}{y^6} = (\dfrac{y^3}{4} + \dfrac{1}{y^3})^2$. Here, $-2 \le y \le -1 < 0$ so that $y^3 < 0$ and the expression in the parentheses is negative. Recalling that $\sqrt{a^2} = |a| = -a$ if $a < 0$, it follows that $\sqrt{1 + g'(y)^2} = -(\dfrac{y^3}{4} + \dfrac{1}{y^3})$. Thus, $L_{-2}^{-1} =$

$-\int_{-2}^{-1} (\dfrac{y^3}{4} + y^{-3})dy = -[\dfrac{y^4}{16} - \dfrac{1}{2y^2}]_{-2}^{-1} = -(-\dfrac{7}{16} - \dfrac{7}{8}) = \dfrac{21}{16}.$

13. Let the line $y = x$ and the given curve intersect when $x = a$. Solving, we find that $2a^{2/3} = 1$ so that $a = (1/2)^{3/2}$. By symmetry then, the desired length is $8\ L_a^1$.
Solving $x^{2/3} + y^{2/3} = 1$ for y, we get $y = (1-x^{2/3})^{3/2}$ and $y' = -(1-x^{2/3})^{1/2}/x^{1/3}$ and $\sqrt{1+y'^2} = \sqrt{1+(1-x^{2/3})/x^{2/3}}$ which simplifies to $x^{-1/3}$. Thus $8\ L_a^1 =$

$8\int_a^1 x^{-1/3}\ dx = 12(1 - a^{2/3}) = 12(1 - 1/2) = 6.$

NOTE. If we had tried to compute L_0^1 directly, we would have obtained $L_0^1 = \int_0^1 1/\sqrt[3]{x}\ dx$ which does not exist because the integrand becomes infinite as

$x \to 0^{+}$. This type of integral is called an "improper" integral and will be studied in Chapter 10.

16. $f(x) = x^{3/2} \Longrightarrow f'(x) = (3/2)x^{1/2} \Longrightarrow \sqrt{1 + f'(x)^2} = \sqrt{1 + 9x/4}$. Thus $s(x) =$

$\int_{1}^{x} \sqrt{1 + 9t/4} \; dt = \frac{8}{27} [(1 + 9t/4)^{3/2}]_{1}^{x} = \frac{8}{27} [(1 + \frac{9}{4}x)^{3/2} - (1 + \frac{9}{4})^{3/2}] =$

$\frac{1}{27} [(4 + 9x)^{3/2} - 13^{3/2}]$. $\Delta s = s(1.1) - s(1) = (\frac{1}{27})[13.9^{3/2} - 13^{3/2}] \approx$

$(51.82 - 46.86)/27 \approx 0.184$.

$ds = \sqrt{1 + f'(x)^2} \; \Delta x$ and with $x = 1$, $\Delta x = 0.1$, $ds = \sqrt{1 + 9/4} \; (.1) = \sqrt{13}(.1)/2$

≈ 0.180.

19. $y = f(x) = 2/x \Longrightarrow f'(x) = -2/x^2 \Longrightarrow L_1^2 = \int_{1}^{2} \sqrt{1 + 4/x^4} \; dx$. Let $g(x)$ denote

the integrand. Then with $a = 1$, $b = 2$, $n = 4$, $L_1^2 = \int_{1}^{2} g(x) dx \approx$

$\frac{2-1}{3(4)}[g(1) + 4g(5/4) + 2g(3/2) + 4g(7/4) + g(2)] = 0.0833[2.2361 + 4(1.6243)$

$+ 2(1.3380) + 4(1.1944) + 1.1180] = 0.0833(17.3047) = 1.4421 \approx 1.44$.

22. $f(x) = x^4/4$, $f'(x) = x^3 \Longrightarrow L_0^2 = \int_{0}^{2} \sqrt{1 + x^6} \; dx$. With $g(x)$ denoting the

integrand, with $a = 0$, $b = 2$, $n = 4$, $L_0^2 \approx \frac{2-0}{3(4)}[g(0) + 4g(0.5) + 2g(1) +$

$4g(1.5) + g(2)] \approx 0.1667[1 + 4(1.0078) + 2(1.4142) + 4(3.5200) + 8.0623] =$

$0.1667(30.0019) = 5.0003 \approx 5.00$.

25. $f'(x) = (8x^3 - 2x^{-3})/8 \Longrightarrow 1 + f'(x)^2 = 1 + (64x^6 - 32 + 4x^{-6})/64 =$

$(64 + 64x^6 - 32 + 4x^{-6})/64 = (64x^6 + 32 + 4x^{-6})/64 =$

$(8x^3 + 2x^{-3})^2/8^2 \Longrightarrow S = 2\pi \int_{1}^{2} \frac{2x^4 + x^{-2}}{8} \frac{8x^3 + 2x^{-3}}{8} dx =$

$(\pi/32) \int_{1}^{2} (16x^7 + 12x + 2x^{-5}) dx = (\pi/32)(2x^8 + 6x^2 - \frac{1}{2}x^{-4})]_{1}^{2} =$

$(\pi/32)[(512 + 24 - 1/32) - (2 + 6 - 1/2) = 16911\pi/1024 \approx 51.88$.

28. $g'(y) = 2y^{-1/2} \Longrightarrow 1 + g'(y)^2 = 1 + 4/y = (y + 4)/y \Longrightarrow S =$

$2\pi \int_{1}^{9} 4\sqrt{y}\sqrt{4 + y}/\sqrt{y} \; dy = 8\pi \int_{1}^{9} (4 + y)^{1/2} dy = \frac{16}{3}\pi (4 + y)^{3/2}]_{1}^{9} =$

$\frac{16}{3}\pi (13^{3/2} - 5^{3/2}) \approx 598$.

34. $f'(x) = x^{-2/3} \Longrightarrow \sqrt{1 + f'(x)^2} = \sqrt{1 + x^{-4/3}} = \sqrt{x^{4/3} + 1/x^{2/3}} \Longrightarrow S =$

$2\pi \int_{1}^{8} x\sqrt{x^{4/3} + 1/x^{2/3}} dx = 2\pi \int_{1}^{8} x^{1/3}(x^{4/3} + 1)^{1/2} dx$. Let $u =$

$x^{4/3} + 1$. Then $du = \frac{4}{3}x^{1/3}dx$, or $x^{1/3}dx = \frac{3}{4}du$. $x = 1$, $8 \Longrightarrow u =$

$2, 17$. Thus $S = 2\pi(\frac{3}{4})\int_{2}^{17} u^{1/2}du = \pi u^{3/2}]_{2}^{17} = \pi(17^{3/2} - 2^{3/2})$.

EXERCISES 6.6

1. If, as in Example 3, x is the number of units that the spring is stretched beyond its natural length, then $f(x) = kx$ and $f(1.5) = 8 \Rightarrow k(3/2) = 8 \Rightarrow k = 16/3$.

    (a) Here x ranges from 0 to 4 (14" is 4" beyond the natural length of 10").

    Thus $W = \int_0^4 (16/3)x \, dx = 128/3$.

    (b) Here, x ranges from 1 to 3 and $W = \int_1^3 (16/3)x \, dx = \frac{64}{3}$.

4. In the formula $f(x) = kx$, x is the amount that the spring has been stretched beyond its natural length. Thus if x_0 is the natural length, then we obtain amounts of stretch of $6-x_0$, $7-x_0$, and $8-x_0$ when the spring is stretched to lengths of 6, 7 and 8 cm., respectively. The data given in the problem, together with (6.10) and (6.11) yield: $\int_{6-x_0}^{7-x_0} kx \, dx = 60$ and $\int_{7-x_0}^{8-x_0} kx \, dx = 120$.

    We seek k and x_0. From the first relation we get: $(k/2)[(7-x_0)^2-(6-x_0)^2]$ $= 60 \Rightarrow k(13-2x_0) = 120 \Rightarrow 13k - 2kx_0 = 120$. From the second relation, we get: $(k/2)[(8-x_0)^2-(7-x_0)^2] = 120 \Rightarrow k(15-2x_0) = 240 \Rightarrow 15k - 2kx_0 = 240$. Subtracting from this equation the equation from the first relation, we get $2k = 120$ or $k = 60$. Using this in either equation, we get $x_0 = 11/2 = 5.5$ cm.

5. $W = 62.5 \int_0^3 8(3-y) \, dy$.

7. By (6.15) and Example 2, the work done in lifting the 1500 kg. elevator 3 m. is $1500g(3)$ joules, where $g \approx 9.81$ m/sec^2. To this we must add the work of lifting the cable. We place the y axis with the origin at the initial position of the bottom of the cable and $y = 3$ at the final position. The center of the winch is at $y = 4$. Partition [0,3] and let $w_k \in [y_{k-1}, y_k]$. When the cable bottom is at w_k, there are $4-w_k$ m still suspended with a mass of 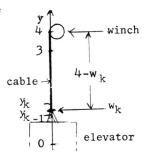 $7(4-w_k)$ kg. The work done in lifting the cable bottom from y_{k-1} to y_k is thus $\approx 7(4-w_k)g\Delta y_k$. The total work in lifting the cable bottom from $y = 0$ to $y = 3$ is approximately the sum of all such terms. In the limit as $\|P\|$

$\rightarrow 0$ we get $\int_0^3 7(4-y)g\, dy = 7g(12 - \frac{9}{2}) = \frac{105}{2}$ g. The total work in lifting

both cable and elevator is then $(\frac{105}{2} + 4500)$ g joules $\approx 4552.5(9.81)$ j

$= 44660$ j.

9. (a) $W = 62.5\pi(9/4) \int_0^6 (6-y)dy$. (b) $W = 62.5\pi(9/4) \int_0^6 (10-y)dy$.

10. Placing the axes as shown we partition the y-interval,
$0 \le y \le 3$, since the tank is only half filled. The kth
slice has volume $V_k = \pi(3/2)^2 \Delta y_k$ and mass $62.5\, V_k =$
$(562.5\pi/4)\Delta y_k$.

(a) The work required to lift this slice to the top is
(the distance lifted) $\cdot$ (mass) $\approx (6-w_k) \cdot$ (mass). The

total work is approximately the sum of these, and,
passing to the limit as $||P|| \to 0$ we obtain

$W = \int_0^3 (6-y)(562.5\pi/4)dy = (562.5\pi/4)(27/2) \approx 5,964$ ft.lbs.

(b) The only difference from (a) is that the slice must be lifted $(10-w_k)$

ft. Thus $W = \int_0^3 (10-y)(562.5\pi/4)dy = (562.5\pi/4)(51/2) \approx 11,266$ ft.lbs.

11. $W = \dfrac{16(62.5)}{\sqrt{3}} \int_0^{\sqrt{3}} y(\sqrt{3} - y)dy$.

13. Here $f(x) = c/d^2$ where c is a constant, and d is
the distance between the electrons.

(a) Partition $0 \le x \le 3$. At $w_k \,\varepsilon\, [x_{k-1}, x_k]$,

$d = 5-w_k$, $f(w_k) = c/(5-w_k)^2$. The work done

in moving the electron from x_{k-1} to x_k is approximately (force) $\cdot$ (distance)

$= \dfrac{c\Delta x_k}{(5-w_k)^2}$. The total work done in moving it from $x = 0$ to $x = 3$ is ap-

proximately the sum of all such terms, and passing to the limit as $||P|| \to$

0, we obtain $W = \int_0^3 \dfrac{c}{(5-x)^2}\, dx = \dfrac{c}{5-x}\Big]_0^3 = c(\frac{1}{2} - \frac{1}{5}) = \dfrac{3c}{10}$.

(b) Begin as in (a). Now, the net force,
f, on the electron at w_k is a sum $f_1 + f_2$
of the force f_1 exerted by the electron at

(-5,0) and the force f_2 exerted by the elec-

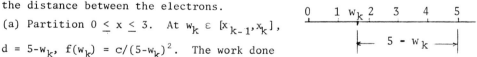

tron at (5,0). Note that f_1 and f_2 must be of opposite signs since f_2 opposes the motion (as in (a)), but f_1 assists it in repelling the moving electron. Thus $f(w_k) = \dfrac{-c}{(5+w_k)^2} + \dfrac{c}{(5-w_k)^2}$ and, as above,

$$W = \int_0^3 [-\frac{c}{(5+x)^2} + \frac{c}{(5-x)^2}]dx = c[\frac{1}{5+x} + \frac{1}{5-x}]_0^3 = \frac{9c}{40} .$$

15. $W = \displaystyle\int_0^{12} (24-y/6)dy.$

16. Since there were 20 lb. of water initially, the bucket will be half-filled when 10 lbs have leaked out. This will happen in 40 sec. since it leaks out at a rate of $0.25 = 1/4$ lb/sec. Since it rises at a rate of 1.5 ft/sec, it will rise $1.5(40) = 60$ ft while half the water leaks out. With $y = 0$ and $y = 60$ as initial and final locations of the bucket, we partition $0 \le y \le 60$ and let $w_k \varepsilon [y_{k-1}, y_k]$. At height w_k, the bucket will have been rising for $w_k/1.5 = 2w_k/3$ sec., and during this time $(0.25)(2w_k/3) = w_k/6$ lbs of water will have leaked out. Since the initial mass of bucket and water was 24 lb., the mass at w_k is thus $24 - w_k/6$ and the work in lifting it from y_{k-1} to y_k is $\approx (24-w_k/6)\Delta y_k$. Summing we obtain $W = \displaystyle\int_0^{60} (24-y/6)dy = 1440-300 = 1140$ ft.lb.

17. $W = 115 \displaystyle\int_{32}^{40} v^{-1.2} dv.$

19. If $K = gm_1m_2$ then $F(s) = Ks^{-2}$ and $W = K\displaystyle\int_{4000}^{4000+h} s^{-2}ds = -Ks^{-}]_{4000}^{4000+h}$

$= -K(\dfrac{1}{4000+h} - \dfrac{1}{4000}) = \dfrac{Kh}{(4000+h)(4000)} .$

22. $a = 1$, $b = 9$, $n = 8 \Longrightarrow (b-a)/2n = 1/2$. Thus $W = \displaystyle\int_1^9 f(x)dx \approx$

$\dfrac{1}{2}[125 + 2(120) + 2(130) + 2(146) + 2(165) + 2(157) + 2(150) + 2(143) + 140]^{-}$

$= \dfrac{2287}{2} = 1143.5$ joules.

EXERCISES 6.7

1. (a) Introduce a coordinate system with the origin at the lower left corner. Then $h(y) = 1 - y$, $L(y) = 1$, so that $F = \rho\int_0^1(1 - y)(1)dy = \rho/2$.   (b) Similar to (a) except that $L(y) = 3$. Then $F = \rho\int_0^1(1 - y)(3)dy = 3\rho/2$.

2. Hint: Let the diagonal be $y = x$. For the upper half, $L(y) = y$, and for the lower half, $L(y) = 1 - y$.

3. (a) $F = \rho \int_0^1 (1 - y)(2\sqrt{3})y\,dy.$ (b) $F = \rho \int_0^{1/2} (1/2 - y)(2\sqrt{3})y\,dy.$

4. The end of the trough and the
   coordinate axes are as shown. (The
   length is immaterial.) The upper
   vertices are at $x = \pm\sqrt{4 - h^2}$. Thus
   the right boundary is $y = (h/\sqrt{4 - h^2})x$
   or $x = (\sqrt{4 - h^2}/h)y$ and, similarly,
   the left-hand boundary is $x = -(\sqrt{4 - h^2}/h)y$. Thus $L(y) = $
   (right) − (left) $= (2\sqrt{4 - h^2}/h)y.$

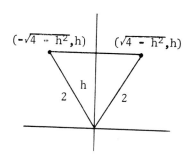

   (a) A full trough means that the surface is at $y = h$ and $h(y) = h - y$. Thus, $F = \rho \int_0^h (h - y)(2\sqrt{4 - h^2}/h)y\,dy = 2\rho(\sqrt{4 - h^2}/h)\int_0^h (h - y)y\,dy.$ The integral is $h^3/6$, and $F = \rho h^2 \sqrt{4 - h^2}/3.$
   (b) If half-full, $h(y) = h/2 - y$ and $2\rho(\sqrt{4 - h^2}/h)\int_0^{h/2}(h/2 - y)y\,dy = \rho h^2 \sqrt{4 - h^2}/24.$

5. $F = 60 \int_{-2}^{0} (0-y)\,2\sqrt{4-y^2}\,dy.$

7. Place the x-axis on a sloping edge of
   the bottom with $x = 0$ at the 3' end and
   $x = \sqrt{40^2 + 6^2} = \sqrt{1636}$ at the 9' end.
   Partition as usual and consider the
   rectangle shown. Above the point w_k,
   the depth d is 3 + L where, by similar

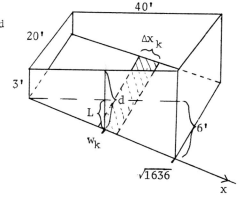

   triangles, $\dfrac{L}{6} = \dfrac{w_k}{\sqrt{1636}}$ so that $d = 3 + \dfrac{6w_k}{\sqrt{1636}}$. The force acting on this rec-

   tangle = (pressure) · (area) $\approx (\rho d)\cdot$

   $(20\Delta x_k)$. Summing over all rectangles and passing to the limit, the total

   force is $F = 20\rho \int_0^{\sqrt{1636}} (3 + \dfrac{6x}{\sqrt{1636}})dx = 120\sqrt{1636}\,\rho \approx 4853.7\rho.$ (Another way

   would be to place the axis on a horizontal edge. With a bit of work, the

   area at the bottom can be shown to be $\sqrt{1636}\Delta x_k/2$ and the depth at w_k is

   $3 + \dfrac{6w_k}{40}$.)

9. $F = \rho \int_0^4 (10-y)[(4-y/2)-(y/2-4)]dy$.   (The y-axis bisects the trapezoid for

   this form of the integral.)

10. Place the coordinate axes with the origin at the
    center of the circle, the water level at $y = 6$.
    The right and left boundaries of the circular
    plate are $f(y) = \sqrt{4-y^2}$ and $g(y) = -\sqrt{4-y^2}$, $L(y)=2\sqrt{4-y^2}$

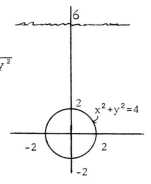

$F = 2\rho \int_{-2}^2 (6-y)\sqrt{4-y^2}\, dy$. To evaluate $\int_{-2}^2 \sqrt{4-y^2}\, dy$,

note that this is the area of half a circle of
radius 2, which is 2π. The other integral is

$\int_{-2}^2 y\sqrt{4-y^2}\, dy = -\frac{1}{3}(4-y^2)^{3/2}\Big]_{-2}^2 = 0$.   Thus $F = 24\pi\rho$.

11. $F = \rho \int_0^4 (4-y)\ 2\sqrt{y}\, dy$.

13. Place the y axis with $y = 0$ at the bottom of the plate.
    Then the top is at $y = 6$, and the surface is at $v = 8$.
    Partition $[0,6]$ and let $w_k \ \epsilon \ [y_{k-1}, y_k]$. Therefore the

    force on the typical rectangle shown is $50(8-w_k)(3\Delta y_k)$

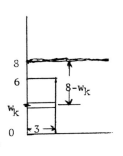

    $= 150(8-w_k)\Delta y_k$. Summing and passing to the limit as

    $\|P\| \to 0$ we get $F = 150 \int_0^6 (8-y)dy = 150(48-18) = 4500$ lbs.

14. HINT. Let the diagonal extend from the lower left to the upper right corner.
    For the left part of the plate, by similar triangles, the area of the typical
    rectangle is $(3/6)w_k\Delta y_k$.

EXERCISES 6.8

1. $m = 2 + 7 + 5 = 14$. $M_x = \sum_{k=1}^3 m_k y_k = 2(-1) + 7(0) + 5(-5) = -27$.

    $M_y = \sum_{k=1}^3 m_k x_k = 2(4) + 7(-2) + 5(-8) = -46$.

    $\bar{x} = \dfrac{M_y}{m} = -\dfrac{46}{14}$, $\bar{y} = \dfrac{M_x}{m} = -\dfrac{27}{14}$.

3. $m = \int_0^1 x^3 dx$, $M_x = (1/2)\int_0^1 x^6 dx$, $M_y = \int_0^1 x^4 dx$.

4. With $\delta = 1$, m = Area = $\int_0^9 x^{1/2} dx = \frac{2}{3}x^{3/2} \,]_0^9 = 18.$

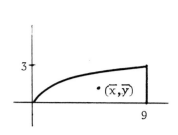

$$M_y = \int_0^9 x \cdot x^{1/2} dx = \frac{2}{5}x^{5/2}\,]_0^9 = \frac{486}{5}. \quad M_x =$$

$$\int_0^9 \frac{1}{2}(x^{1/2})^2 dx = \frac{x^2}{4}\,]_0^9 = \frac{81}{4}, \quad \bar{x} = \frac{486}{5(18)} = \frac{27}{5},$$

$$\bar{y} = \frac{81}{4(18)} = \frac{9}{8}.$$

5. $m = \delta \int_{-2}^2 (4 - x^2)\,dx.$

7. The curves intersect when $y^2 = 2y \Rightarrow y = 0,2 \Rightarrow x = 0,4.$ $y^2 = x$ $\Rightarrow y = \sqrt{x}$ in this region (see figure in answer section of text). Thus, $f(x) = \sqrt{x}$ and $g(x) = x/2$, $f(x) - g(x) = \sqrt{x} - x/2$. With the density = 1, $m = \int_0^4 (\sqrt{x} - x/2)dx = \frac{2}{3}x^{3/2} - \frac{x^2}{4}\,]_0^4 = \frac{4}{3}.$
$M_y = \int_0^4 x(\sqrt{x} - x/2)dx = \frac{2}{5}x^{5/2} - \frac{x^3}{6}\,]_0^4 = \frac{32}{15}.$
$M_x = \frac{1}{2}\int_0^4 (f^2(x) - g^2(x))dx = \frac{1}{2}\int_0^4 (x - \frac{x^2}{4})dx = \frac{1}{2}(\frac{x^2}{2} - \frac{x^3}{12})\,]_0^4 = \frac{4}{3}.$
$\bar{x} = \frac{32/15}{4/3} = \frac{8}{5}, \quad \bar{y} = \frac{4/3}{4/3} = 1.$

9. $m = \delta \int_{-2}^1 [(1 - x^2) - (x - 1)]dx$

10. Here $f(x) = 2 - x$, $g(x) = x^2$. The curves intersect when $x^2 = 2 - x \Rightarrow$ $x^2 + x - 2 = 0 \Rightarrow x = -2,1.$ Since $f(x) - g(x) = 2 - x - x^2$, $m =$ $\int_{-2}^1 (2 - x - x^2)dx = 9/2.$ $M_y =$ $\int_{-2}^1 x(2 - x - x^2)dx = -\frac{9}{4}.$ $M_x =$ $\frac{1}{2}\int_{-2}^1 (f^2(x) - g^2(x))dx =$

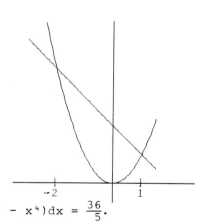

$\frac{1}{2}\int_{-2}^1 [(2 - x)^2 - x^4]dx = \frac{1}{2}\int_{-2}^1 (4 - 4x + x^2 - x^4)dx = \frac{36}{5}.$
$\bar{x} = \frac{-9/4}{9/2} = -\frac{1}{2}. \quad \bar{y} = \frac{36/5}{9/2} = \frac{8}{5}.$

11. $m = \delta \int_{-1}^2 (y + 2 - y^2)dy.$

13. With $\delta = 1$, m = Area = $\pi a^2/4$. Since the region is homogeneous (constant density) and symmetric about the line $y = x$, $\bar{x} = \bar{y}$ and we need calculate only M_x or M_y. Solving $x^2 + y^2 = a^2$ for y, we

obtain $y = \sqrt{a^2 - x^2}$, $0 \le x \le a$. $M_x = \frac{1}{2}\int_0^a (\sqrt{a^2 x^2})^2 dx =$

$\frac{1}{2}\int_0^a (a^2 - x^2) dx = \frac{a^3}{3}$. Thus $\bar{x} = \bar{y} = \frac{a^3/3}{\pi a^2/4} = \frac{4a}{3\pi}$. ($M_y = \int_0^a x\sqrt{a^2 - x^2}\, dx$

can be evaluated using the substitution $u = a^2 - x^2$.)

16. The described region may be considered as
being formed by cutting a semicircular region of radius a from a semicircular region of radius b, both centered at the
origin and lying above the x-axis.

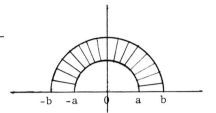

Thus $m = A = \frac{1}{2}\pi b^2 - \frac{1}{2}\pi a^2 = \pi(b^2 - a^2)/2$.

Because of symmetry about the y-axis, $\bar{x} = 0$.

To compute $\bar{y}$, we need M_x. Let M_{xa} and M_{xb} denote the moments of the semi-

circular regions of radii a and b, respectively, with respect to the x-axis.

By Example 5, $M_{xa} = \frac{2}{3}a^3$, $M_{xb} = \frac{2}{3}b^3$. Because moments are additive $M_x + M_{xa}$

$= M_{xb}$ since the union of the region and the semicircular region of radius a

is the entire semicircular region of radius b. Thus $M_x = M_{xb} - M_{xa} =$

$\frac{2}{3}(b^3 - a^3) = \frac{2}{3}(b-a)(b^2 + ab + a^2)$. Finally, $\bar{y} = \frac{M_x}{m} = \frac{(2/3)(b-a)(b^2 + ab + a^2)}{(\pi/2)(b^2 - a^2)} =$

$\frac{4(b^2 + ab + a^2)}{3\pi(b+a)}$.

EXERCISES 6.9

1. Since v is in mph, we need t in hrs. Thus 12 min $= 12/60 = 1/5$
hrs. Using the Trapezoidal Rule, $(b - a)/2n = (1/5)/2(12) =$
$1/120$, we obtain $\int_0^{1/5} v(t)dt \approx \frac{1}{120}[40 + 2(45) + 2(40) + 2(50) +$
$2(55) + 2(65) + 2(60) + 2(53) + 2(53) + 2(58) + 2(63) + 2(63) +$
$53] = 1303/120 \approx 10.9$ miles.

4. Using the Trapezoidal Rule with $a = 0$, $b = 10$, $n = 5$, $\int_0^{10} v(t)dt \approx$
$\frac{10 - 0}{2(5)}[24 + 2(22) + 2(16) + 2(10) + 2(2) + 0] = 124$ feet. (Note:
n is odd, thus Simpson's Rule cannot be used.)

7. $F = Q_0/\int_0^T c(t)dt$. Here, $Q_0 = 5$, $T = 12$. Using Simpson's Rule with
$a = 0$, $b = T = 12$, and $n = 12$, $\int_0^{12} c(t)dt \approx \frac{12 - 0}{3(12)}[0 + 4(0) +$
$2(0.15) + 4(0.48) + 2(0.86) + 4(0.72) + 2(0.48) + 4(0.26) +$
$2(0.15) + 4(0.09) + 2(0.05) + 4(0.01) + 0] = \frac{9.62}{3} \approx 3.21$. Thus,
$F \approx 5/3.21 \approx 1.56$ liters/min.

10. Over [0,5], the amount of capital formation is $\int_0^5 2t(3t+1)dt = \int_0^5 (6t^2+2t)dt$

= 275 thousand dollars. Over [5,10], we obtain

$\int_5^{10} (6t^2+2t)dt = 2100-275 = 1825$ thousands of dollars.

13. (a) 1-item time is $f(1) \approx 18.16$. (b) 4-item time $\approx \int_0^4 (20(x+1)^{-.4}+3)dx =$

$\frac{20}{.6}(x+1)^{.6}\Big]_0^4 +12 = \frac{20}{.6}(5^{.6}-1)+12 \approx 66.22$. (c) 8-item time $\approx \frac{20}{.6}(9^{.6}-1)+32 \approx$

115.24.   (d) 16-item time $\approx \frac{20}{.6}(17^{.6}-1)+48 \approx 197.12$ min.

EXERCISES 6.10 (Review)

1. The curves intersect at $(\pm 2,4)$. Thus for x integration the interval is
    $-2 \le x \le 2$, the upper boundary is $y = -x^2$, the lower is $y = x^2 - 8$, and

$A = \int_{-2}^2 [(-x^2)-(x^2-8)]dx = \int_{-2}^2 (8-2x^2)dx = 64/3$. For the y integration the

interval is $-8 \le y \le 0$. For $-8 \le y \le -4$, the right and left boundaries are

$x = \pm(y+8)^{1/2}$, and for $-4 \le y \le 0$, the boundaries are $x = \pm(-y)^{1/2}$. Thus

$A = \int_{-8}^{-4} 2(y+8)^{1/2}\,dy + \int_{-4}^0 2(-y)^{1/2}\,dy = \frac{32}{3} + \frac{32}{3} = \frac{64}{3}$.

3. $A = \int_a^b [(1-y)-y^2]dy$ where $a = (-1 - \sqrt{5})/2$, $b = (-1 + \sqrt{5})/2$.

4. Find point A by substituting $y = -x^3$ into the
    line's equation to obtain $-3x^3 + 7x - 10 = 0$
    which has $x = -2$ as the only real solution.
    For point B, put $y = \sqrt{x}$, or $y^2 = x$, into the
    linear equation to obtain $3y + 7y^2 - 10 = 0$
    with positive solution $y = 1$. Thus the x-
    interval is $-2 \le x \le 1$ with the obvious change
    in the lower boundary at $x = 0$. Thus

$A = \int_{-2}^0 [(-\frac{7}{3}x + \frac{10}{3}) - (-x^3)]dx +$

$\int_0^1 [(-\frac{7}{3}x + \frac{10}{3}) - x^{1/2}]dx = \frac{1}{3}\int_{-2}^1 (-7x+10)dx + \int_{-2}^0 x^3\,dx - \int_0^1 x^{1/2}\,dx =$

$\frac{27}{2} - 4 - \frac{2}{3} = \frac{53}{6}$.

5. $A = \int_{-2}^{0} (x^3 - x^2 - 6x)\,dx + \int_{0}^{3} (-x^3 + x^2 + 6x)\,dx.$

7. Using the disc method, $V = \pi \int_{0}^{2} (\sqrt{4x + 1})^2\,dx = \pi \int_{0}^{2} (4x + 1)\,dx.$

9. $V = 2\pi \int_{0}^{1} x(2 - (x^3 + 1))\,dx.$

10. The graphs intersect when $x = 0$ and 1. $\sqrt[3]{x} \geq \sqrt{x}$ on $[0,1]$

   $\implies$ (using the washer method) $V = \pi \int_{0}^{1} (x^{2/3} - x)\,dx = $

   $\pi(3/5 - 1/2) = \pi/10.$

11. (a) $V = \pi \int_{-2}^{1} [(-4x+8)^2 - (4x^2)^2]\,dx.$

   (b) $V = 2\pi \int_{-2}^{1} (1-x)[(-4x+8) - 4x^2]\,dx.$

   (c) $V = 2\pi \int_{-2}^{1} [(16 - 4x^2)^2 - (4x + 8)^2]\,dx.$

13. $y = 1 + (1/2)(x+3)^{2/3} = f(x) \implies f'(x) = (1/3)(x+3)^{-1/3} \implies 1 + f'(x)^2 = 1 +$

   $(1/9)(x+3)^{-2/3} = [(x+3)^{2/3} + 1/9](x+3)^{-2/3} \implies L_{-2}^{5} = \int_{-2}^{5} \sqrt{(x+3)^{2/3} + 1/9}\;\cdot$

   $(x+3)^{-1/3}\,dx = [(x+3)^{2/3} + 1/9]^{3/2}]_{-2}^{5} = (4 + 1/9)^{3/2} - (1 + 1/9)^{3/2} =$

   $(37^{3/2} - 10^{3/2})/27.$

   (The substitution $u = (x+3)^{2/3} + 1/9$, $du = \frac{2}{3}(x+3)^{-1/3}\,dx$ leads to the anti-derivative above.)

15. $W = 36\pi(62.5) \int_{0}^{4} (5-y)\,dy.$

16. Position the y-axis vertically so that $y = 0$ is the bottom and $y = 30$ is the top of the well. The bucket loses 8 pounds of water in the 30 feet for an average loss of 8/30 pounds per foot. Partition $[0,30]$, and select w_k in $[y_{k-1}, y_k]$. When the bucket bottom is at w_k, $(8/30)w_k$ of water has leaked out, and the total mass of bucket and water is $4 + (24 - (8/30)w_k)$. The work done in lifting the bucket from y_{k-1} to y_k, a distance of Δy_k feet, is $(28 - (8/30)w_k)\Delta y_k$. Summing for the total work approximation and passing to the limit as $\|P\| \to 0$, we obtain $W = \int_{0}^{30} (28 - (8/30)y)\,dy = 840 - 120 = 720.$

17. $F = \rho \int_0^{2\sqrt{2}} (6-y)2(2\sqrt{2} - y)dy + \rho \int_{-2\sqrt{2}}^0 (6-y)2(2\sqrt{2} + y)dy.$ (The y-axis bisects

the plate for this form of the answer.)

19. $f(x) = 1 + x^3$, $g(x) = -x - 1$ whose graphs intersect at $x = -1$.
Since $f(x) - g(x) = x^3 + x + 2$, $m = \int_{-1}^1 (x^3 + x + 2)dx = 4$,

$M_y = \int_{-1}^1 x(x^3 + x + 2)dx = \frac{16}{15}$, $M_x = \frac{1}{2}\int_{-1}^1 [f^2(x) - g^2(x)]dx =$

$\frac{1}{2}\int_{-1}^1 [(1 + x^3)^2 - (-1 - x)^2]dx = \frac{1}{2}\int_{-1}^1 [(1 + 2x^3 + x^6) - (1 + 2x + x^2)]dx$

$= -\frac{4}{21}$. $\bar{x} = \frac{16/15}{4} = \frac{4}{15}$, $\bar{y} = \frac{-4/21}{4} = -\frac{1}{21}$.

22. We introduce a coordinate system with the origin at the vertex
and the positive y-axis along the axis of the parabola. Then the
equation of the parabola is $y = dx^2$. The opening 4 ft. wide one
foot from the vertex means that $x = \pm2 \Rightarrow y = 1$. Thus, $a(\pm2)^2 = 1$
or $a = 1/4$ and $y = x^2/4$. To generate the surface, we revolve the
right half of the parabola, $0 \le x \le 2$, about the y-axis. Since
$y' = x/2$, $S = 2\pi \int_0^2 x\sqrt{1 + x^2/4}\ dx$.   Let $u = 1 + x^2/4$, $du = (x/2)dx$
so that $x\ dx = 2du$. $x = 0, 2 \Rightarrow u = 1,2$, and $S = 4\pi \int_1^2 u^{1/2}\ du =$
$\frac{8}{3}\pi u^{3/2}]_1^2 = \frac{8}{3}\pi(2^{3/2} - 1)$.

25. The limit is $\int_0^1 \pi x^4\ dx = \pi/5$.

28. If the force function is $f(x) = \pi x^4$ for $0 \le x \le 1$, then the limit is the
work done in moving an object from $x = 0$ to $x = 1$.

CHAPTER 7
EXPONENTIAL AND LOGARITHMIC FUNCTIONS

EXERCISES 7.1

1. $(f \circ g)(x) = 7g(x) + 5 = 7(x - 5)/7 + 5 = (x - 5) + 5 = x;$
    $(g \circ f)(x) = (f(x) - 5)/7 = ((7x - 5) + 5)/7 = 7x/7 = x.$

4. $(f \circ g)(x) = g(x)^3 + 1$
    $= \sqrt[3]{x - 1}^3 + 1$
    $= (x - 1) + 1 = x;$
    $(g \circ f)(x) = \sqrt[3]{f(x) - 1}$
    $= \sqrt[3]{(x^3 + 1) - 1} = \sqrt[3]{x^3} = x.$

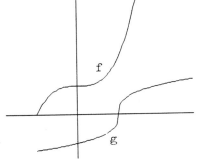

7. First, f is one-to-one if $x < -5/2$ since it is decreasing there
    $(f'(x) < 0)$. Next, to find f^{-1}, we solve $f(x) = y$ for x to
    obtain $x = f^{-1}(y)$. $f(x) = y \Rightarrow 1/(2x + 5) = y \Rightarrow 2x + 5 = 1/y \Rightarrow$
    $2x = 1/y - 5 = (1 - 5y)/y \Rightarrow x = (1 - 5y)/2y = f^{-1}(y)$. Since
    $f(x)$ is positive for $x > -5/2$, y in this formula must also be
    positive. Replacing y by x, we finally obtain $f^{-1}(x) =$
    $(1 - 5x)/2x$ for $x > 0$.

10. $f(x) = 4x^2 + 1$ is increasing if $x \geq 0$ $(f'(x) \geq 0)$ and $f(x) \geq 1$.
    $f(x) = y \Rightarrow 4x^2 + 1 = y \Rightarrow 4x^2 = y - 1 \Rightarrow x = \sqrt{y - 1}/2$, rejecting
    the negative square root which would have made $x \leq 0$. Thus,
    $f^{-1}(x) = \sqrt{x - 1}/2$ for $x \geq 1$.

13. $\sqrt{3x - 5} = y \Rightarrow y \geq 0$ and $3x - 5 = y^2 \Rightarrow x = (y^2 + 5)/3 = f^{-1}(y)$
    for $y \geq 0$. Thus, $f^{-1}(x) = (x^2 + 5)/3$, $x \geq 0$.

16. f is one-to-one for $-\infty < x < \infty$ since it is an increasing function
    $(f'(x) \geq 0)$. $f(x) = y \Rightarrow (x^3 + 1)^5 = y \Rightarrow x^3 + 1 = y^{1/5} \Rightarrow$
    $x^3 = y^{1/5} - 1 \Rightarrow x = (y^{1/5} - 1)^{1/3} = f^{-1}(y) \Rightarrow f^{-1}(x) =$
    $(x^{1/5} - 1)^{1/3}$, all x.

22. (a) (a,b) on the graph of $f \Rightarrow b = f(a) \Rightarrow a = f^{-1}(b) \Rightarrow (b,a)$
    is on the graph of f^{-1}. (b) The midpoint of PQ is
    $((a+b)/2, (a+b)/2)$, i.e. $y = x$. (c) $m_{PQ} = -1 \Rightarrow$ PQ and the line
    $y = x$ (with slope 1) are perpendicular.

EXERCISES 7.2

1. $f'(x) = \dfrac{D_x(9x+4)}{9x+4} = \dfrac{9}{9x+4}$.

4. $f(x) = \ln(5x^2+1)^3 = 3 \ln(5x^2+1)$. Thus $f'(x) = 3 \dfrac{D_x(5x^2+1)}{5x^2+1} = \dfrac{3(10x)}{5x^2+1}$.

7. $f'(x) = \dfrac{D_x(3x^2-2x+1)}{(3x^2-2x+1)} = \dfrac{6x-2}{3x^2-2x+1}$.

10. $f'(x) = (\ln x)D_x \ln(x+5) + D_x(\ln x) \ln(x+5) = \dfrac{\ln x}{x+5} + \dfrac{\ln(x+5)}{x}$.

13. $f(x) = \ln\sqrt{x} + \sqrt{\ln x} = (1/2)\ln x + (\ln x)^{1/2} \implies f'(x) = 1/2x + (1/2)(\ln x)^{-1/2}$
    $D_x(\ln x) = 1/2x + 1/2x\sqrt{\ln x}$.

16. $f(x) = \ln\sqrt{\dfrac{4+x^2}{4-x^2}} = \dfrac{1}{2} \ln \dfrac{4+x^2}{4-x^2} = \dfrac{1}{2}[\ln(4+x^2) - \ln(4-x^2)] \implies f'(x) =$
    $\dfrac{1}{2}(\dfrac{2x}{4+x^2} - \dfrac{-2x}{4-x^2}) = \dfrac{8x}{16-x^4}$.

19. $f(x) = \ln \dfrac{\sqrt{x^2+1}}{(9x-4)^2} = \dfrac{1}{2} \ln(x^2+1) - 2 \ln(9x-4) \implies f'(x) = (\dfrac{1}{2})2x/(x^2+1) -$
    $(2)9/(9x-4) = x/(x^2+1) - 18/(9x-4)$.

22. $f'(x) = \dfrac{D_x \ln x}{\ln x} = \dfrac{1}{x \ln x}$.

25. $f'(x) = 2 \ln\sqrt{x^2+1} \, D_x \ln\sqrt{x^2+1}$. Using $\ln\sqrt{a} = (1/2)\ln a$, we obtain $f'(x) =$
    $\ln(x^2+1)D_x((1/2)\ln(x^2+1)) = \ln(x+1)\dfrac{(1/2)2x}{x^2+1}$.

28. $x > 0 \implies |x| = x$, and $f'(x) = D_x \ln|x| = D_x \ln x = 1/x$. $x < 0 \implies |x| = -x$,
    and $f'(x) = D_x \ln |x| = D_x \ln(-x) = \dfrac{D_x(-x)}{-x} = \dfrac{-1}{-x} = \dfrac{1}{x}$. Thus, for all $x \neq 0$,
    $D_x \ln |x| = 1/x$.

31. $x \ln y - y \ln x = 1 \implies (x\dfrac{y'}{y} + \ln y) - (\dfrac{y}{x} + y' \ln x) = 0 \implies x^2 y' + xy \ln y$
    $- y^2 - y'xy \ln x) = 0$ (having multiplied through by xy) $\implies (x^2 - xy \ln x)y' =$
    $y^2 - xy \ln y \implies y' = (y^2 - xy \ln y)/(x^2 - xy \ln x)$.

34. $\ln x \ln y = xy - 1 \implies (\ln x) \dfrac{y'}{y} + \dfrac{1}{x} \ln y = xy' + y \implies x(\ln x)y' + y \ln y$
    $= x^2 yy' + xy^2 \implies (x \ln x - x^2 y)y' = xy^2 - y \ln y \implies y' =$
    $y(xy - \ln y)/x(\ln x - xy)$.

37. The given line has slope 6, and $y' = 2x + 4/x$. $y' = 6 \implies$
    $2x + 4/x = 6 \implies 2x^2 + 4 = 6x \implies 2x^2 - 6x + 4 = 0 \implies$
    $2(x - 1)(x - 2) = 0$. Thus $x = 1,2$ at which $y = 1 + 4 \ln 1 = 1$
    and $y = 4 + 4 \ln 4$.

40. $y = \ln(x^2 + 1) \implies y' = \dfrac{2x}{x^2 + 1} \implies y'' = \dfrac{2(1 - x^2)}{(x^2 + 1)^2} = 0$ if $x = \pm 1$.
    $y'' > 0$ if $|x| < 1$, and $y'' < 0$ if $|x| > 1$. Thus, there are PI's
    at $x = \pm 1$ at which $y = \ln 2$.

43. Writing $T = f(L) = -2.57 \ln \frac{87 - L}{63}$, recall that the error is ΔT
     $\simeq dT = f'(L)\Delta L$. We want ΔT when $L = 80$ and $\Delta L = 2$. Now,
     $f'(L) = -2.57 \frac{(-1/63)}{(87 - L)/63} = \frac{2.57}{87 - L}$. Thus $f'(80) = 2.57/7 \simeq 0.367$,
     and $\Delta T \simeq f'(80)\Delta L = 0.367(2) \simeq 0.73$ yrs.

46. (a) $S(0) = a \ln 1 = 0$.   (b) $S(1) = a \ln(\frac{a + 1}{a}) = a \ln(1 + 1/a)$
     $= \ln[(1 + 1/a)^a]$.   (c) $S'(N) = a/(a + N) > 0 \Rightarrow S$ is increasing.
     $S''(N) = -a/(a + N)^2 < 0 \Rightarrow$ CD. Except for a scale change, the
     graph is the same as that of $\ln(1 + x)$, $x \geq 0$.

52. Hint: $\ln x = \int_1^x \frac{1}{t}dt = -\int_x^1 \frac{1}{t}dt$.

55. $D_x(x \ln x - x + C) = x \cdot \frac{1}{x} + \ln x - 1 + 0 = \ln x$ verifies the formula.

EXERCISES 7.3

1. $f'(x) = e^{-5x}D_x(-5x) = -5e^{-5x}$

4. $f'(x) = e^{1-x^3}D_x(1-x^3) = -3x^2e^{1-x^3}$

7. $f'(x) = e^{\sqrt{x+1}} D_x\sqrt{x+1} = (e^{\sqrt{x+1}})/2\sqrt{x+1}$.

10. $f'(x) = \frac{1}{2}(e^{2x}+2x)^{-1/2}D_x(e^{2x} + 2x) = \frac{1}{2}(e^{2x}+2x)^{-1/2}(2e^{2x}+2) = (2e^{2x}+2)/2\sqrt{e^{2x}+2x}$.

13. $f'(x) = 3(e^{4x}-5)^2D_x(e^{4x}-5) = 3(e^{4x}-5)^2(4e^{4x})$.

16. $f'(x) = D_x(e^{x^{1/2}} + e^{x/2}) = e^{\sqrt{x}}D_x(x^{1/2}) + e^{x/2}D_x(x/2) = e^{\sqrt{x}}/2\sqrt{x} + \frac{1}{2}e^{x/2}$.

19. $f'(x) = e^{-2x}D_x(\ln x) + \ln x\, D_x(e^{-2x}) = \frac{e^{-2x}}{x} + \ln x(-2)e^{-2x}$.

22. $f'(x) = \frac{\ln(e^x-1)D_x \ln(e^x+1) - \ln(e^x+1)D_x \ln(e^x-1)}{[\ln(e^x-1)]^2} =$

     $\{\frac{[\ln(e^x-1)]e^x}{e^x + 1} - \frac{[\ln(e^x+1)]e^x}{e^x - 1}\}/[\ln(e^x-1)]^2$.

25. $e^{xy} - x^3 + 3y^2 = 11 \Rightarrow e^{xy}(xy'+y) - 3x^2 + 6yy' = 0 \Rightarrow (xe^{xy}+6y)y' +$
     $(ye^{xy}-3x^2) = 0$.

28. $xe^y - ye^x = 2 \Rightarrow (xe^yy' + e^y) - (ye^x + e^xy') = 0 \Rightarrow (xe^y - e^x)y' =$
     $ye^x - e^y \Rightarrow y' = (ye^x - e^y)/(xe^y - e^x)$.

31. (a) $y' = e^{-x}/x - e^{-x} \ln x = 0 \Rightarrow 1/x - \ln x = 0 \Rightarrow 1 - x \ln x = 0$
$\Rightarrow x \ln x = 1.$ (b) See the text answer for explanation of
uniqueness. Let $f(x) = x \ln x - 1.$ Since $f(1) = -1$ and
$f(e) = e - 1 > 0$, the solution lies between 1 and $e \simeq 2.718.$
Since $f'(x) = 1 + \ln x$, the Newton recursion formula is
$$x_{n+1} = x_n - \frac{x_n \ln x_n - 1}{1 + \ln x_n}, \text{ and our initial guess will be } x_1 = 2.$$
We obtain:

n	x_n	$f(x_n)$
1	2.000	0.386
2	1.772	0.014
3	1.763	0.000
4	1.763	0.000

Thus the solution is $x = 1.76$ to two places.

34. $f'(x) = (-2x^2 + 2x)e^{-2x} = 2x(1 - x)e^{-2x} = 0$
at $x = 0$ and $x = 1.$ $f''(x) = (4x^2 - 8x + 2)e^{-2x}.$
$f''(0) = 2 \Rightarrow f(0) = 0$ is a local minimum;
$f''(1) = -2e^{-2} \Rightarrow f(1) = e^{-2}$ is a local
maximum. $f''(x) = 0$ if $4x^2 - 8x + 2 = 0$ or
$x = 1 \pm 1/\sqrt{2}.$ $f''(x) > 0$ on $(-\infty, 1-1/\sqrt{2})$
and $(1+1/\sqrt{2}, \infty)$ and the graph is CU there;
$f''(x) < 0$ on $(1-1/\sqrt{2}, 1+1/\sqrt{2})$ and the graph is CD there. The PI's
are at $x = 1 \pm 1/\sqrt{2}.$ f is decreasing for $(-\infty, 0]$ and $[1, \infty)$ and
increasing for $[0, 1].$

37. $f'(x) = 1 + \ln x = 0 \Rightarrow \ln x = -1 \Rightarrow x = e^{-1} = 1/e.$ $0 < x < 1/e$
$\Rightarrow f'(x) < 0 \Rightarrow f$ decreasing; and $x > 1/e \Rightarrow f'(x) > 0 \Rightarrow f$
increasing. Thus $f(1/e)$ is a local min. $f''(x) = 1/x > 0$ for
$x > 0$ and the graph is CU.

38. $f'(x) = e^x - e^{-x} = e^{-x}(e^{2x} - 1).$ Thus $f'(x) = 0$ if $x = 0$, $f'(x) > 0$ if
$x > 0$, $f'(x) < 0$ if $x < 0.$ Thus f is increasingly on $[0, \infty)$, decreasing on
$(-\infty, 0]$ and $f(0) = 2$ is a local minimum. $f''(x) = e^x + e^{-x} > 0$ for all $x \Rightarrow$
graph is CU on $(-\infty, \infty).$

40. $I'(t) = I_o e^{-Rt/L} D_t(-Rt/L) = (-RI_o/L)e^{-Rt/L} = (-R/L)I(t).$

43. $h = f(x) = 79.041 + 6.39x - e^{3.261-0.993x} \Rightarrow f'(x) =$
$6.39 + 0.993e^{3.261-0.993x}$ is the rate of growth at age x,
$1/4 \le x \le 6.$ (a) For a one-year-old baby, $x = 1$, and $f(1) \simeq$
75.77 cm $\simeq$ 30 inches, and the growth rate is $f'(1) \simeq$ 15.98 cm/yr
$\simeq 6\frac{1}{4}$ inches/yr.   (b) $\frac{d}{dx}f'(x) = -(0.993)^2 e^{3.261-0.993} < 0 \Rightarrow$
$f'(x)$ decreases, i.e., the child's growth rate decreases with

age. Thus the maximum growth rate is at the youngest age for which this model applies, $x = 1/4$ yr (3 months) and the minimum growth rate is at $x = 6$ yrs.

46. To simplify notation, let $g(v) = -mv^2/2KT$ so that $g'(v) = -mv/KT$ and $F(v) = cv^2 e^{g(v)}$. Then $F'(v) = c(v^2 e^{g(v)} g'(v) + 2v e^{g(v)})$ $= cve^{g(v)}(vg'(v) + 2) = 0$ only if $vg'(v) = -2$ since all other factors are positive. Substituting for $g'(v)$ we obtain $-mv^2/KT = -2 \Rightarrow v^2 = 2KT/m \Rightarrow v = \sqrt{2KT/m} = v_0$. $0 < v < v_0 \Rightarrow$ $F' > 0$ and $v_0 < v \Rightarrow F' < 0$. Thus a local max, and since there is only one critical number, it is the absolute max.

49. First write R as $R = rx(\ln K - \ln x) = (r \ln K)x - r x \ln x$. Then $\frac{dR}{dx} = r \ln K - r(x/x + \ln x) = r(\ln K - 1 - \ln x) = 0$ if $\ln x = \ln K - 1 \Rightarrow x = e^{\ln K - 1} = e^{\ln K} e^{-1} = Ke^{-1}$.

52. With $a = 0$, $b = 1$, $n = 10$, we obtain

k	x_k	$f(x_k)$	m	$mf(x_k)$
0	0.0000	1.0000	1	1.0000
1	0.1000	0.9900	2	1.9801
2	0.2000	0.9608	2	1.9216
3	0.3000	0.9139	2	1.8279
4	0.4000	0.8521	2	1.7043
5	0.5000	0.7788	2	1.5576
6	0.6000	0.6977	2	1.3954
7	0.7000	0.6126	2	1.2253
8	0.8000	0.5273	2	1.0546
9	0.9000	0.4449	2	0.8897
10	1.0000	0.3679	1	0.3679
				14.9242

Since $(b - a)/2n = 0.0500$, we obtain 0.7462.

EXERCISES 7.4

1. Let $u = x^2+1$. Then $du = 2x\,dx$, $x\,dx = (1/2)du$. $\int \frac{x}{x^2+1}\,dx = \frac{1}{2}\int\frac{du}{u} = \frac{1}{2}\ln|u|$ $+ C = \frac{1}{2}\ln|x^2+1| + C = \frac{1}{2}\ln(x^2+1) + C$ since $x^2+1 > 0$. ($\ln\sqrt{x^2+1} + C$ is another correct form of the answer.)

4. Let $u = x^4-5$, $du = 4x^3 dx$, $x^3 dx = (1/4)du$. $\int \frac{x^3}{x^4-5}\,dx = \frac{1}{4}\int\frac{du}{u} = \frac{1}{4}\ln|u| + C =$ $\frac{1}{4}\ln|x^4-5| + C$.

7. With $u = x^3+1$, $du = 3x^2 dx$, $\int \frac{x^2}{x^3+1}\,dx = \frac{1}{3}\int\frac{du}{u} = (1/3)\ln|u| + C = (1/3)\ln|x^3+1| + C$.

10. Let $u = 4-5x$, $du = -5dx$. When $x = -1$, $u = 9$ and when $x = 0$, $u = 4$. Thus $\int_{-1}^{0}\frac{dx}{4-5x} = -\frac{1}{5}\int_{9}^{4}\frac{du}{u} = -\frac{1}{5}\ln|u|]_{9}^{4} = -\frac{1}{5}(\ln 4 - \ln 9) = \frac{1}{5}\ln(9/4)$.

13. $\int (x + e^{5x})dx = \frac{x^2}{2} + \int e^{5x}dx$. Let $u = 5x$, $du = 5\,dx$, $dx = (1/5)du$. Using

   this in the remaining integral, we get $\frac{x^2}{2} + \frac{1}{5}\int e^u du = \frac{x^2}{2} + \frac{e^u}{5} + C = \frac{x^2}{2} + \frac{e^{5x}}{5} + C$.

16. With $u = \ln x$, $du = \frac{1}{x}dx$, $\int \frac{1}{x(\ln x)^2}dx = \int \frac{1}{u^2}du = -\frac{1}{u} + C = -\frac{1}{\ln x} + C$.

19. With $u = \sqrt{x}$, $du = \frac{1}{2\sqrt{x}}dx$, $\int \frac{e^{\sqrt{x}}}{\sqrt{x}}dx = 2\int e^u du = 2e^u + C = 2e^{\sqrt{x}} + C$.

22. With $u = e^x+1$, $du = e^x dx$, $\int \frac{e^x}{(e^x+1)^2}dx = \int \frac{1}{u^2}du = -\frac{1}{u} + C = -1/(e^x+1) + C$.

25. Since $x^2 + 2x + 1 = (x+1)^2$, we let $u = (x+1)$, $du = dx$ so that $\int \frac{1}{x^2+2x+1}dx$

   $= \int \frac{1}{(x+1)^2}dx = \int \frac{1}{u^2}du = \int u^{-2}du = -u^{-1} + C = -1/(x+1) + C$.

28. $\int \frac{x^2+3x+1}{x}dx = \int (x + 3 + \frac{1}{x})dx = \frac{x^2}{2} + 3x + \ln|x| + C$.

31. $A = \int_1^2 (\frac{1}{x} - e^{-x})dx = \ln|x| + e^{-x}]_1^2 = (\ln 2 + e^{-2}) - (\ln 1 + e^{-1})$

   $= \ln 2 + e^{-2} - e^{-1} \approx 0.460$. (Recall that $\ln 1 = 0$.)

34. $V = \pi\int_1^4 (\frac{1}{\sqrt{x}})^2 dx = \pi\int_1^4 \frac{1}{x}dx = \pi \ln x]_1^4 = \pi \ln 4 \approx 1.39\pi$.

37. $y = (x^2 + 3)^5/(x + 1)^{1/2} \Rightarrow \ln y = \ln(x^2 + 3)^5 - \ln(x + 1)^{1/2} =$
   $5\ln(x^2 + 3) - \frac{1}{2}\ln(x + 1)$. Thus $\frac{y'}{y} = \frac{5(2x)}{x^2 + 3} - \frac{1}{2}\cdot\frac{1}{x + 1}$

   $= \frac{10x(2x + 2) - (x^2 + 3)}{(x^2 + 3)(2x + 2)} = \frac{20x^2 + 20x - x^2 - 3}{2(x^2 + 3)(x + 1)} = \frac{19x^2 + 20x - 3}{2(x^2 + 3)(x + 1)}$

   $\Rightarrow y' = \frac{19x^2 + 20x - 3}{2(x^2 + 3)(x + 1)}$ $y = \frac{19x^2 + 20x - 3}{2(x^2 + 3)(x + 1)}\cdot\frac{(x^2 + 3)^5}{(x + 1)^{1/2}}$

   $= \frac{(19x^2 + 20x - 3)(x^2 + 3)^4}{2(x + 1)^{3/2}}$.

40. $\ln y = \frac{2}{3}\ln(x^2 + 3) + 4\ln(3x - 4) - \frac{1}{2}\ln x$.
   $\frac{y'}{y} = \frac{2(2x)}{3(x^2 + 3)} + \frac{4(3)}{3x - 4} - \frac{1}{2x}$.
   $y' = [\frac{4x}{3(x^2 + 3)} + \frac{12}{3x - 4} - \frac{1}{2x}]\frac{(x^2 + 3)^{2/3}(3x - 4)^4}{\sqrt{x}}$.

43. $\Delta S = \int_{T_1}^{T_2} (c/T)dT = c \ln T]_{T_1}^{T_2} = c(\ln T_2 - \ln T_1) = c \ln(T_2/T_1)$.

46. Since R is the rate of consumption, the amount consumed between
   $t = 0$ and $t = T$ is $\int_0^T R\,dt$. Thus we seek T such that $\int_0^T R\,dt =$

   $50 \Rightarrow 6.5\int_0^T e^{0.02t}dt = 50 \Rightarrow \frac{6.5}{0.02}e^{0.02t}]_0^T = 50 \Rightarrow 6.5(e^{0.02T} - 1)$

$= 1 \Rightarrow e^{0.02T} = 1 + 1/6.5 = 7.5/6.5 \Rightarrow 0.02T = \ln(7.5/6.5) \Rightarrow T$
$= 50 \ln(7.5/6.5) \approx 7.2$ yrs.

49. $\int_0^3 f(x)dx = c\int_0^3 \dfrac{x}{x^2 + 4}dx = \dfrac{c}{2} \ln(x^2 + 4)]_0^3 = \dfrac{c}{2}(\ln 13 - \ln 4)$

$= \dfrac{c}{2} \ln(\dfrac{13}{4}) = \dfrac{c}{2} \ln 3.25 = 1$ if $c = 2/\ln 3.25$.

EXERCISES 7.5

1. $f'(x) = 7^x \ln 7$.

4. $f'(x) = 9^{\sqrt{x}} \ln 9\, D_x\sqrt{x} = (9^{\sqrt{x}} \ln 9)/2\sqrt{x}$.

7. $f'(x) = 5^{3x-4} \ln 5\, D_x(3x - 4) = 5^{3x-4}(\ln 5) \cdot 3$.

10. $f'(x) = 10(10^x + 10^{-x})^9 D_x(10^x + 10^{-x}) =$

$10(10^x + 10^{-x})^9(10^x \ln 10 - 10^{-x} \ln 10)$.

13. $f'(x) = D_x 5 \log(3x^2 + 2) = \dfrac{5(6x)}{(\ln 10)(3x^2+2)}$.

16. $f'(x) = D_x(\log |1-x^2| - \log |2-5x^3|) = \dfrac{-(2x)}{(\ln 10)(1-x^2)} - \dfrac{(-15x^2)}{(\ln 10)(2-5x^3)}$.

19. $f'(x) = D_x(x^e) + D_x(e^x) = ex^{e-1} + e^x$ since the first function is a <u>power</u>

<u>function</u>; the second is the familiar exponential.

22. Method 1. $D_x(x^{x^2+4}) = D_x e^{(x^2+4)\ln x} = e^{(x^2+4)\ln x} D_x(x^2+4)\ln x =$

$x^{x^2+4}[(x^2+4)/x + 2x \ln x]$.

Method 2. $y = x^{x^2+4} \Longleftrightarrow \ln y = (x^2+4) \ln x \Longrightarrow \dfrac{y'}{y} = \dfrac{x^2+4}{x} + 2x \ln x \Longrightarrow$
$y' = y((x^2+4)/x + 2x \ln x)$.

25. Let $u = -x^2$, $du = -2x\, dx$, $x\, dx = -(1/2)du$. Then $\int x3^{-x^2}dx = -\dfrac{1}{2}\int 3^u du =$

$-\dfrac{1}{2}\dfrac{3^u}{\ln 3} + C = -\dfrac{3^{-x^2}}{2 \ln 3} + C$.

28. Let $u = 3^x+4$, $du = 3^x \ln 3\, dx$. Then $3^x dx = (1/\ln 3)du$ and $\int \dfrac{3^x}{\sqrt{3^x+4}} dx =$

$\dfrac{1}{\ln 3}\int \dfrac{du}{\sqrt{u}} = \dfrac{1}{\ln 3}\int u^{-1/2}du = \dfrac{2}{\ln 3} u^{1/2} + C = \dfrac{2}{\ln 3}\sqrt{3^x+4} + C$.

31. With $u = x^3$, $du = 3x^2 dx$, $\int x^2 2^{x^3}dx = \dfrac{1}{3}\int 2^u du = 2^u/3 \ln 2 + C = 2^{x^3}/3 \ln 2 + C$.

34. $u = 4^x + 1 \Longrightarrow du = 4^x \ln 4\, dx \Longrightarrow 4^x dx = (1/\ln 4)du$. $\int 4^x(4^x + 1)^3 dx =$

$\dfrac{1}{\ln 4}\int u^3 du = \dfrac{u^4}{4 \ln 4} + C = \dfrac{(4^x + 1)^4}{4 \ln 4} + C$.

37. (a) $B'(t) = 0.95^t \ln 0.95 \Rightarrow B'(2) = 0.95^2 \ln(0.95) \approx -\$0.05$.
    (b) $(1/2)\int_0^2 0.95^t dt = (1/2 \ln 0.95)(0.95^2 - 1) \approx \$0.95$.

40. $\log P = a + b/(c + T) \Rightarrow P'/P \ln 10 = -b/(c + T)^2$. Since
     $P \ln 10 > 0$, $P' > 0$ if $b < 0$.

43. (a) $R(x_0) = a \log 1 = 0$.   (b) $S = dR/dx = (a/\ln 10)/x = K/x$.
     $S(2x) = K/2x = (1/2)S(x)$, or $S(x) = 2S(2x)$.

EXERCISES 7.6

1. The number of bacteria after t hours is $y = q(t) = y_0 e^{ct}$. We
    are given that $y_0 = 5,000$ and that $y = 15,000$ when $t = 10$. Thus
    $y = 5,000e^{ct}$ and, setting $t = 10$, we obtain $5000e^{10c} = 15000 \Rightarrow$
    $e^{10c} = 3 = (e^{10c})^{1/10} = 3^{1/10} \Rightarrow e^c = 3^{1/10}$. Thus $y = 5000(e^c)^t$
    $= 5000(3^{1/10})^t = 5000 \cdot 3^{t/10}$. (Alternatively, $e^{10c} = 3 \Rightarrow$
    $10c = \ln 3 \Rightarrow c = (\ln 3)/10 \Rightarrow e^{ct} = e^{(\ln 3/10)t} = 3^{t/10}$ or
    $e^{0.11t}$.) After 20 hours, $t = 20$ and $q(20) = (5000)3^2 = 45,000$.
    To find when $q(t)$ is 50,000 we solve $(5000)3^{t/10} = 50,000 \Leftrightarrow$
    $3^{t/10} = 10 \Leftrightarrow (t/10)\ln 3 = \ln 10 \Leftrightarrow t = (10 \ln 10)/\ln 3 \approx 23/1.1$
    ≈ 21 hours.

4. If $P(t) =$ the population t years from the present, then $\frac{dP}{dt} = .05P$ and $P(t) =$
    $P(0)e^{.05t} = 500,000 \, e^{.05t}$. In 10 years $t = 10$ and $p(10) = 500,000e^{.5} \approx$
    $500,000(1.65) = 825,000$.

7. Let $y(t)$ be the temperature of the thermometer t minutes after it's brought
    in. (Assume that the reading is the same as the actual temperature.) Then
    $\frac{dy}{dt} = c(y-70)$, and $y(10) = 40$ which yields $\frac{dy}{y-70} = c \, dt \Rightarrow \ln|y-70| = ct + b$
    $\Rightarrow |y-70| = ke^{ct}$. Since the instrument is cooler than 70, $y-70 < 0$ and the
    last formula becomes $70-y = ke^{ct}$. Setting $t = 0$, $y(0) = 40$ and $k = 30$.
    Using $y(5) = 60$ we obtain $10 = 30e^{5c}$ or $c = (-1/5)\ln 3$. To find when it
    registers 65, we solve $70-65 = 30e^{(-1/5)(\ln 3)t} \Leftrightarrow (-1/5)(\ln 3)t = \ln(1/6)$
    $= -\ln 6 \Rightarrow t = (5 \ln 6)/\ln 3 \approx 5(1.792)/1.099 \approx 8.2$ minutes.

10. Here we are given that $y = y_0 e^{ct}$, $y_0 = 2000$, $y = 1500$ when
     $t = 10$ days where y is the number of counts after t days. We
     seek the time t at which $y = y_0/2 = 1000$, the half-life. Setting
     $t = 10$ and using y_0, we obtain $2000e^{10c} = 1500 \Rightarrow e^{10c} = 3/4 \Rightarrow$
     $e^c = (3/4)^{1/10} \Rightarrow y = 2000(3/4)^{t/10}$. (Alternatively, $e^{10c} = 3/4$
     $\Rightarrow c = (\ln (3/4))/10 \approx -0.0288$ and $y = 2000e^{-0.0288t}$.) Now set
     $y = y_0/2$ to find the half-life: $2000(3/4)^{t/10} = 1000 \Rightarrow$
     $(3/4)^{t/10} = 1/2 \Rightarrow (t/10)\ln(3/4) = \ln(1/2) \Rightarrow t =$
     $10 \ln(0.5)/\ln(0.75) \approx 24.09$ days.

13. Let y_0 be the amount of ^{90}Sr in the field now. Then $y_0 = 2.5S$ and we seek t at which $y = S$. The half-life of 29 years means that $y(t + 29) = y(t)/2$ or $y_0 e^{c(t+29)} = y_0 e^{ct}/2 \Rightarrow e^{29c} = 1/2$ $\Rightarrow e^c = (1/2)^{1/29}$ so that $y = y_0(1/2)^{t/29}$. Substituting $y_0 = 2.5S$ we then solve $y = S$ for t: $2.5S(1/2)^{t/29} = S \Rightarrow$ $(1/2)^{t/29} = 0.4 \Rightarrow (t/29)\ln(0.5) = \ln(0.4) \Rightarrow t = 29 \ln(0.4)/\ln(0.5) \approx 38.3$ years.

16. $\frac{dh}{dt} = -\frac{V}{Q}\frac{h}{k + h} \Rightarrow \frac{k + h}{h}dh = -\frac{V}{Q}dt \Rightarrow (\frac{k}{h} + 1)dh + \frac{V}{Q}dt = 0 \Rightarrow$ $k \ln h + h + \frac{V}{Q}t = C.$

19. With y the amount of ^{14}C in the fossil, we know that $y = y_0 e^{ct}$, that the half-life is 5700 years, and we seek t at which $y = 20\%$ of y_0, or $y = 0.2y_0$. At $t = 5700$, $y = y_0/2 \Rightarrow e^{5700c} = 1/2 \Rightarrow c = \ln(0.5)/5700 \approx -1.216 \times 10^{-4}$. Finally, $y = .2y_0 \Rightarrow$ $y_0 e^{ct} = 0.2y_0 \Rightarrow e^{ct} = 0.2 \Rightarrow ct = \ln(0.2) \Rightarrow t = \ln(0.2)/c \approx$ 13,235 years.

22. Since $f(n) = 3 + 20(1 - e^{-0.1n})$ is approximately the number of items produced after n days, in (a) we must round off our answers to the nearest integer. (a) $f(5) = 3 + 20(1 - e^{-0.5}) \approx 10.87 \approx 11$ items; $f(9) = 3 + 20(1 - e^{-0.9}) \approx 15$; $f(24) \approx 21$; $f(30) \approx 22$.

(b) and (c) $f(x) = 3 + 20(1 - e^{-0.1x}) \Rightarrow f'(x) = 2e^{-0.1x} > 0 \Rightarrow$ f is increasing. $f''(x) = -0.2e^{-0.1x} < 0 \Rightarrow CD$. $\lim_{x \to \infty} f(x) = 23$. These yield the graph shown.

25. (a) Writing $G'(t) = ABke^{-Bt}e^{-Ae^{-Bt}}$, we see that if $t = (\ln A)/B$ then $e^{Bt} =$ $e^{\ln A} = A$, $e^{-Bt} = 1/A$, and $G'((\ln A)/B) = ABk(1/A)e^{-A/A} = Bke^{-1}$. That this is a maximum for G' follows from the first derivative test. All factors in $G''(t)$ are positive except $(-1 + Ae^{-Bt})$ which is < 0 if $t > (\ln A)/B$ and > 0 if $t < (\ln A)/B$.

(b) $B > 0 \Rightarrow e^{-Bt} \to 0$ as $t \to \infty$ and $e^{-Ae^{-Bt}} \to e^0 = 1$. Thus $\lim_{t \to \infty} G'(t) = ABk(0)1 = 0.$

(c) $G(t) = ke^{-Ae^{-Bt}} \to ke^{-(A)(0)} = k$ as $t \to \infty$.

EXERCISES 7.7

1. $f'(x) = 1/\sqrt{2x+3} > 0 \Longrightarrow$ f is increasing on $[1,11]$

    and has an inverse with domain $[f(1),f(11)] =$

    $[\sqrt{5},5]$. As in Sec. 7.1, we find f^{-1} by solving $y =$

    $\sqrt{2x+3}$ to obtain $y^2 = 2x+3$ and $x = (y^2-3)/2$. Thus

    $f^{-1}(x) = (x^2-3)/2$. Directly, $D_x f^{-1}(x) = x$. Using

    (7.29) with $g = f^{-1}$, $D_x f^{-1}(x) = 1/f'(f^{-1}(x)) =$

    $\sqrt{2f^{-1}(x) + 3}$ (from the formula for f' in the first

    line) $= \sqrt{(x^2-3) + 3} = x$ since $x > 0$.

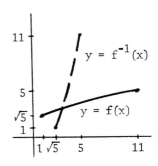

4. $f'(x) = 2x-4 < 0 \Longrightarrow$ f is decreasing on $[-1,1]$

    and has an inverse with domain $[f(1),f(-1)] =$

    $[2,10]$. Setting $y = x^2-4x+5$ and solving

    for x, we get $x = 2 - \sqrt{y-1}$ by the quadratic

    formula. (The negative sign must be se-

    lected for x to lie in $[-1,1]$.) Thus $f^{-1}(x)$

    $= 2 - \sqrt{x-1}$. Using (7.29), $D_x f^{-1}(x) =$

    $1/(2f^{-1}(x) - 4) = -1/2\sqrt{x-1}$, the same as

    computing it directly.

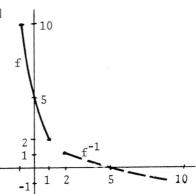

7. $f'(x) = -2xe^{-x^2} < 0$ if $x > 0$. Thus f is decreas-

    ing on $[0,\infty)$ and has an inverse with domain equal

    to the range of f. Since $f(0) = 1$ and $f(x) \to 0$

    as $x \to \infty$, f^{-1} has domain $(0,1]$. Solving $y =$

    e^{-x^2} we get $-x^2 = \ln y \Longrightarrow x = \sqrt{-\ln y}$. Thus

    $f^{-1}(x) = \sqrt{-\ln x}$ and $D_x f^{-1}(x) = -1/2\sqrt{-\ln x}$

    by either method.

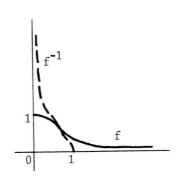

10. $f'(x) = e^x - e^{-x} > 0$ if $x > 0$. Thus f is in-

    creasing on $[0,\infty)$ and has an inverse with do-

    main equal to the range of f. Since $f(0) = 2$

    and $f(x) \to \infty$ as $x \to \infty$, the domain of f^{-1} is $[2,\infty)$.

    Next, $y = e^x + e^{-x} \Longrightarrow e^x - y + e^{-x} = 0$

    $\Longrightarrow e^{2x} - ye^x + 1 = 0$. Now solve for e^x by

    the quadratic formula! $e^x = (y \pm \sqrt{y^2-4})/2$.

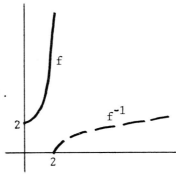

The + sign must be chosen so that $x = \ln[(y + \sqrt{y^2-4})/2]$. Thus $f^{-1}(x) = \ln[(x + \sqrt{x^2-4})/2]$ and $Df^{-1}(x) = [1 + x/\sqrt{x^2-4}]/(x + \sqrt{x^2-4}) = 1/\sqrt{x^2-4}$.

(NOTE: The quadratic formula can be used since $e^{2x} = (e^x)^2$. Also, if the - sign had been selected, we would have $x = \ln[(y - \sqrt{y^2-4})/2]$. But as $y \to \infty$, $y - \sqrt{y^2-4} \to 0$ and x would tend to $-\infty$ contrary to the earlier observation that $y = f(x)$ and x go to ∞ simultaneously.)

13. $f'(x) = 4e^{2x}/(e^{2x}+1)^2 > 0 \implies$ f has an inverse. Using 7.29, if g is the inverse function, $g'(0) = 1/f'(g(0)) = 1/f'(0) = 1$.

16. Since $|x-2| = -(x-2)$, if $x \le 2$ and $|x-2| = x-2$ if $x \ge 2$, f is decreasing on $(-\infty, 2]$, increasing on $[2,\infty)$ and, thus, has no inverse. On any subset of either of these intervals, f will have an inverse.

EXERCISES 7.8 (Review)

1. $f'(x) = (1-2x)D_x \ln|1-2x| + \ln|1-2x| \cdot D_x(1-2x) = \frac{(1-2x)(-2)}{(1-2x)} + (-2)\ln|1-2x| = -2(1 + \ln|1-2x|)$.

4. $f(x) = \log|2-9x| - \log|1-x^2| \implies f'(x) = \frac{1}{\ln 10}(\frac{-9}{2-9x} - \frac{-2x}{1-x^2})$.

7. $f(x) = e^{\ln(x^2+1)} = x^2+1$, and $f'(x) = 2x$.

10. $f(x) = \frac{1}{4}\ln\frac{x}{3x+5} = \frac{1}{4}(\ln x - \ln(3x+5)) \implies f'(x) = \frac{1}{4}(\frac{1}{x} - \frac{3}{3x+5}) = \frac{5}{4x(3x+5)}$.

13. $f'(x) = x^2 e^{1-x^2}(-2x) + 2xe^{1-x^2} = 2x(1-x^2)e^{1-x^2}$.

16. $f'(x) = 5^{3x}(\ln 5)(3) + 5(3x)^4 \cdot 3$.

19. $f'(x) = 10^{\ln x}(\ln 10)D_x \ln x = 10^{\ln x}(\ln 10)/x$.

22. $f'(x) = 7^{\ln|x|}\ln 7\, D_x(\ln|x|) = 7^{\ln|x|}(\ln 7)/x$.

25. $1 + xy = e^{xy} \implies xy' + y = e^{xy}(xy' + y) \implies x(1 - e^{xy})y' = y(e^{xy} - 1) \implies y' = y(e^{xy} - 1)/(1 - e^{xy}) = -y/x$ if $xy \ne 0$.

28. $\int_0^1 e^{-3x+2}dx = -\frac{1}{3}e^{-3x+2}\Big]_0^1 = -\frac{1}{3}[e^{-1} - e^2] = (e^2 - e^{-1})/3$.

31. With $u = 1 - \ln x$, $du = -\frac{1}{x}dx$, $\int\frac{1}{x - x\ln x}dx = \int\frac{1}{x(1 - \ln x)}dx = -\int\frac{1}{u}du = -\ln|u| + C = -\ln|1 - \ln x| + C$.

34. With $u = \frac{1}{x}$, $du = -\frac{1}{x^2}dx$, $\int\frac{e^{1/x}}{x^2}dx = -\int e^u du = -e^u + C = -e^{1/x} + C$.

37. With $u = \log x$, $du = \frac{1}{x \ln 10}dx$, $\frac{dx}{x} = \ln 10\, du$, $\int\frac{1}{x\sqrt{\log x}}dx = \ln 10\int u^{-1/2}du = 2\ln 10\, u^{1/2} + C = 2\ln 10\sqrt{\log 10} + C$.

40. With $u = 1 + e^x$, $du = e^x dx$, $\int \frac{2e^x}{1 + e^x} dx = 2\int \frac{du}{u} = 2 \ln|u| + C = 2 \ln|1 + e^x| + C = 2 \ln(1 + e^x) + C$ since $e^x + 1 > 0$.

43. This is just the integral of the power function, x^e. Thus the answer is simply $x^{e+1}/(e + 1) + C$.

46. $f'(x) = x + 2x \ln x = x(1 + 2 \ln x) = 0$ if $\ln x = -1/2 \Rightarrow x = e^{-1/2}$. (Recall $x > 0$ here, so $x = 0$ is not a critical number.) $f''(x) = 1 + 2 + 2 \ln x = 3 + 2 \ln x$. $f''(e^{-1/2}) = 3 + 2 \ln(e^{-1/2}) = 3 + 2(-1/2) = 2 \Rightarrow f(e^{-1/2}) = -1/2e$ is a local minimum. Next, $f''(x) = 0$ if $\ln x = -3/2 \Rightarrow x = e^{-3/2}$. $0 < x < e^{-3/2} \Rightarrow \ln x < -3/2 \Rightarrow f''(x) < 0 \Rightarrow$ CD on $(0, e^{-3/2})$. Also, $x > e^{-3/2} \Rightarrow \ln x > -3/2 \Rightarrow f''(x) > 0 \Rightarrow$ CU on $(e^{-3/2}, \infty)$, and a PI at $x = e^{-3/2}$.

49. $V = \int_{-3}^{-2} \pi(e^{4x})^2 = (\pi/8)e^{8x}\Big]_{-3}^{-2} = (\pi/8)(e^{-16} - e^{-24})$.

52. Let $f(x) = -8310 \ln x$ so that $T = f(x)$. (a) $T = f(0.04) \approx 24{,}000$ yrs. (b) With $x = 0.04$ and $\triangle x = 0.005$, $\triangle T \approx dT = f'(x)\triangle x = -\frac{8310}{x}\triangle x \Rightarrow |\triangle T| = \frac{8310}{0.04}\,0.005 \approx 1040 \approx 1000$ yrs.

55. Let y be the number of cells measured in hundreds of thousands. Then $y_0 = 1$, so that $y = e^{ct}$, and $y = 2$ when $t = 1/3 \Rightarrow e^{c/3} = 2 \Rightarrow e^c = 2^3 = 8 \Rightarrow y = 8^t$. When $t = 2$, $y = 8^2 = 64$ hundred thousand or 6.4 million.

58. (a) $f'(x) = 2e^{2x} + 2e^x > 0 \Rightarrow f$ increasing $\Rightarrow f^{-1}$ exists. $f(0) = 4$ and $\lim_{x \to \infty} f(x) = \infty$ imply that the range of f is $[4, \infty)$, the same as the domain of f^{-1}. (b) Noting that $f(x) = (e^x + 1)^2$, we solve $y = (e^x + 1)^2$, $y \geq 4$ for x: $e^x + 1 = \sqrt{y} \Rightarrow x = \ln(\sqrt{y} - 1) = f^{-1}(y)$, and $f^{-1}(x) = \ln(\sqrt{x} - 1)$. $D_x f^{-1}(x) = \frac{D_x(\sqrt{x} - 1)}{\sqrt{x} - 1} = \frac{1/2\sqrt{x}}{\sqrt{x} - 1}$. (c) $f'(0) = 4$, $D_x f^{-1}(x)\Big|_{x=4} = \frac{1/2\sqrt{4}}{\sqrt{4} - 1} = \frac{1/4}{1} = 1/4$.

EXERCISES 8.1

1. $\frac{\pi}{6}$ rad. $= \frac{\pi}{6} \cdot \frac{180}{\pi} = 30°$; $\frac{\pi}{4}$ rad. $= \frac{\pi}{4} \cdot \frac{180}{\pi} = 45°$; $\frac{\pi}{3}$ rad. $= \frac{\pi}{3} \cdot \frac{180}{\pi} = 60°$; $\frac{\pi}{2}$ rad

   $= \frac{\pi}{2} \cdot \frac{180}{\pi} = 90°$; $\frac{2\pi}{3}$ rad. $= \frac{2\pi}{3} \cdot \frac{180}{\pi} = \frac{360°}{3} = 120°$; $\frac{3\pi}{4}$ rad. $= \frac{3\pi}{4} \cdot \frac{180}{\pi} =$

   $\frac{540°}{4} = 135°$; $\frac{5\pi}{6}$ rad. $= \frac{5\pi}{6} \cdot \frac{180}{\pi} = \frac{5}{6} \cdot 180° = 150°$; π rad. $= \pi \cdot \frac{180}{\pi} = 180°$.

4. (a) $\sin t = -4/5$, $\cos t = 3/5 \Longrightarrow \tan t = \frac{\sin t}{\cos t} = -\frac{4/5}{3/5} = -\frac{4}{3}$; $\csc t =$

   $\frac{1}{\sin t} = -\frac{1}{4/5} = -\frac{5}{4}$; $\sec t = \frac{1}{\cos t} = \frac{1}{3/5} = \frac{5}{3}$; $\cot t = \frac{\cos t}{\sin t} = \frac{1}{\tan t} = -\frac{3}{4}$.

   (b) $\csc t = \sqrt{13}/2$, $\cot t = -3/2 \Longrightarrow \sin t = 1/\csc t = 2/\sqrt{13}$; $\cos t =$

   $\frac{\cos t}{\sin t} \cdot \sin t = \cot t \sin t = -\frac{3}{2} \cdot \frac{2}{\sqrt{13}} = -\frac{3}{\sqrt{13}}$; $\tan t = \frac{1}{\cot t} = -\frac{2}{3}$; $\sec t =$

   $\frac{1}{\cos t} = -\frac{\sqrt{13}}{3}$.

7. $\frac{9\pi}{2}$ rad. $= \frac{9\pi}{2} \cdot \frac{180}{\pi} = \frac{9}{2} \cdot 180° = 810°$; $-\frac{2\pi}{3} = -\frac{2\pi}{3} \cdot \frac{180}{\pi} = -\frac{360°}{3} = -120°$;

   $\frac{7\pi}{4} = \frac{7\pi}{4} \cdot \frac{180}{\pi} = (7)45° = 315°$; $5\pi = 5\pi \cdot \frac{180}{\pi} = 900°$; $\frac{\pi}{5} = \frac{\pi}{5} \cdot \frac{180}{\pi} = \frac{180°}{5} = 36°$.

NOTE: As you may recall, the concept of "reference angle" is very useful. If θ is any angle, in standard position, its reference angle, θ', is the (positive) acute angle between the terminal side of θ and either the positive or negative x-axis. (E.g. $\theta = 225° \Longrightarrow \theta' = 45°$; $\theta = 5\pi/6 \Longrightarrow \theta' = \pi/6$; $\theta = 358° \Longrightarrow \theta' = 2°$.) In trigonometry courses it is shown that if f is any trigonometric function, then $f(\theta) = \pm f(\theta')$, the correct sign determined from the quadrant in which the terminal side of θ lies. (E.g. $\sin 225° = -\sin 45° = -1/\sqrt{2}$ since in quadrant III, $\sin \theta < 0$; $\cos 5\pi/6 = -\cos \pi/6 = -\sqrt{3}/2$ since, in quadrant II, $\cos \theta < 0$; etc.) In all of the following problems, we will use the results of Examples 2 and 3 of the text, when appropriate.

10. (a) $225°$, in quadrant III where $\cos \theta < 0$, has reference angle $45°$. Thus, $\cos 225° = -\cos 45° = -1/\sqrt{2}$. (b) $150°$, in quadrant II where $\tan \theta < 0$, has reference angle $30°$. Thus $\tan 150° = -\tan 30° = -1/\sqrt{3}$. (c) $\sin(-\pi/6) = -\sin(\pi/6) = -1/2$. (d) $\sec(4\pi/3) = -\sec(\pi/3) = -1/\cos(\pi/3) = -2$. (e) $\cot(7\pi/4) = -\cot(\pi/4) = -1/\tan(\pi/4) = -1$. (f) $\csc 300° = -\csc 60° = -1/\sin 60° = -2/\sqrt{3}$.

13. $\sin \theta \sec \theta = \sin \theta \cdot \frac{1}{\cos \theta} = \frac{\sin \theta}{\cos \theta} = \tan \theta$.

16. $\cot \beta \sec \beta = \frac{\cos \beta}{\sin \beta} \cdot \frac{1}{\cos \beta} = \frac{1}{\sin \beta} = \csc \beta$.

19. First, $\sin^2 t + \cos^2 t = 1 \Longrightarrow \sin^2 t = 1 - \cos^2 t$. Now, $\cos^2 t - \sin^2 t = \cos^2 t - (1-\cos^2 t) = \cos^2 t - 1 + \cos^2 t = 2\cos^2 t - 1$.

22. As in #19, $1 - 2\sin^2 x = 1 - 2(1 - \cos^2 x) = 1 - 2 + 2\cos^2 x = 2\cos^2 x - 1$.

25. $\sec\beta - \cos\beta = \dfrac{1}{\cos\beta} - \cos\beta = \dfrac{1 - \cos^2\beta}{\cos\beta} = \dfrac{\sin^2\beta}{\cos\beta} = \dfrac{\sin\beta}{\cos\beta} \cdot \sin\beta = \tan\beta\sin\beta$.

28. $\sin x + \cos x\cot x = \sin x + \cos x\,\dfrac{\cos x}{\sin x} = \dfrac{\sin^2 x + \cos^2 x}{\sin x} = \dfrac{1}{\sin x} = \csc x$.

31. $2\cos^3\theta - \cos\theta = 0 \iff \cos\theta(2\cos^2\theta - 1) = 0$. To solve, we set each factor equal to 0, and find the values of θ in $[0, 2\pi)$ that make each factor 0, using reference angles if necessary. $\cos\theta = 0 \implies \theta = \pi/2$ and $3\pi/2$ (90° and 270°). $2\cos^2\theta - 1 = 0 \implies \cos^2\theta = 1/2 \implies \cos\theta = \pm 1/\sqrt{2}$. The acute angle (reference angle) whose cosine is $1/\sqrt{2}$ is $\pi/4$. Since $\cos\theta > 0$ in quadrants I and IV, < 0 in quadrants II and III, we obtain: $\cos\theta = 1/\sqrt{2} \implies \theta = \pi/4$ and $7\pi/4$ (45° and 315°); $\cos\theta = -1/\sqrt{2} \implies \theta = \pi - \pi/4 = 3\pi/4$ and $\theta = \pi + \pi/4 = 5\pi/4$ (135° and 225°).

34. $\csc^5\theta - 4\csc\theta = 0 \iff \csc\theta(\csc^4\theta - 4) = 0 \iff \csc\theta(\csc^2\theta + 2)(\csc^2\theta - 2) = 0$. Since $|\csc\theta| \geq 1$, the first 2 factors are never 0. Thus we must have $\csc^2\theta - 2 = 0 \implies \csc\theta = \pm\sqrt{2} \implies \sin\theta = \pm 1/\sqrt{2}$. The reference angle is $\pi/4$. $\sin\theta = 1/\sqrt{2} \implies \theta = \pi/4, 3\pi/4$ (45°, 135°), and $\sin\theta = -1/\sqrt{2} \implies \theta = 5\pi/4, 7\pi/4$ (225°, 315°).

37. Using $\cos^2\beta = 1 - \sin^2\beta$ we obtain: $\sin\beta + 2\cos^2\beta = 1 \iff \sin\beta + 2 - 2\sin^2\beta = 1 \iff 2\sin^2\beta - \sin\beta - 1 = 0 \iff (2\sin\beta + 1)(\sin\beta - 1) = 0$. From $\sin\beta - 1 = 0$ we get $\sin\beta = 1 \implies \beta = \pi/2$ (90°). From $2\sin\beta + 1 = 0$ we get $\sin\beta = -1/2$. With $\pi/6$ as reference angle and $\sin\theta < 0$ in quadrants III and IV, we obtain $\beta = \pi + \pi/6 = 7\pi/6$ and $\beta = 2\pi - \pi/6 = 11\pi/6$ (210°, 330°).

40. $\sin 2u = \sin u \iff 2\sin u\cos u = \sin u \iff \sin u(2\cos u - 1) = 0$. $\sin u = 0 \implies u = 0, \pi$ (0°, 180°). $2\cos u - 1 = 0 \implies \cos u = 1/2 \implies u = \pi/3$ and $2\pi - \pi/3 = 5\pi/3$ (60°, 300°).

NOTE: Before working #43-51, observe that $\csc\theta = 5/3 \implies \sin\theta = 3/5$, $\cos\theta = \sqrt{1 - \sin^2\theta} = \sqrt{1 - 9/25} = \sqrt{16/25} = 4/5$, $\tan\theta = \sin\theta/\cos\theta = 3/4$. Also $\cos\phi = 8/17 \implies \sin\phi = \sqrt{1 - 64/289} = \sqrt{225/289} = 15/17$, $\tan\phi = 15/8$.

43. $\sin(\theta + \phi) = \sin\theta\cos\phi + \cos\theta\sin\phi = \dfrac{3}{5} \cdot \dfrac{8}{17} + \dfrac{4}{5} \cdot \dfrac{15}{17} = \dfrac{24+60}{85} = \dfrac{84}{85}$.

46. $\sin(\phi - \theta) = \sin\phi\cos\theta - \cos\phi\sin\theta = \dfrac{15}{17} \cdot \dfrac{4}{5} - \dfrac{8}{17} \cdot \dfrac{3}{5} = \dfrac{60-24}{85} = \dfrac{36}{85}$.

49. $\tan 2\theta = \dfrac{2\tan\theta}{1 - \tan^2\theta} = \dfrac{2(3/4)}{1 - 9/16} = \dfrac{3/2}{7/16} = \dfrac{3}{2} \cdot \dfrac{16}{7} = \dfrac{24}{7}$.

52. Grouping $\alpha + \beta + \gamma$ as $(\alpha + \beta) + \gamma$, we have, $\cos(\alpha + \beta + \gamma) = \cos(\alpha + \beta)\cos\gamma - \sin(\alpha + \beta)\sin\gamma = [\cos\alpha\cos\beta - \sin\alpha\sin\beta]\cos\gamma - [\sin\alpha\cos\beta + \cos\alpha\sin\beta]\sin\gamma$.

55. (a) $f(g(\pi)) = f(\pi/4) = \cos(\pi/4) = 1/\sqrt{2}$.

(b) $g(f(\pi)) = g(\cos \pi) = g(-1) = -1/4$.

EXERCISES 8.2

1. $\lim\limits_{x\to 0} \dfrac{x}{\sin x} = \lim\limits_{x\to 0} \dfrac{1}{(\sin x/x)} = \dfrac{1}{1} = 1$.

4. $\lim\limits_{\theta\to 0} \dfrac{3\theta + \sin \theta}{\theta} = \lim\limits_{\theta\to 0} \left(\dfrac{3\theta}{\theta} + \dfrac{\sin \theta}{\theta}\right) = \lim\limits_{\theta\to 0} 3 + \lim\limits_{\theta\to 0} \dfrac{\sin \theta}{\theta} = 3 + 1 = 4$.

7. $\lim\limits_{\theta\to 0} \dfrac{2 \cos \theta - 2}{3\theta} = \lim\limits_{\theta\to 0} \dfrac{2}{3} \dfrac{\cos \theta - 1}{\theta} = -\dfrac{2}{3} \lim\limits_{\theta\to 0} \dfrac{1 - \cos \theta}{\theta} = -\dfrac{2}{3} \cdot 0 = 0$.

10. $\lim\limits_{x\to 0} \dfrac{x \sin x}{x^2 + 1} = \lim\limits_{x\to 0} \dfrac{x}{x^2+1} \lim\limits_{x\to 0} \sin x = 0 \cdot 0 = 0$.

13. $\lim\limits_{t\to 0} \dfrac{4t^2 + 3t \sin t}{t^2} = \lim\limits_{t\to 0} \left(4 + 3 \dfrac{\sin t}{t}\right) = 4 + 3 \cdot 1 = 7$.

16. $\lim\limits_{t\to 0} \dfrac{\sin t}{1 + \cos t} = \dfrac{0}{1+1} = 0$.

19. $\lim\limits_{x\to 0} \dfrac{x + \tan x}{\sin x} = \lim\limits_{x\to 0} \left(\dfrac{x}{\sin x} + \dfrac{\tan x}{\sin x}\right) = \lim\limits_{x\to 0} \left(\dfrac{1}{(\sin x/x)} + \dfrac{\sin x}{\cos x} \cdot \dfrac{1}{\sin x}\right) = \dfrac{1}{1} +$

$\lim\limits_{x\to 0} \dfrac{1}{\cos x} = 1 + \dfrac{1}{1} = 2$.

22. $\lim\limits_{x\to 0} \dfrac{\csc 2x}{\cot x} = \lim\limits_{x\to 0} \dfrac{1}{\sin 2x} \tan x = \lim\limits_{x\to 0} \dfrac{1}{2 \sin x \cos x} \dfrac{\sin x}{\cos x} = \lim\limits_{x\to 0} \dfrac{1}{2 \cos^2 x} =$

$\dfrac{1}{2 \cdot 1^2} = \dfrac{1}{2}$.

25. $\lim\limits_{v\to 0} \dfrac{\cos(v + \pi/2)}{v} = \lim\limits_{v\to 0} \dfrac{\cos v \cos \pi/2 - \sin v \sin \pi/2}{v} = \lim\limits_{v\to 0} \dfrac{0 - \sin v}{v} = -1$.

28. $\lim\limits_{x\to 0} \dfrac{1 - \cos ax}{bx} = \dfrac{a}{b} \lim\limits_{x\to 0} \dfrac{1 - \cos ax}{ax}$. With $t = ax$, $t \to 0$ as $x \to 0$. Thus the

limit is $\dfrac{a}{b} \lim\limits_{t\to 0} \dfrac{1 - \cos t}{t} = \dfrac{a}{b} \cdot 0 = 0$.

EXERCISES 8.3

1. $f'(x) = D_x(4 \cos x) = 4D_x(\cos x) = 4(-\sin x) = -4 \sin x$.

4. $f'(x) = -\sin(4 - 3x)D_x(4 - 3x) = -(-3)\sin(4 - 3x) = 3 \sin(4 - 3x)$.

7. $f'(x) = x^3 D_x \cos(1/x) + D_x(x^3) \cos(1/x) = -x^3 \sin(1/x) D_x(1/x) +$

   $3x^2 \cos(1/x) = -x^3 \sin(1/x)(-1/x^2) + 3x^2 \cos(1/x) = x \sin(1/x) +$

   $3x^2 \cos(1/x)$.

10. $f'(x) = -\csc^2(x/2) D_x(x/2) = -(1/2)\csc^2(x/2)$.

13. $f'(x) = -\csc^2(x^3-2x) D_x(x^3-2x) = -[\csc^2(x^3-2x)](3x^2-2)$, or, better,

   $-(3x^2-2)\csc^2(x^3-2x)$.

16. $f'(x) = 3(\tan^2 6x) D_x \tan 6x = 3 \tan^2 6x \sec^2 6x D_x(6x) = 18 \tan^2 6x \sec^2 6x$.

19. $f'(x) = x^2 D_x \csc 5x + (D_x x^2)\csc 5x = -x^2 \csc 5x \cot 5x D_x(5x) + 2x \csc 5x$

   $= -5x^2 \csc 5x \cot 5x + 2x \csc 5x$.

22. $f'(x) = x^2 \cdot 3 \sec^2 4x \cdot D_x \sec 4x + D_x(x^2)\sec^3 4x = 3x^2 \sec 4x(\sec 4x \tan 4x)$

   $\cdot D_x(4x) + 2x \sec^3 4x = 12x^2 \sec^3 4x \tan 4x + 2x \sec^3 4x$.

25. $f'(x) = 3 \cot^2(3x+1) D_x \cot(3x+1) = 3 \cot^2(3x+1)(-\csc^2(3x+1)) \cdot 3$.

28. $f'(x) = \dfrac{[(\tan 2x + 1)(\sec 2x \tan 2x)(2) - \sec 2x(\sec^2 2x)(2)]}{(\tan 2x + 1)^2}$.

31. $f'(x) = e^{-3x} D_x \tan \sqrt{x} + \tan \sqrt{x} D_x e^{-3x} = (e^{-3x} \sec^2 \sqrt{x})/2\sqrt{x} - 3(\tan \sqrt{x})e^{-3x}$.

34. $f'(x) = -\csc(\cot 4x)\cot(\cot 4x)D_x \cot 4x$

   $= (\csc(\cot 4x))(\cot(\cot 4x))(\csc^2 4x)(4)$.

37. $f'(x) = [(x^3+1) D_x \csc 3x - \csc 3x D_x(x^3+1)]/(x^3+1)^2$

   $= [-(x^3+1)\csc 3x \cot 3x D_x(3x) - 3x^2 \csc 3x]/(x^3+1)^2$

   $= -3[(x^3+1)\csc 3x \cot 3x + x^2\csc 3x]/(x^3+1)^2$.

40. Using logarithmic differentiation, $y = (\tan x)^{3x} \Rightarrow \ln y = 3x \ln(\tan x) \Rightarrow$

   $\dfrac{y'}{y} = 3x \dfrac{D_x(\tan x)}{\tan x} + 3 \ln(\tan x) = 3x \dfrac{\sec^2 x}{\tan x} + 3 \ln(\tan x) \Rightarrow y' =$

   $(3x \dfrac{\sec^2 x}{\tan x} + 3 \ln(\tan x))(\tan x)^{3x}$.

43. $\dfrac{dy}{dx} = \cos x - x(-\sin x) - \cos x = x \sin x$

   $\dfrac{d^2 y}{dx^2} = x \cos x + \sin x$.

46. $\dfrac{dy}{dx} = \dfrac{(\cos x + 1)(-\sin x) - (\cos x - 1)(-\sin x)}{(\cos x + 1)^2} = (-2 \sin x)/(\cos x + 1)^2$.

   $\dfrac{d^2 y}{dx^2} = \dfrac{(\cos x + 1)^2(-2 \cos x) + 2 \sin x(2)(\cos x + 1)(-\sin x)}{(\cos x + 1)^4}$

   $= (-2 \cos x(\cos x + 1) - 4 \sin^2 x)/(\cos x + 1)^3$.

49. $e^x \cot y = xe^{2y} \implies -e^x(\csc^2 y)y' + e^x \cot y = 2xe^{2y}y' + e^{2y} \implies$

$-(e^x \csc^2 y + 2xe^{2y})y' = (e^x \cot y - e^{2y})$.

52. $f'(x) = -\sin x - \cos x = 0$ if $\sin x = -\cos x$, or $\tan x = -1 \implies x = 3\pi/4$, $7\pi/4$. Tabulating as before:

Interval	$[0, 3\pi/4)$	$(3\pi/4, 7\pi/4)$	$(7\pi/4, 2\pi]$
k	$\pi/2$	π	$11\pi/6$
$f'(k)$	-1	1	$(1-\sqrt{3})/2$
$f'(x)$	$-$	$+$	$-$
Variation of f	decreasing on $[0, 3\pi/4]$	increasing on $[3\pi/4, 7\pi/4]$	decreasing on $[7\pi/4, 2\pi]$

Thus $f(3\pi/4) = -2/\sqrt{2} = -\sqrt{2}$ is a local minimum and $f(7\pi/4) = \sqrt{2}$ is a local maximum.

55. $f'(x) = -2 \sin x + 2 \cos 2x = -2 \sin x + 2(1 - 2 \sin^2 x) = -4 \sin^2 x - 2 \sin x + 2 = -4(\sin x - 1/2)(\sin x + 1) = 0$ if $\sin x = 1/2$ at $x = \pi/6$, $5\pi/6$ and if $\sin x = -1$ at $x = 3\pi/2$. Since $(\sin x + 1) \geq 0$ always, the sign of $f'(x)$ is the same as that of $-4(\sin x - 1/2)$ and, thus, opposite to that of $\sin x - 1/2$. (This follows from the final factored form of $f'(x)$ above.) On $(\pi/6, 5\pi/6)$, $\sin x > 1/2$, $\sin x - 1/2 > 0$ and, thus $f'(x) < 0$. On the remainder of $[0, 2\pi]$, namely on $[0, \pi/6]$ and $(5\pi/6, 2\pi]$, $\sin x < 1/2$ and $f'(x) > 0$. Thus f is decreasing on $[\pi/6, 5\pi/6]$ and increasing on $[0, \pi/6]$ and $[5\pi/6, 2\pi]$. Thus $f(\pi/6)$ is a local maximum, and $f(5\pi/6)$ is a local minimum. (There is no extremum at $x = 3\pi/2$, only a horizontal tangent.)

58. $f'(x) = e^{-x} \sec x \tan x - e^{-x} \sec x = e^{-x} \sec x(\tan x - 1) = 0$ only if $\tan x = 1$ since e^{-x} and $\sec x$ are never 0. Thus the critical numbers are $x = \pi/4$ and $5\pi/4$. Now, $f''(x) = e^{-x} \sec^3 x + e^{-2x} \sec^2 x (\tan x - 1)^2$. $f''(\pi/4) = e^{-\pi/4} \sqrt{2}^3 > 0 \implies f(\pi/4)$ is a local minimum. $f(5\pi/4) = e^{-5\pi/4}(-\sqrt{2})^3 < 0 \implies f(5\pi/4)$ is a local maximum.

61. $y' = \frac{3}{8} \csc^2 x \, D_x \csc x = -\frac{3}{8} \csc^3 x \cot x$. $x = \pi/6 \implies y' = -\frac{3}{8}(2)^2\sqrt{3} = -3\sqrt{3}$. Thus the tangent line is $y-1 = -3\sqrt{3} (x - \pi/6)$. Since $-1/y' = 1/3\sqrt{3}$, the normal line is $y-1 = (1/3\sqrt{3})(x-\pi/6)$.

63. Let x, y, z and θ be as shown. We
    are given that at a certain instant,
    z = 60,000 ft, θ 30° = π/6 radians
    and $\frac{d\theta}{dt} = \frac{1}{2}$ degree/sec = $\frac{\pi}{360}$
    radians/sec. We seek y and $\left|\frac{dx}{dt}\right|$.

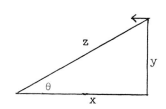

sin θ = y/z $\Rightarrow$ y = z sin θ =
60,000 sin π/6 = 30,000. tan θ =
y/x $\Rightarrow$ x = y cot θ = 30,000 cot θ
$\Rightarrow \frac{dx}{dt} = -30,000 \csc^2\theta \frac{d\theta}{dt} \Rightarrow \left|\frac{dx}{dt}\right| =$
$30,000(\csc^2 \frac{\pi}{6})\frac{\pi}{360} = 120,000\pi/360 = 1047.2$ ft/sec ≈ 714 mph.

65. Refer to my sketch for Exercise 49 of Sec. 8.6. If A and B are the angles
    shown between the horizontal and the top and bottom, respectively, of the
    billboard, then θ = A-B is to be maximized. Note that 0 < B < A < π/2 so
    that 0 < θ = A-B < π/2. If the viewer is x feet from the building, then by
    elementary trigonometry, tan A = 80/x and tan B = 60/x. Thus tan θ =

$\tan(A-B) = \frac{\tan A - \tan B}{1 + \tan A \tan B} = \frac{80/x - 60/x}{1 + 4800/x^2} = \frac{20x}{x^2+4800}$. Differentiating, we

obtain: $\sec^2\theta \frac{d\theta}{dx} = \frac{20(x^2+4800) - 40x^2}{(x^2+4800)^2}$, or $\frac{d\theta}{dx} = \frac{96,000 - 20x^2}{(x^2+4800)^2} \cos^2\theta$. Since

0 < θ < π/2, cos θ can not be zero. Thus the critical numbers are solutions
of 96000 - 20x² = 0 (and x > 0, of course). Solving, we obtain x² = 4800
$\Rightarrow$ x = $\sqrt{4800}$ = 40$\sqrt{3}$ ≈ 69.3 ft.

66. Let θ = angle AIP, d_1 = d(I,P) = the distance swum = k sec θ,
    and d_2 = the distance between the camp and P = the distance
    walked. d - d_2 = k tan θ $\Rightarrow$ d_2 = d - k tan θ.

    (a) Then c = $c_1 d_1$ + $c_2 d_2$ = c_1k sec θ + c_2(d - k tan θ). To
    minimize c, θ must be restricted to the interval
    0 ≤ θ ≤ $\tan^{-1}$(d/k).

    (b) $\frac{dc}{d\theta}$ = c_1k sec θ tan θ - c_2k $\sec^2\theta$ = 0 if c_1 sec θ   tan θ =
    $c_2 \sec^2 θ \Rightarrow c_1$ tan θ = c_2 sec θ $\Rightarrow c_1 \frac{\sin θ}{\cos θ} = \frac{c_2}{\cos θ} \Rightarrow$

    sin θ = c_2/c_1, or θ = $\sin^{-1}(c_2/c_1)$. However, if
    $\sin^{-1}(c_2/c_1)$ > $\tan^{-1}$(d/k), the minimum of c is at the right
    end point, $\tan^{-1}$(d/k). (This will happen if c_2 is only a little
    larger than c_1. Then c_2/c_1 ≈ 1 and $\sin^{-1}(c_2/c_1)$ ≈ π/2.)

70. Let r, h, V be the base radius, height, and volume of the cylinder.
    Then V = πr²h and, from the text figure, h = L sin θ and the base
    circumference is 2πr = L cos θ or r = (L/2π)cos θ. As a function

of θ, $V = K \cos^2\theta \sin\theta$ where $K = L^3/4\pi^2$.

$V' = K(\cos^3\theta - 2\cos\theta\sin^2\theta) = K\cos\theta(\cos^2\theta - 2\sin^2\theta) = 0$ when $\cos\theta = 0$ at $\theta = \pi/2$ or when $\cos^2\theta - 2\sin^2\theta = 0$. We disregard $\theta = \pi/2$ at which $V = 0$. The second condition is satisfied if $\sin^2\theta = \frac{1}{2}\cos^2\theta \Rightarrow \tan^2\theta = \frac{1}{2} \Rightarrow \tan\theta = 1/\sqrt{2} \Rightarrow \theta \approx 35.3°$.

75. Strangely, this maximum problem is solved by considering a minimum problem. Let $L = L_1 + L_2$ be, as shown, the distance between points on the walls measured on a line touching the inner corner. As $\theta \to 0$ or $\theta \to \pi/2$, $L \to \infty$, and it is clear that there is an angle θ_0 which makes L a minimum. Let L_0 be the minimum value of L. A rod of length L_0 <u>will</u> just fit around the corner. It will touch both walls when the turning angle is θ_0, but there is excess room for any other angle. However, any rod of length $> L_0$ will not fit around the corner. It will be jammed tight touching the corner and the walls, at some angle $\theta \neq \theta_0$, and further rotation will be impossible. So, our problem is to minimize L as a function of θ. By elementary trigonometry, $L = L_1 + L_2 = \dfrac{3}{\cos\theta} + \dfrac{4}{\sin\theta}$. $\dfrac{dL}{d\theta} = \dfrac{3\sin\theta}{\cos^2\theta} - \dfrac{4\cos\theta}{\sin^2\theta} = $

$\dfrac{3\sin^3\theta - 4\cos^3\theta}{\cos^2\theta\,\sin^2\theta} = 0$ when $3\sin^3\theta = 4\cos^3\theta$ or $\tan^3\theta = \dfrac{4}{3} \Rightarrow \tan\theta = \sqrt[3]{4/3}$.

Using tables or a calculator, we find that $\theta \approx 47°45'$ and $L_0 \approx 9.87$ ft.

76. In the text figure we introduce a coordinate system with the x-axis along the air-water interface and the y-axis passing through the point P, so that $P = (0,a)$. Let $A = (x,0)$ be the point at which the light ray strikes the surface, and let $Q = (c,-b)$ where $c > x$. Then $d_1 = d(P,A) = \sqrt{a^2 + x^2}$, $d_2 = d(A,Q) = \sqrt{(c-x)^2 + b^2}$. From trigonometry, $\sin\theta_1 = x/d_1$ and $\sin\theta_2 = (c-x)/d_2$. The total time of travel, T, is (time in air) + (time in water) $= d_1/v_1 + d_2/v_2$. Thus, $T = \sqrt{a^2 + x^2}/v_1 + $

$\sqrt{(c-x)^2 + b^2}/v_2 \Rightarrow \dfrac{dT}{dx} = \dfrac{x}{v_1\sqrt{a^2 + x^2}} - \dfrac{c-x}{v_2\sqrt{(c-x)^2 + b^2}}$

$= \dfrac{x}{v_1 d_1} - \dfrac{c-x}{v_2 d_2} = \dfrac{\sin\theta_1}{v_1} - \dfrac{\sin\theta_2}{v_2} = 0 \Rightarrow \dfrac{\sin\theta_1}{v_1} = \dfrac{\sin\theta_2}{v_2}$,

Snell's Law.

78. $s(t) = 4\sin\pi t \Rightarrow$ amplitude $= |4| = 4$, period $= 2\pi/\pi = 2$, frequency $= \pi/2\pi = 1/2$. Note that $v(t) = 4\pi\cos\pi t < 0$ if $\pi/2 < \pi t < 3\pi/2$ or $1/2 < t$

< 3/2. Similarly, v(t) > 0 if 0 ≤ t < 1/2 or 3/2 < t ≤ 2. Thus the motion
is in the "+" direction (from s = 0 to s = 4) during the time interval
0 ≤ t ≤ 1/2; then in the "-" direction (from s = 4 to s = -4) during the
interval 1/2 ≤ t ≤ 3/2, and, finally in the "+" direction (from s = -4 to
s = 0) in the interval 3/2 ≤ t ≤ 2. This motion repeats itself every 2
seconds, and the motion is oscillatory.

86. Preliminary analysis: writing the equation as cos x + x - 2 = 0, let f(x)
= cos x + x - 2. Then $f(\pi/2) = \pi/2 - 2 < 0$ and $f(\pi) = -1 + \pi - 2 > 0$.
Thus the root lies between $\pi/2 \approx 1.57$ and $\pi \approx 3.14$. Our initial guess will
be x_1 = 3.

i	x_i	$f(x_i)$
1	3.0000	0.0100
2	2.9883	0.0001
3	2.9883	0.0000

Thus, the root is 2.9883 to 4 places.

89. Here $x_{n+1} = x_n - \dfrac{\sin x_n}{\cos x_n} = x_n - \tan x_n$.

(a) x_1 = 3

n	x_n
1	3.00000
2	3.14255
3	3.14159
4	3.14159

(b) x_1 = 6

n	x_n
1	6.00000
2	6.29181
3	6.28319
4	6.28319

In (b), $x_n \to 2\pi$.

EXERCISES 8.4

1. With u = 4x, du = 4 dx, dx = (1/4)du and $\int \cos 4x\, dx = \frac{1}{4}\int \cos u\, du = $
$\frac{1}{4}\sin u + C = \frac{1}{4}\sin 4x + C$.

4. With $u = x^3$, $\int x^2 \cot x^3 \csc x^3 dx = (1/3)\int \cot u \csc u\, du = -(1/3)\csc u + C$
$= -(1/3)\csc x^3 + C$.

7. $\int \frac{1}{\cos 2x}\, dx = \int \sec 2x\, dx = \frac{1}{2}\ln|\sec 2x + \tan 2x| + C$, (u = 2x could be
used as in #1).

10. $\int (x + \csc 8x)dx = x^2/2 + (1/8)\ln|\csc 8x - \cot 8x| + C$.

13. With $u = \sin 3x$, $du = 3 \cos 3x \, dx$, $\cos 3x \, dx = (1/3)du$ and

$$\int \cos 3x \sqrt[3]{\sin 3x} \, dx = \frac{1}{3} \int u^{1/3} \, du = \frac{1}{3} \cdot \frac{3}{4} u^{4/3} + C = \frac{1}{4}(\sin 3x)^{4/3} + C.$$

16. With $u = \sin x$, $du = \cos x \, dx$ and $\int \sin^3 x \cos x \, dx = \int u^3 \, du =$

$$\frac{1}{4} u^4 + C = \frac{1}{4} \sin^4 x + C.$$

19. With $u = \tan x$, $du = \sec^2 x \, dx$, $\displaystyle\int_0^{\pi/4} \tan x \sec^2 x \, dx = \int_0^1 u \, du = 1/2.$

22. $\displaystyle\int_{\pi/6}^{\pi/2} \frac{\cos^2 x}{\sin x} \, dx = \int_{\pi/6}^{\pi/2} \frac{1 - \sin^2 x}{\sin x} \, dx = \int_{\pi/6}^{\pi/2} (\csc x - \sin x)dx = \ln|\csc x - \cot x|$

$$+ \cos x]_{\pi/6}^{\pi/2} = 0 - (\ln|2-\sqrt{3}| + \sqrt{3}/2).$$

25. With $u = x + \cos x$, $du = (1 - \sin x)dx$ and $\displaystyle\int \frac{1 - \sin x}{x + \cos x} \, dx = \int \frac{du}{u} = \ln|u| + C$

$= \ln|x + \cos x| + C.$

28. $\displaystyle\int_0^{\pi/4} (1 + \sec x)^2 dx = \int_0^{\pi/4} (1 + 2 \sec x + \sec^2 x)dx = [x + 2 \ln|\sec x + \tan x|$

$$+ \tan x]_0^{\pi/4} = \pi/4 + 2 \ln|\sqrt{2}+1| + 1.$$

31. Since $\sin x = \dfrac{1}{\csc x}$, $\displaystyle\int \frac{e^{\cos x}}{\csc x} \, dx = \int e^{\cos x} \sin x \, dx.$ Let $u = \cos x$, $du =$

$-\sin x$ and the integral is $-\displaystyle\int e^u \, du = -e^u + C = -e^{\cos x} + C.$

34. With $u = 1 + 3 \sec x$, $du = 3 \sec x \tan x \, dx$, and $\displaystyle\int \frac{\sec x \tan x}{1 + 3 \sec x} \, dx = \frac{1}{3} \int \frac{du}{u} =$

$\frac{1}{3} \ln|u| + C = \frac{1}{3} \ln|1 + 3 \sec x| + C.$

37. For $x \in [-\pi/4, \pi/4]$, $x \le \pi/4 < 1 \le \sec x$. Thus $A = \displaystyle\int_{-\pi/4}^{\pi/4} (\sec x - x)dx =$

$\ln|\sec x + \tan x| - \dfrac{x^2}{2}]_{-\pi/4}^{\pi/4} = \ln|\sqrt{2} + 1| - \ln|\sqrt{2} - 1|.$

40. Using the shell method, $V = 2\pi \displaystyle\int_0^{\sqrt{\pi}/2} x \tan x^2 \, dx.$ With $u = x^2$, $du = 2x \, dx$,

$x = 0 \Rightarrow u = 0$, $x = \sqrt{\pi}/2 \Rightarrow u = \pi/4$, $V = \pi \displaystyle\int_0^{\pi/4} \tan u \, du = \pi \ln|\sec u|]_0^{\pi/4}$

$= \pi(\ln\sqrt{2} - \ln 1) = \pi \ln\sqrt{2}$ or $(\pi \ln 2)/2.$

43. First way: $u = \tan x \Rightarrow du = \sec^2 x \, dx \Rightarrow I = \int u \, du = u^2/2 + C =$

$\tan^2 x/2 + C.$ Second way: $u = \sec x \Rightarrow du = \sec x \tan x \, dx \Rightarrow$

$I = \int u \, du = u^2/2 + D = \sec^2 x/2 + D.$ The answers can be reconciled

because each answer is an antiderivative of f(x) = tan x sec²x and
hence one is equal to the other plus a constant, i.e., they differ
by a constant. Using the relation sec²x = tan² x + 1, the second
answer is tan² x/2 + D + 1/2, and their difference is the constant
C - D - 1/2.

46. With $f(x) = \sin \sqrt{x}$ and $(b - a)/n = \pi/4$ we obtain:

(a) $I \approx \frac{\pi}{8}[f(0) + 2f(\pi/4) + 2f(\pi/2) + 2f(3\pi/4) + f(\pi)] \approx$

$\frac{\pi}{8}[0 + 1.5494 + 1.9000 + 1.9987 + 0.9797] \approx 2.52.$

(b) $I \approx \frac{\pi}{12}[f(0) + 4f(\pi/4) + 2f(\pi/2) + 4f(3\pi/4) + f(\pi)] \approx$

$\frac{\pi}{12}[0 + 3.0988 + 1.9000 + 3.9974 + 0.9797] \approx 2.61.$

49. (a) The sign of q'(t) is that of sin 2πt since k and q are > 0.
Now, sin 2πt > 0 if 0 < 2πt < π or 0 < t < 1/2. Since t = 0 is
the start of spring and t = 1/4 the start of summer, etc., this
means that q' > 0 and q is increasing during spring and summer.
Similarly, q' < 0 and q is decreasing for 1/2 < t < 1, during
fall and winter.

(b) Integrating the given relation yields $\ln q(t) = -\frac{k}{2\pi} \cos 2\pi t + C.$
Set t = 0, $q = q_0$ to obtain $\ln q_0 = -\frac{k}{2\pi} + C$ or $C = \ln q_0 + k/2\pi.$

Thus, $\ln q(t) = \frac{k}{2\pi}(1 - \cos 2\pi t) + \ln q_0$ or $q(t) =$

$q_0 e^{k(1-\cos 2\pi t)/2\pi}.$

51. Let x be angle PBC and f(x) = d(P,B). Then f(x) = a sec x,
$0 \leq x \leq \theta$, and the average of f is $(1/\theta)\int_0^\theta a \sec x \, dx =$

$\frac{a}{\theta} \ln|\sec \theta + \tan \theta|.$

52. If the shape is a square of area A, then each side has length
$s = \sqrt{A}$. The distance a, the perpendicular distance from the
center to the side, is then $\sqrt{A}/2$. Since $\theta = \pi/4$, the average
distance of P from C is $\frac{\sqrt{A}/2}{\pi/4} \ln|\sec \frac{\pi}{4} + \tan \frac{\pi}{4}| =$

$\frac{2}{\pi} \ln(\sqrt{2} + 1)\sqrt{A} \approx 0.56 \sqrt{A}.$

If it is a regular hexagon of area A, it is made up of 6 equilateral triangles each of area A/6, side s, and altitude a.
Each of these is the union of 2 right triangles (such as pictured
in the preceding text problem), with horizontal leg a, vertical
leg s/2, interior angle $\theta = \pi/6$, and area A/12. Thus, $\frac{1}{2} a \frac{s}{2} = \frac{A}{12}$

and $\frac{s}{2} = a \tan \frac{\pi}{6} = a/\sqrt{3}$. Thus, $\frac{A}{12} = \frac{a^2}{2\sqrt{3}}$, and $a^2 = \frac{\sqrt{3}}{6} A \Rightarrow a =$

$\frac{\sqrt[4]{3}}{\sqrt{6}} \sqrt{A}$. Thus, the average distance of P from C is

$$\frac{a}{\theta} \ln|\sec\theta + \tan\theta| = \frac{(\sqrt{3}/\sqrt{6})\sqrt{A}}{\pi/6} \ln|\sec\frac{\pi}{6} + \tan\frac{\pi}{6}| =$$

$$\frac{\sqrt{3}\sqrt{6}}{\pi} \ln(\frac{2}{\sqrt{3}} + \frac{1}{\sqrt{3}})\sqrt{A} \approx 0.564\sqrt{A}.$$

(These results should not be surprising to you if you think of
the limiting case as the number of sides increases, and the
region approaches a circle of area A and radius r. The actual
(not <u>average</u>) distance of each point on the circle from C is r.
$\pi r^2 = A \Rightarrow r = \sqrt{A}/\sqrt{\pi} \approx 0.564\sqrt{A}!$)

EXERCISES 8.5

1. $\sin^{-1}(\pm\sqrt{3}/2) = \pm\pi/3$ since $\sin(\pm\pi/3) = \pm\sqrt{3}/2$ and both angles are in $[-\pi/2, \pi/2]$.

4. (a) $\arcsin(-1) = -\pi/2$ since $\sin(-\pi/2) = -1$ and $-\pi/2 \in [-\pi/2, \pi/2]$.

    (b) $\cos^{-1}(-1) = \pi$ since $\cos\pi = -1$ and $\pi \in [0,\pi]$.

7. $\sin(\cos^{-1}(\sqrt{3}/2)) = \sin\pi/6 = 1/2$.

10. $\tan(\tan^{-1}10) = 10$ since $\tan x$, $\tan^{-1}x$ are inverse functions.

13. Let $a = \sin^{-1}3/5$. Then $\sin a = 3/5$, $\cos a = \sqrt{1-9/25} = 4/5$. Let $b = \tan^{-1}4/3$.
    Then $\sin b = 4/5$, $\cos b = 3/5$. $\cos(\sin^{-1}3/5 + \tan^{-1}4/3) = \cos(a+b) =$

    $\cos a \cos b - \sin a \sin b = (\frac{4}{5})(\frac{3}{5}) - (\frac{3}{5})(\frac{4}{5}) = 0.$

16. Let $a = \sin^{-1}8/17$. Then $\sin a = \frac{8}{17}$, $\cos a = \sqrt{1-(8/17)^2} = \frac{15}{17}$. Then

    $\cos(2\sin^{-1}8/17) = \cos(2a) = \cos^2 a - \sin^2 a = (\frac{15}{17})^2 - (\frac{8}{17})^2 = \frac{161}{289}.$

19. Two equivalent methods can be used : let $y = \tan^{-1}x$ so that $x =$
    $\tan y = \sin y/\sqrt{1 - \sin^2 y}$, and now solve for $\sin y$. An alternate method
    is to sketch a right triangle with y as interior
    angle whose tangent is x. We can do this by mak-
    ing the opposite side of length x and the adjacent
    side of length 1. Then the hypotenuse is $\sqrt{x^2+1}$
    and $\sin(\tan^{-1}x) = \sin y = $ opposite/hypotenuse
    $= x/\sqrt{x^2+1}$.

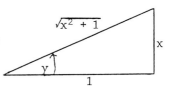

22. Let $y = \tan^{-1}x$. Then $\cos 2y = \cos^2 y - \sin^2 y = (1/\sqrt{x^2+1})^2 - (x/\sqrt{x^2+1})^2 =$
    $(1-x^2)/(x^2+1)$. (Refer to the sketch for #19.)

25. Let $y = \arctan x$. Then the right side of the desired identity is $2y$. Now,
    from #19 above, $\sin y = \sin(\arctan x) = x/\sqrt{x^2+1}$. By the figure there, $\cos y$
    $= 1/\sqrt{x^2+1}$. Thus $2\sin y \cos y = 2x/(x^2+1)$, and the left side of the identity
    is $\arcsin(2x/(x^2+1)) = \arcsin(2\sin y \cos y) = \arcsin(\sin 2y) = 2y$, the same
    as the right side.

28. cos(arccos(-x)) = -x and, from the formula cos(π-θ) = -cos θ we obtain
    cos(π-arccos x) = -cos(arccos x) = -x. The identity is thus established
    since cos x is one-to-one on [0,π].

34.   37.

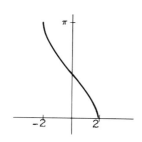   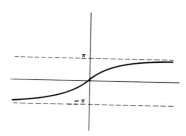

40. Let a = arccos x. Then x = cos a and y = sin a. Since $0 \le$ arccos x $\le$ π,
    $0 \le$ sin a $\le$ 1 and hence, $0 \le y \le$ 1. y = sin a = $\sqrt{1-\cos^2 a}$ = $\sqrt{1-x^2}$ $\Rightarrow$ y^2 =
    $1-x^2$ $\Rightarrow$ $x^2 + y^2$ = 1. Since $y \ge$ 0, the graph is the upper half of the unit
    circle centered at (0,0).

43. Set x = 1. The left side is $\tan^{-1} 1$ = π/4 $\approx$ 0.785. The right side is
    1/tan 1 $\approx$ 1/1.557 $\approx$ 0.642; definitely not π/4.

46. Using the quadratic formula, sin t = $(-7 \pm \sqrt{49 - 36})/6$ =
    $(-7 \pm \sqrt{13})/6$. The "-" sign yields a number not in the range of
    the sine function, and it must be rejected. Thus, sin t =
    $(-7 + \sqrt{13})/6$ $\Rightarrow$ t = $\sin^{-1}[(-7 + \sqrt{13})/6]$ $\approx$ -0.6013.

49. Let z = sin θ so that the equation becomes $6z^3 + 18z^2 - 5z - 15 = 0$,
    or $6z^2(z + 3) - 5(z + 3) = (6z^2 - 5)(z + 3) = 0$ $\Rightarrow$ z = -3,
    $\pm\sqrt{5/6}$. Rejecting the root -3, which is not in the range of sine,
    we have the solutions z = sin θ = $\pm\sqrt{5/6}$ $\Rightarrow$ θ = ±1.1503. (The
    text answer is obtained by multiplying numerator and denominator
    of $\sqrt{5/6} = \sqrt{5}/\sqrt{6}$ by $\sqrt{6}$ to obtain $\sqrt{5}\sqrt{6}/(\sqrt{6})^2 = \sqrt{30}/6$.)

EXERCISES 8.6

1. f'(x) = $\dfrac{D_x(3x-5)}{1 + (3x-5)^2}$ = $\dfrac{3}{9x^2-30x+26}$.

2. f'(x) = $\dfrac{D_x(x/3)}{\sqrt{1 - (x/3)^2}}$ = $\dfrac{1/3}{\sqrt{1 - x^2/9}}$ = $\dfrac{1}{\sqrt{9 - x^2}}$.

4. f'(x) = $2x/(1 + (x^2)^2) = 2x/(1+x^4)$.

7. f'(x) = $x^2 \cdot \dfrac{2x}{1+x^4}$ + 2x arctan(x^2).

10. $f'(x) = x^2 \cdot \dfrac{5}{5x\sqrt{(5x)^2-1}} + 2x \sec^{-1}5x = x/\sqrt{25x^2-1} + 2x \sec^{-1}5x.$

13. $f'(x) = D_x(\sin^{-1}x)^{-1} = -(\sin^{-1}x)^{-2} D_x\sin^{-1}x = -1/\sqrt{1-x^2}\ (\sin^{-1}x)^2.$

16. $f'(x) = 4(\dfrac{1}{x} - \arcsin\dfrac{1}{x})^3 D_x(\dfrac{1}{x} - \arcsin\dfrac{1}{x}).$ The last derivative is

$-\dfrac{1}{x^2} - \dfrac{D_x(1/x)}{\sqrt{1-1/x^2}} = -\dfrac{1}{x^2}(1 - \dfrac{1}{\sqrt{1-1/x^2}}).$

19. $f'(x) = \sqrt{x}\, D_x \sec^{-1}\sqrt{x} + \sec^{-1}\sqrt{x}\, D_x\, \sqrt{x} = \dfrac{\sqrt{x}\, D_x\,\sqrt{x}}{\sqrt{x}\,\sqrt{(\sqrt{x})^2-1}} + \dfrac{\sec^{-1}\sqrt{x}}{2\sqrt{x}}$

$= \dfrac{1/2\sqrt{x}}{\sqrt{x-1}} + \dfrac{\sec^{-1}\sqrt{x}}{2\sqrt{x}}.$

22. $f'(x) = x\, D_x \arccos\sqrt{4x+1} + \arccos\sqrt{4x+1}\, D_x\, x = \dfrac{-x\, D_x\sqrt{4x+1}}{\sqrt{1 - \sqrt{4x+1}^2}} + \arccos\sqrt{4x+1}$

$= \dfrac{-x(4/2\sqrt{4x+1})}{\sqrt{-4x}} + \arccos\sqrt{4x+1} = -2x/\sqrt{4x+1}\,\sqrt{-4x} + \arccos\sqrt{4x+1}.$

(Note that the domain of f is [-1/4,0], so there's no problem with $\sqrt{-4x}$.)

25. $x^2 + x \sin^{-1}y = ye^x \Longrightarrow 2x + xy'/\sqrt{1-y^2} + \sin^{-1}y = ye^x + y'e^x \Longrightarrow$

$(x/\sqrt{1-y^2} - e^x)y' = ye^x - 2x - \sin^{-1}y$ which yields the given answer.

28. With $u = e^x$, $du = e^x dx$, $\displaystyle\int_0^1 \dfrac{e^x}{1+e^{2x}}\,dx = \int_1^e \dfrac{du}{1+u^2} = \tan^{-1}u]_1^e = \tan^{-1}e - \tan^{-1}1$

$= \tan^{-1}e - \pi/4.$

31. With $u = \cos x$, $du = -\sin x\, dx$, $\displaystyle\int \dfrac{\sin x}{\cos^2x + 1}\,dx = -\int \dfrac{du}{u^2+1} = -\tan^{-1}u + C$

$= -\tan^{-1}(\cos x) + C.$

34. First, $\displaystyle\int \dfrac{dx}{e^x\sqrt{1-e^{-2x}}} = \int \dfrac{e^{-x}}{\sqrt{1 - e^{-2x}}}\,dx.$ Let $u = e^{-x}$ so that this becomes

$-\displaystyle\int \dfrac{du}{\sqrt{1-u^2}} = -\sin^{-1}u + C = -\sin^{-1}(e^{-x}) + C.$

37. First multiply and divide by x^2 so that $\displaystyle\int \dfrac{1}{x\sqrt{x^6-4}}\,dx = \int \dfrac{x^2}{x^3\sqrt{x^6-4}}\,dx.$

Let $u = x^3$, $du = 3x^2\,dx$, $x^2dx = (1/3)du.$ The integral becomes $\dfrac{1}{3}\displaystyle\int \dfrac{1}{u\sqrt{u^2-4}}\,du$

$= \dfrac{1}{3} \cdot \dfrac{1}{2} \sec^{-1}\dfrac{u}{2} + C = \dfrac{1}{6}\sec^{-1}\dfrac{x^3}{2} + C.$

40. We use a slightly different substitution, or change of variable, technique.
Let $u = \sqrt{x}$. Then $x = u^2$ and $dx = 2u\,du$, and

$\displaystyle\int \dfrac{1}{x\sqrt{x-1}}\,dx = \int \dfrac{2u}{u^2\sqrt{u^2-1}}\,du = 2\int \dfrac{1}{u\sqrt{u^2-1}}\,du = 2\sec^{-1}u + C = 2\sec^{-1}\sqrt{x} + C.$

43. $A = \int_{-2}^{2} \dfrac{4}{\sqrt{16-x^2}}\, dx = 4 \sin^{-1} \dfrac{x}{4}\Big]_{-2}^{2} = 4[\dfrac{\pi}{6} - (-\dfrac{\pi}{6})] = \dfrac{4\pi}{3}$.

46. With $f(x) = \arcsin x$, $x = .25$, $\Delta x = .01$, $\Delta f \approx df = f'(.25)(.01) =$

$\dfrac{1}{\sqrt{1-.25^2}}\,(.01) = \dfrac{.01}{\sqrt{1 - 1/16}} = .04/\sqrt{15}$.

49. If A and B are the angles shown, between the
horizontal and the top and bottom, respectively,
of the billboard, then $\theta = A-B$ is to be
maximized. If the viewer is x feet from the
base of the building then $\theta = \tan^{-1}(80/x)$
$- \tan^{-1}(60/x)$ and $\dfrac{d\theta}{dx} = \dfrac{-80x^2}{1+(80/x)^2} - \dfrac{-60/x^2}{1+(60/x)^2}$

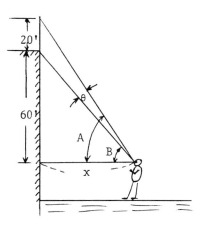

$= \dfrac{60}{x^2 + 3600} - \dfrac{80}{x^2 + 6400}$; $\dfrac{d\theta}{dx} = 0$ if $60(x^2+6400)$

$- 80(x^2 + 3600) = 0$ or $96,000 - 20x^2 = 0$. Thus
$x = \sqrt{4800} = 40\sqrt{3} \approx 69.3$ ft.

52. $y' = 2/(1 + 4x^2)$. The given line has slope 2/13. $y' = 2/13 \Rightarrow$
$2/(1 + 4x^2) = 2/13 \Rightarrow 1 + 4x^2 = 13 \Rightarrow x^2 = 3 \Rightarrow x = \pm\sqrt{3}$.

55. We use the disc method. Since $e^x \geq (1/\sqrt{x^2 + 1}$ on $[0,1]$, we have
$V = \pi \int_{0}^{1} [(e^x)^2 - (1/\sqrt{x^2 + 1})^2] dx = \pi \int_{0}^{1} [e^{2x} - (1/x^2 + 1)] dx =$
$\pi[(1/2)e^{2x} - \arctan x]_{0}^{1} = \pi[(1/2)e^2 - \arctan 1]$
$- \pi[(1/2)(1) - \arctan 0] = \pi[(e^2/2) - (\pi/4)] - \pi[1/2]$
$= (\pi/2)[e^2 - (\pi/2) - 1]$.

59. $\int_{0}^{1} 4/(1 + x^2) dx = 4 \tan^{-1} x]_{0}^{1} = 4(\tan^{-1} 1 - \tan^{-1} 0) =$
$4(\pi/4 - 0) = \pi$. With $a = 0$, $b = 1$, and $n = 10$, the approxima-
tions to π are, to 4 places, 3.1416 by Simpson's Rule and 3.1599
by the Trapezoidal Rule.

EXERCISES 8.7

1. and 4. These are immediate from the definitions and some easy calculations.

7. $2 \sinh x \cosh x = (2/4)(e^x - e^{-x})(e^x + e^{-x}) = (1/2)(e^{2x} - e^{-2x}) = \sinh 2x$.

10. $\tanh 2x = \dfrac{\sinh 2x}{\cosh 2x} = \dfrac{2 \sinh x \cosh x}{\cosh^2 x + \sinh^2 x}$ (by #7,8)

$= \dfrac{2 \tanh x}{1 + \tanh^2 x}$ (dividing numerator and denominator by $\cosh^2 x$).

13. $2 \sinh \frac{x+y}{2} \cosh \frac{x-y}{2} = 2 \frac{(e^{\frac{x+y}{2}} - e^{\frac{-x-y}{2}})}{2} \frac{(e^{\frac{x-y}{2}} + e^{\frac{-x-y}{2}})}{2}$. When the first terms

in each parentheses are multiplied together, we get: $e^{\frac{x+y}{2}} e^{\frac{x-y}{2}} = e^{\frac{x+y+x-y}{2}} =$

$e^{\frac{2x}{2}} = e^x$. Similarly, the other three products can be simplified, and we ob-

tain for the entire expression: $(e^x + e^y - e^{-y} - e^{-x})/2 =$

$(e^x - e^{-x})/2 + (e^y - e^{-y})/2 = \sinh x + \sinh y$, as desired.

16. $f'(x) = \sinh\sqrt{4x^2+3}\ D_x\sqrt{4x^2+3} = \frac{(\sinh\sqrt{4x^2+3})(8x)}{2\sqrt{4x^2+3}}$.

19. $f'(x) = \frac{(x^2+1)(-\operatorname{sech} x^2 \tanh x^2)D_x x^2 - (\operatorname{sech} x^2)D_x(x^2+1)}{(x^2+1)^2}$

   $= \frac{-[(x^2+1)(\operatorname{sech} x^2 \tanh x^2)2x + (\operatorname{sech} x^2)(2x)]}{(x^2+1)^2}$.

22. $f'(x) = \cosh(x^2+1)\ D_x(x^2+1) = 2x \cosh(x^2+1)$.

25. $f'(x) = \frac{D_x \sinh 2x}{\sinh 2x} = \frac{2 \cosh 2x}{\sinh 2x} = 2 \coth 2x$.

28. $f'(x) = (1/2)(\operatorname{sech} 5x)^{-1/2}\ D_x \operatorname{sech} 5x = -(5 \operatorname{sech} 5x \tanh 5x)/2\sqrt{\operatorname{sech} 5x}$.

31. $\sinh xy = ye^x \Longrightarrow (\cosh xy)(xy' + y) = ye^x + y'e^x \Longrightarrow (x \cosh xy - e^x)y' = y(e^x - \cosh xy)$.

34. With $u = \ln x$, $du = (1/x)dx$, $\displaystyle\int \frac{\cosh \ln x}{x}\ dx = \int \cosh u\ du = \sinh u + C$
   $= \sinh \ln x + C$.

37. With $u = \sinh x$, $\displaystyle\int \sinh x \cosh x\ dx = \int u\ du = u^2/2 + C = (\sinh^2 x)/2 + C$.

40. Let $u = \cosh x$, $du = \sinh x\ dx$. Then $\displaystyle\int \sinh x\sqrt{\cosh x}\ dx = \int \sqrt{u}\ du = \frac{2}{3}u^{3/2}$
   $+ C = \frac{2}{3}(\cosh x)^{3/2} + C$.

43. Let $u = 1 - 2 \tanh x$, $du = -2 \operatorname{sech}^2 x\ dx$. Then $\displaystyle\int \frac{\operatorname{sech}^2 x}{1 - 2 \tanh x}\ dx = -\frac{1}{2}\int \frac{du}{u}$
   $= -\frac{1}{2} \ln|u| + C = -\frac{1}{2} \ln|1 - 2 \tanh x| + C$.

46. $L_0^1 = \displaystyle\int_0^1 \sqrt{1 + (D_x \cosh x)^2}\ dx = \int_0^1 \sqrt{1 + \sinh^2 x}\ dx = \int_0^1 \cosh x\ dx = \sinh x\Big]_0^1$
   $= \sinh 1 \approx 1.175$.

52.

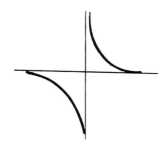

55. By symmetry we can double the area of the right half from $x = 0$
to $x = 315$. Thus, $A = 2\int_0^{315}(-127.7\ \cosh\frac{x}{127.7} + 757.7)dx =$

$2(-127.7^2\ \sinh\frac{x}{127.7} + 757.7x)]_0^{315} =$

$2(-127.7^2\ \sinh\frac{315}{127.7} + 757.7 \cdot 315) \approx 2(143,287) = 286,574\ ft^2$.

(b) $y' = \sinh\frac{x}{127.7} \implies 1 + (y')^2 = 1 + \sinh^2\frac{x}{127.7} = \cosh^2\frac{x}{127.7}$

$\implies L = 2\int_0^{315} \cosh\frac{x}{127.7}dx = 2(127.7\ \sinh\frac{315}{127.7}) \approx 1,494\ ft$.

58. $y = A\cosh Bx \implies y' = AB\sinh Bx \implies y'' = AB^2\cosh Bx$.
$yy'' = (y')^2 \iff A^2B^2\cosh^2 Bx = 1 + A^2B^2\sinh^2 Bx \iff$
$A^2B^2(1 + \sinh^2 Bx) = 1 + A^2B^2\sinh^2 Bx$ since $\cosh^2 = 1 + \sinh^2$.
Now cancel $A^2B^2 \sinh^2 Bx$ from both sides to obtain $A^2B^2 = 1 \iff$
$(AB)^2 = 1 \iff AB = 1$ since both A and B are positive.

EXERCISES 8.8

1. $y = \cosh^{-1}x \implies x = \cosh y = (e^y + e^{-y})/2,\ (x \geq 1),\ \implies e^{2y} - 2xe^y + 1 = 0$

$\implies e^y = (2x \underset{-}{+} \sqrt{4x^2-4})/2 \implies e^y = x + \sqrt{x^2-1}$. (The + sign must be chosen to
guarantee that $x \to \infty \iff y \to \infty$.)

4. $D_x(\tanh^{-1}u) = D_u(\frac{1}{2} \ln \frac{1+u}{1-u})\ D_xu$ 10.

$= \frac{1}{2}\ D_u(\ln(1+u) - \ln(1-u)D_xu$

$= \frac{1}{2}(\frac{1}{1+u} + \frac{1}{1-u})D_xu = \frac{1}{1-u^2}\ D_xu$.

13. $f'(x) = \dfrac{D_x\sqrt{x}}{\sqrt{(\sqrt{x})^2 - 1}} = 1/2\ \sqrt{x}\ \sqrt{x-1}$.

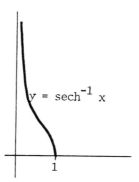

$y = \text{sech}^{-1}\ x$

1

16. $f'(x) = \dfrac{D_x\sin 3x}{1 - \sin^2 3x} = \dfrac{3\cos 3x}{\cos^2 3x} = 3\sec 3x$.

19. $f'(x) = \dfrac{D_x\cosh^{-1}4x}{\cosh^{-1}4x} = 4/(\sqrt{16x^2-1}\ \cosh^{-1}4x)$.

22. With $u = 4x$, $\displaystyle\int \frac{1}{\sqrt{16x^2-9}}\,dx = \frac{1}{4}\int \frac{du}{\sqrt{u^2-9}} = (1/4)\cosh^{-1}\left(\frac{u}{3}\right) + C =$

$(1/4)\cosh^{-1}(4x/3) + C.$

25. With $u = e^x$, $\displaystyle\int \frac{e^x}{\sqrt{e^{2x}-16}} = \int \frac{du}{\sqrt{u^2-16}} = \cosh^{-1}(u/4) + C = \cosh^{-1}(e^x/4) + C.$

28. First multiply and divide the integrand by e^x. Then let $u = e^x$, $du = e^x dx$

and $\displaystyle\int \frac{1}{\sqrt{5-e^{2x}}}\,dx$ and $\displaystyle\int \frac{e^x dx}{e^x\sqrt{5-e^{2x}}} = \int \frac{du}{u\sqrt{5-u^2}} = -\frac{1}{\sqrt{5}}\,\text{sech}^{-1}\frac{|u|}{5} + C =$

$-\dfrac{1}{\sqrt{5}}\,\text{sech}^{-1}\dfrac{e^x}{5} + C.$ ($|e^x| = e^x$ since $e^x > 0$.)

31. If at time t, the point is at $P(1,y)$, then the motion is vertical along the line $x = 1$ so that the velocity is dy/dt and $d(0,P) = \sqrt{1 + y^2}$. Thus, $\dfrac{dy}{dt} = k\sqrt{1 + y^2}$. When $t = 0$, $y = 0$ and $\dfrac{dy}{dt} = 3$. Thus, $k = 3$ and $\dfrac{dy}{dt} = 3\sqrt{1 + y^2} \Rightarrow \dfrac{dy}{\sqrt{1 + y^2}} = 3dt \Rightarrow$

$\sinh^{-1} y = 3t + C \Rightarrow y = \sinh(3t + C).$ Set $t = 0$, $y = 0$ to obtain $0 = \sinh C \Rightarrow C = 0$ so that $y = \sinh 3t$.

EXERCISES 8.9 (Review)

1. $\displaystyle\lim_{x\to 0}\frac{x^2}{\sin x} = \lim_{x\to 0} x \cdot \frac{x}{\sin x} = \lim_{x\to 0} x\,\frac{1}{(\sin x)/x} = 0 \cdot \frac{1}{1} = 0.$

4. $\displaystyle\lim_{x\to 0}\frac{2 - \cos x}{1 + \sin x} = \frac{2 - 1}{1 + 0} = 1.$

7. $f'(x) = -\sin\sqrt{3x^2+x}\; D_x\sqrt{3x^2+x} = (-\sin\sqrt{3x^2+x})\left(\frac{1}{2}\right)(3x^2+x)^{-1/2} D_x(3x^2+x)$

$= (-\sin\sqrt{3x^2+x})(6x+1)/2\sqrt{3x^2+x}.$

10. $f'(x) = D_x(x^3 + \csc 6x)^{1/3} = (1/3)(x^3 + \csc 6x)^{-2/3} D_x(x^3 + \csc 6x) =$

$(1/3)(x^3 + \csc 6x)^{-2/3}(3x^2 - 6\csc 6x \cot 6x)$ or

$(3x^2 - 6\csc 6x \cot 6x)/3\sqrt[3]{(x^3 + \csc 6x)^2}.$

13. $f'(x) = (\sin^{-1}5x\; D_x(3x+7)^4 - (3x+7)^4 D_x \sin^{-1}5x)/(\sin^{-1}5x)^2.$

$= [(\sin^{-1}5x)4(3x+7)^3 3 - (3x+7)^4 \cdot \dfrac{5}{\sqrt{1-25x^2}}]/(\sin^{-1}5x)^2.$

16. $f'(x) = 5^{\tan 2x}(\ln 5)\, D_x \tan 2x = 5^{\tan 2x}(\ln 5)(2\sec^2 2x).$

19. $f(x) = \ln(\csc^3 2x) = 3\ln\csc 2x \Rightarrow f'(x) = 3 \cdot \dfrac{D_x \csc 2x}{\csc 2x} =$

$3 \cdot \dfrac{-(\csc 2x \cot 2x)(2)}{\csc 2x} = -6\cot 2x.$

22. $f(x) = \cot(1/x) + 1/\cot x = \cot(1/x) + \tan x \Longrightarrow f'(x) = -\csc^2(1/x)D_x(1/x)$
     $+ \sec^2 x = (1/x^2)\csc^2(1/x) + \sec^2 x.$

25. $f'(x) = \sinh e^{-5x} D_x e^{-5x} = (\sinh e^{-5x})(-5e^{-5x}).$

28. $f'(x) = \sec(\sec x)\tan(\sec x)D_x \sec x = \sec(\sec x)\tan(\sec x)(\sec x \tan x).$

31. $f'(x) = 3 \sin^2 e^{-2x} D_x \sin e^{-2x} = 3 \sin^2 e^{-2x} \cos e^{-2x} D_x(e^{-2x}) =$
     $3 \sin^2 e^{-2x} \cos e^{-2x} (-2e^{-2x}).$

34. $f'(x) = \sec 5x D_x \tan 5x + \tan 5x D_x \sec 5x = 5 \sec^3 5x + 5 \tan^2 5x \sec 5x.$

37. $f'(x) = \dfrac{D_x\sqrt{1-x^2}}{\sqrt{1-(\sqrt{1-x^2})^2}} = \dfrac{-x}{\sqrt{1-x^2}\,\sqrt{x^2}} = \pm\,\dfrac{1}{\sqrt{1-x^2}},$ $+$ if $x < 0,$ $-$ if $x > 0.$

40. $f'(x) = \dfrac{D_x\sqrt{\tan 2x}}{1 + (\sqrt{\tan 2x})^2} = \dfrac{D_x(\tan 2x)/2\sqrt{\tan 2x}}{1 + \tan 2x}$
     $= \dfrac{2 \sec^2 2x}{2\sqrt{\tan 2x}(1 + \tan 2x)} = \dfrac{\sec^2 2x}{\sqrt{\tan 2x}(1 + \tan 2x)}$

43. $f'(x) = \dfrac{D_x\tan^{-1}x}{1+(\tan^{-1}x)^2} = 1/(1+x^2)\,[1 + (\tan^{-1}x)^2].$

46. $f'(x) = \dfrac{D_x\tanh(5x+1)}{\tanh(5x+1)} = \dfrac{5 \operatorname{sech}^2(5x+1)}{\tanh (5x+1)}$

49. $f'(x) = (D_x x^2)/\sqrt{(x^2)^2+1} = 2x/\sqrt{x^4+1}.$

52. With $u = x/2$, $2\,du = dx$, $\displaystyle\int \csc(x/2)\cot(x/2)\,dx = 2\int \csc u \cot u\,du =$
     $-2 \csc u + C = -2 \csc(x/2) + C.$

55. With $u = 9x$, $dx = (1/9)du$, $\displaystyle\int (\cot 9x + \csc 9x)\,dx = (1/9)\int (\cot u + \csc u)\,du$
     $= (1/9)(\ln|\sin u| + \ln|\csc u - \cot u|) + C,$ where $u = 9x.$

58. $\displaystyle\int \cot 2x \csc 2x\,dx = -(1/2)\csc 2x + C.$

61. $\displaystyle\int \dfrac{\sin 4x}{\tan 4x}\,dx = \int \cos 4x\,dx = (1/4)\sin 4x + C.$

64. $u = \cos 2x \Longrightarrow du = -2 \sin 2x\,dx$ and $\sin 2x\,dx = -(1/2)du.$ $x = 0$ or $\pi/4$
     $\Longrightarrow u = 0$ or $1.$ Thus $\displaystyle\int_0^{\pi/4} \sin 2x \cos^2 2x\,dx = -\frac{1}{2}\int_1^0 u^2 du = +\frac{1}{2}\int_0^1 u^2 du =$
     $\frac{1}{2}\cdot\frac{1}{3} = \frac{1}{6}.$

67. $u = \sin 3x \Longrightarrow du = 3 \cos 3x\,dx,$ and $\cos 3x\,dx = (1/3)du.$
     $\displaystyle\int \cos 3x(\sin 3x)^{-3}dx = \frac{1}{3}\int u^{-3}du = \frac{1}{3}\frac{u^{-2}}{-2} + C = -\frac{1}{6}(\sin 3x)^{-2} + C =$
     $-\frac{1}{6}\csc^2 3x + C.$

70. $\int \dfrac{1}{4 + 9x^2}\, dx = \dfrac{1}{9} \int \dfrac{dx}{(4/9) + x^2} = \dfrac{1}{9} \cdot \dfrac{3}{2} \tan^{-1} \dfrac{x}{2/3} + C = \dfrac{1}{6}\tan^{-1}\left(\dfrac{3x}{2}\right) + C.$

73. With $u = x^2$, $\int \dfrac{x}{\text{sech } x^2}\, dx = \dfrac{1}{2} \int \cosh u\, du = \dfrac{1}{2} \sinh u + C = \dfrac{1}{2} \sinh(x^2) + C.$

76. $\displaystyle\int_0^{\pi/2} \dfrac{\cos x}{1 + \sin^2 x}\, dx = \tan^{-1}(\sin x)\Big]_0^{\pi/2} = \tan^{-1}1 - \tan^{-1}0 = \dfrac{\pi}{4}.$

79. $u = 2 + \cot x \Rightarrow du = -\csc^2 x\, dx.$ $\int \dfrac{\csc^2 x}{2 + \cot x}\, dx = -\int \dfrac{du}{u} = -\ln|u| + C =$

$-\ln|2 + \cot x| + C.$

82. $u = 1 - 2x \Rightarrow dx = -(1/2)\,du.$ $\int \text{sech}^2(1-2x)\,dx = -(1/2)\int \text{sech}^2 u\, du =$

$-(1/2)\tanh u + C = -(1/2)\tanh(1-2x) + C.$

85. $u = 2x \Rightarrow dx = (1/2)\,du$ and $x = u/2.$ $\int \dfrac{1}{x\sqrt{9-4x^2}}\, dx = \int \dfrac{(1/2)\,du}{(u/2)\sqrt{9-u^2}} =$

$\int \dfrac{1}{u\sqrt{3^2-u^2}}\, du = -\dfrac{1}{3} \text{sech}^{-1} \dfrac{|u|}{3} + C = -\dfrac{1}{3} \text{sech}^{-1} \dfrac{|2x|}{3} + C.$

88. Instead of changing variables, we proceed as follows. $\int \dfrac{1}{\sqrt{25x^2+36}}\, dx =$

$\int \dfrac{1}{\sqrt{25(x^2+36/25)}}\, dx = \dfrac{1}{5} \int \dfrac{1}{\sqrt{x^2+(6/5)^2}}\, dx = \dfrac{1}{5} \sinh^{-1} \dfrac{x}{6/5} + C = \dfrac{1}{5} \sinh^{-1} \dfrac{5x}{6} + C.$

91. $f'(x) = 8 \sec x \tan x - \csc x \cot x = \dfrac{8 \sin x}{\cos^2 x} - \dfrac{\cos x}{\sin^2 x} = \dfrac{8 \sin^3 x - \cos^3 x}{\sin^2 x \cos^2 x} = 0$

$\Rightarrow 8 \sin^3 x = \cos^3 x \Rightarrow 2 \sin x = \cos x \Rightarrow \tan x = 1/2 \Rightarrow x = \tan^{-1} 1/2.$
If $0 < x < \tan^{-1} 1/2$, we find, by retracing our steps above with the "=" replaced by "<", that $f'(x) < 0$ and f is decreasing on $(0, \tan^{-1} 1/2)$. Similarly, f is increasing on $(\tan^{-1} 1/2, \pi/2)$, and $f(\tan^{-1} 1/2) = 5\sqrt{5}$ is a local minimum.

94. With $u = x^2$, $A = \displaystyle\int_0^1 \dfrac{x}{1+x^4}\, dx = \dfrac{1}{2} \int_0^1 \dfrac{du}{1+u^2} = \dfrac{1}{2} \tan^{-1}u\Big]_0^1 = \dfrac{\pi}{8}.$

97. All distances will be in meters. Thus the observer is 500 m (1/2 km) from the point below the balloon. With y the distance of the balloon above ground and θ as shown, we are given that $\dfrac{dy}{dt} = 2$ m/sec and seek $\dfrac{d\theta}{dt}$ when $y = 100.$ $\tan \theta = y/500 \Rightarrow \theta = \arctan(y/500)$

$\Rightarrow \dfrac{d\theta}{dt} = \dfrac{D_y(y/500)}{1+(y/500)^2} \dfrac{dy}{dt} = \dfrac{1}{500} \dfrac{1}{1+(y/500)^2} \cdot \dfrac{dy}{dt}.$ Substituting $y = 100$, dy/dt

$= 2$, we obtain: $\dfrac{d\theta}{dt} = \dfrac{1}{500} \dfrac{1}{1+(1/5)^2} (2) = \dfrac{1}{500} \cdot \dfrac{25}{26} (2) = \dfrac{1}{260}$ rad/sec.

CHAPTER 9

EXERCISES 9.1

1. $u = x$, $dv = e^{-x}dx \Longrightarrow du = dx$, $v = -e^{-x}$. $\int xe^{-x}dx = -xe^{-x} - \int (-e^{-x})dx =$
    $-(x+1)e^{-x} + C$.

4. $u = x^2$, $dv = \sin 4x\,dx \Longrightarrow du = 2x\,dx$, $v = -(1/4)\cos 4x$, $I = \int x^2 \sin 4x\,dx$
    $= x^2(-1/4)\cos 4x - \int (-1/4)\cos 4x(2x)dx = -(x^2/4)\cos 4x + (1/2)\int x \cos 4x\,dx$.
    Another integration by parts is necessary. $u = x$, $dv = \cos 4x\,dx \Longrightarrow du = dx$,
    $dv = (1/4)\sin 4x$, and $\int x \cos 4x\,dx = (x/4)\sin 4x - (1/4)\int \sin 4x\,dx =$
    $(x/4)\sin 4x + (1/16)\cos 4x$. Combining results we obtain:
    $I = -(x^2/4)\cos 4x + (1/2)[(x/4)\sin 4x + (1/16)\cos 4x] + C$.

7. $u = x$, $dv = \sec x \tan x\,dx \Longrightarrow du = dx$, $v = \sec x$. $\int x \sec x \tan x\,dx =$
    $x \sec x - \int \sec x\,dx = x \sec x - \ln|\sec x + \tan x| + C$.

10. $u = x^3$, $dv = e^{-x}dx \Longrightarrow du = 3x^2dx$, $v = -e^{-x}$. $I = \int x^3 e^{-x}dx = -x^3 e^{-x} +$
    $3\int x^2 e^{-x}dx$. Now, $u = x^2$, $dv = e^{-x}dx \Longrightarrow du = 2x\,dx$, $v = -e^{-x}$. So, $I =$
    $-x^3 e^{-x} + 3[-x^2 e^{-x} + 2\int xe^{-x}dx]$. The last integral was done in #1. So, com-
    bining all, $I = -(x^3 + 3x^2 + 6x + 6)e^{-x} + C$.

13. $u = \ln x$, $dv = \sqrt{x}\,dx \Longrightarrow du = \frac{1}{x}dx$, $v = \frac{2}{3}x^{3/2}$. $\int \sqrt{x} \ln x\,dx = \frac{2}{3}x^{3/2} \ln x -$
    $\frac{2}{3}\int \frac{x^{3/2}}{x}\,dx = \frac{2}{3}x^{3/2} \ln x - \frac{4}{9}x^{3/2} + C$. N.B. $\int \frac{x^{3/2}}{x}dx = \int x^{1/2}dx = \frac{2}{3}x^{3/2} + D$.

16. $u = \tan^{-1}x$, $dv = x\,dx \Longrightarrow du = \frac{1}{1+x^2}\,dx$, $v = \frac{x^2}{2}$. $\int x \tan^{-1}x\,dx = \frac{x^2}{2}\tan^{-1}x -$
    $\frac{1}{2}\int \frac{x^2}{1+x^2}\,dx = \frac{x^2}{2}\tan^{-1}x - \frac{1}{2}\int (1 - \frac{1}{1+x^2})dx = \frac{x^2}{2}\tan^{-1}x - \frac{x}{2} + \frac{1}{2}\tan^{-1}x + C$.
    N.B. $\frac{x^-}{1+x^2} = 1 - \frac{1}{1+x^2}$ by ordinary long division.

19. With $y = \cos x$, $dy = -\sin x\,dx$, $\int \sin x \ln \cos x\,dx = -\int \ln y\,dy = -(y \ln y - y)$
    $+ C$ (by Example 3) $= -\cos x \ln \cos x + \cos x + C$.

22. $u = \sec^3 x\,dx$, $dv = \sec^2 x\,dx \Longrightarrow du = 3\sec^2 x \tan x\,dx$,
    $v = \tan x$, and $I = \int \sec^5 x\,dx = \sec^3 x \tan x - 3\int \sec^3 x \tan^2 x\,dx$.
    Using $\tan^2 x = \sec^2 x - 1$, $I = \sec^3 x \tan x -$
    $3\int \sec^3 x(\sec^2 x - 1)dx = \sec^3 x \tan x - 3\int \sec^5 x\,dx + 3\int \sec^3 x\,dx$.

The second last term is $-3I$, and the final integral was
evaluated in Example 6. Substituting this result, and adding
$3I$ to both sides yields
$$4I = \sec^3 x \tan x + \frac{3}{2}(\sec x \tan x + \ln|\sec x + \tan x|).$$
Now divide by 4 and add C.

25. $u = x$, $dv = \sin 2x \, dx \implies du = dx$, $v = -\frac{1}{2}\cos 2x$ and $\int_0^{\pi/2} x \sin 2x \, dx =$

$-\frac{x}{2}\cos 2x \,]_0^{\pi/2} + \frac{1}{2}\int_0^{\pi/2} \cos 2x \, dx = (\frac{\pi}{4} - 0) + \frac{1}{4}\sin 2x \,]_0^{\pi/2} = \frac{\pi}{4} + 0 = \frac{\pi}{4}$.

28. This is a tricky one to get started correctly. We have to break up the

integrand as follows: $u = x^3$, $dv = \dfrac{x^2}{\sqrt{1-x^3}} \, dx \implies du = 3x^2 dx$, $v = \int \dfrac{x^2}{\sqrt{1-x^3}} \, dx$.

To evaluate v, let $y = 1-x^3$, $dy = -3x^2 dx$, $x^2 dx = -(1/3)dy$. Then $v =$

$-\frac{1}{3}\int y^{-1/2} \, du = -(2/3)y^{1/2} = -(2/3)\sqrt{1-x^3}$. Using these results, $I = \int \dfrac{x^5}{\sqrt{1-x^3}} dx$

$= -\frac{2}{3}x^3\sqrt{1-x^3} - \frac{2}{3}\int \sqrt{1-x^3} \, 3x^2 dx$. This last integral can be done with the same

substitution, $y = 1-x^3$, as above to obtain $I = -\frac{2}{3}x^3\sqrt{1-x^3} - \frac{2}{3} \cdot \frac{2}{3}(1-x^3)^{3/2} +$

C. By factoring out $\sqrt{1-x^3}$, and simplifying, this reduces to $-(2/9)\sqrt{1-x^3}(x^3+2)$
$+$ C.

31. $u = (\ln x)^2$, $dv = dx \implies du = (2 \ln x)\frac{1}{x} \, dx$, $v = x$. $\int (\ln x)^2 dx = x(\ln x)^2 -$

$2\int (\ln x)\frac{1}{x} x \, dx = x(\ln x)^2 - 2\int \ln x \, dx = x(\ln x)^2 - 2(x \ln x - x) + C$ by

the result of Example 3.

34. $u = x+4$, $dv = \cosh 4x \, dx \implies du = dx$, $v = (1/4)\sinh 4x$. $\int (x+4)\cosh 4x \, dx =$

$(1/4)(x+4)\sinh 4x - (1/4)\int \sinh 4x \, dx = (1/4)(x+4)\sinh 4x - (1/16)\cosh 4x + C$.

37. $u = \cos^{-1} x$, $dv = dx \implies du = -1/\sqrt{1-x^2} \, dx = -(1-x^2)^{-1/2} dx$, $v = x$. $\int \cos^{-1} x \, dx$

$= x \cos^{-1} x - \int (1-x^2)^{-1/2}(-x) dx = x \cos^{-1} x - (1-x^2)^{1/2} + C$.

40 . Let $u = x^m$, $dv = \sin x \, dx$, and the formula follows immediately.

43. $\int x^5 e^x dx = x^5 e^x - 5\int x^4 e^x dx = x^5 e^x - 5[x^4 e^x - 4\int x^3 e^x dx] = x^5 e^x - 5x^4 e^x +$

$20[x^3 e^x - 3\int x^2 e^x dx] = x^5 e^x - 5x^4 e^x + 20x^3 e^x - 60[x^2 e^x - 2\int x \, e^x] =$

$x^5 e^x - 5x^4 e^x + 20x^3 e^x - 60x^2 e^x + 120[x \, e^x - \int e^x]$. The last integral is

just $e^x + C$.

45. $A = \int_0^{\pi^2} \sin\sqrt{x} \, dx$. First, $y = \sqrt{x}$, $dy = (1/2\sqrt{x})dx \implies dx = 2\sqrt{x} \, dy = 2y \, dy$, and

$A = 2\int_0^{\pi} y \sin y \, dy$. Now, $u = y$, $dv = \sin y \, dy \implies du = dy$, $v = -\cos y$, and

$$A = 2(-y \cos y]_0^\pi + \int_0^\pi \cos y \, dy) = 2(\pi + \sin y]_0^\pi) = 2\pi.$$

46. By the disc method, $V = \pi \int_0^{\pi/2} x^2 \sin x \, dx$. Using #40, $\int x^2 \sin x \, dx =$

$-x^2 \cos x + 2\int x \cos x \, dx$. In the last integral, $u = x$, $dv = \cos x \, dx \implies du$

$= dx$, $v = \sin x$, and $\int x \cos x \, dx = x \sin x - \int \sin x \, dx = x \sin x + \cos x$.

Combining these results: $V = \pi(-x^2 \cos x + 2x \sin x + 2 \cos x)]_0^{\pi/2}$

$= \pi[(0 + 2(\frac{\pi}{2}) + 0) - (0 + 0 + 2)] = \pi(\pi - 2).$

49. Because of symmetry, we can compute the force
on the right half and double it to get the
total force. The left boundary of this half
is $x = 3\pi/2$. The right boundary is $y =$
$\sin x$, $3\pi/2 \le x \le y$. We must express this
equation in terms of y as the independent
variable. We obtain for such x, $\sin^{-1} y =$
$x-2\pi$, or $x = \sin^{-1} y + 2\pi$, since, by the
definition, $-\pi/2 \le \sin^{-1} y \le \pi/2$, and here,

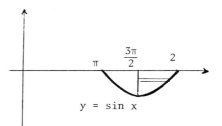

$y = \sin x$

$3\pi/2 \le x \le 2\pi$. Then by (6.19), if F is the force on the right half, F =

$$\rho \int_{-1}^0 (0-y)[(\sin^{-1}y + 2\pi) - 3\pi/2] \, dy = -\rho \int_{-1}^0 y(\sin^{-1}y + \pi/2) \, dy = \rho \int_{-1}^0 y \sin^{-1}y \, dy$$

$- \frac{\pi\rho}{2} \int_{-1}^0 y \, dy$. Let $I = \int_{-1}^0 y \sin^{-1}y \, dy$. $u = y$, $dv = \sin^{-1}y \, dy \implies du = dy$,

$v = y \sin^{-1}y + \sqrt{1-y^2}$ (by #12). So, $I = y^2 \sin^{-1}y + y\sqrt{1-y^2}]_{-1}^0 - \int_{-1}^0 (y \sin^{-1}y + \sqrt{1-y^2}) \, dy$

$= (0 - \sin^{-1}(-1)) - I - \int_{-1}^0 \sqrt{1-y^2} \, dy$. We cannot evaluate the last integral

by antidifferentiation yet. However, it is the area of a quarter circle of
radius one. (It's the area of the region in the 4th quadrant enclosed by
$x^2 + y^2 = 1$.) Thus its value is $\pi/4$, and, using $\sin^{-1}(-1) = -\pi/2$ and trans-
posing the I term, we obtain $2I = -(-\pi/2) - \pi/4 = \pi/4$, or $I = \pi/8$. Since the

value of $\int_{-1}^0 y \, dy$ is $-1/2$, we obtain: $F = -\rho(\pi/8) - (\pi\rho/2)(-1/2) =$

$(\pi\rho/4)(1-1/2) = \pi\rho/8$. The total force is thus $\pi\rho/4 = 62.5 \, \pi/4$.

EXERCISES 9.2

1. $\int \cos^3 x \, dx = \int \cos^2 x \cos x \, dx = \int (1 - \sin^2 x) \cos x \, dx = \int (1-u^2) \, du$ (with

$u = \sin x) = u - \dfrac{u^3}{3} + C = \sin x - \dfrac{\sin^3 x}{3} + C.$

4. $\int \cos^7 x \, dx = \int (1-\sin^2 x)^3 \cos x \, dx = \int (1-u^2)^3 du = \int (1 - 3u^2 + 3u^4 - u^6) du$

(with $u = \sin x) = u - u^3 + \dfrac{3}{5}u^5 - \dfrac{1}{7}u^7 + C = \sin x - \sin^3 x + \dfrac{3}{5}\sin^5 x -$

$\dfrac{1}{7}\sin^7 x + C.$

7. $\int \sin^6 x \, dx = \int [\dfrac{1}{2}(1 - \cos 2x)]^3 dx = \dfrac{1}{8}\int (1 - 3\cos 2x + 3\cos^2 2x - \cos^3 2x) dx$

$= \dfrac{1}{8}(A - B + C - D)$ where $A = \int 1 \, dx = x$, $B = 3\int \cos 2x \, dx = \dfrac{3}{2}\sin 2x$,

$C = 3\int \cos^2 2x \, dx = \dfrac{3}{2}\int (1 + \cos 4x) dx = \dfrac{3}{2}(x + (\sin 4x)/4),$

$D = \int \cos^3 2x \, dx = \dfrac{1}{2}\int \cos^3 t \, dt$ (with $t = 2x) = \dfrac{1}{2}(\sin t - \dfrac{\sin^3 t}{3})$ (by Exercise 1)

$= \dfrac{1}{2}(\sin 2x - \dfrac{\sin^3 2x}{3})$. Now combine these to obtain the answer in the text.

10. $\int \sec^6 x \, dx = \int \sec^4 x \sec^2 x \, dx = \int (1 + \tan^2 x)^2 \sec^2 x \, dx = \int (1 + u^2)^2 \, du = (u = \tan x)$

$\int (1 + 2u^2 + u^4) du = u + \dfrac{2}{3}u^3 + \dfrac{1}{5}u^5 + C = \tan x + \dfrac{2}{3}\tan^3 x + \dfrac{1}{5}\tan^5 x + C.$

11. $\int \tan^3 x \sec^3 x \, dx = \int \tan^2 x \sec^2 x \sec x \tan x \, dx = \int (\sec^2 x - 1)\sec^2 x \sec x \cdot$

$\tan x \, dx = \int (u^2-1)u^2 du = \dfrac{1}{5}u^5 - \dfrac{1}{3}u^3 + C$ $(u = \sec x)$ $= \dfrac{1}{5}\sec^5 x - \dfrac{1}{3}\sec^3 x + C.$

13. $\int \tan^6 x \, dx = \int \dfrac{\tan^6 x}{\sec^2 x} \sec^2 x \, dx = \int \dfrac{\tan^6 x}{\tan^2 x + 1} \sec^2 x \, dx = \int \dfrac{u^6}{u^2+1} du$ $(u = \tan x)$

$= \int (u^4 - u^2 + 1 - \dfrac{1}{u^2+1}) du$ (by long division) $= \dfrac{1}{5}u^5 - \dfrac{1}{3}u^3 + u - \tan^{-1}u + C =$

$\dfrac{1}{5}\tan^5 x - \dfrac{1}{3}\tan^3 x + \tan x - x + C.$

16. $\int \dfrac{\cos^3 x}{\sqrt{\sin x}} dx = \int \sin^{-1/2} x (1-\sin^2 x) \cos x \, dx = \int (u^{-1/2} - u^{3/2}) du = 2u^{1/2} -$

$\dfrac{2}{5}u^{5/2} + C$ $(u = \sin x)$ $= 2\sqrt{\sin x} - \dfrac{2}{5}\sqrt{\sin^5 x} + C.$

19. $\displaystyle\int_0^{\pi/4} \sin^3 x \, dx = \int_0^{\pi/4} (1-\cos^2 x)\sin x \, dx = -\cos x + \dfrac{1}{3}\cos^3 x \,]_0^{\pi/4} =$

$(-1 + \dfrac{1}{6})/\sqrt{2} - (-\dfrac{2}{3}) = 2/3 - 5/6\sqrt{2}.$

22. By the product formula, $\cos x \cos 5x = \dfrac{1}{2}(\cos 4x + \cos 6x)$. So,

$$\int_0^{\pi/4} \cos x \cos 5x \, dx = \frac{1}{2} \int_0^{\pi/4} (\cos 4x + \cos 6x) dx = \frac{1}{2}[\frac{\sin 4x}{4} + \frac{\sin 6x}{6}]_0^{\pi/4}$$

$$= \frac{1}{2}[(0 - \frac{1}{6}) - (0-0)] = -1/12.$$

25. $\int \csc^4 x \cot^4 x \, dx = \int \csc^2 x \cot^4 x \csc^2 x \, dx = \int (1 + \cot^2 x)\cot^4 x \csc^2 x \, dx =$

   $-\int (1 + u^2)u^4 du \quad (u = \cot x) \quad = -u^5/5 - u^7/7 + C = -(\cot^5 x)/5 - (\cot^7 x)/7 + C.$

28. $\cos x = 1/\sec x$, $\tan x = \sin x/\cos x \implies \frac{\tan^2 x - 1}{\sec^2 x} = \sin^2 x - \cos^2 x = -\cos 2x.$

   Thus $\int \frac{\tan^2 - 1}{\sec^2 x} \, dx = -\int \cos 2x \, dx = -\frac{1}{2} \sin 2x + C = -\sin x \cos x + C.$

   (Note that if $\tan x$ and $\sec x$ had been retained, we would have obtained:

   $\int \frac{\tan^2 x - 1}{\sec^2 x} \, dx = \int \frac{\tan^2 x - 1}{\sec^4 x} \sec^2 x \, dx = \int \frac{(\tan^2 x - 1)\sec^2 x}{(\tan^2 x + 1)^2} dx = \int \frac{u^2 - 1}{(u^2 + 1)^2} \, du,$

   an integral which we can not yet evaluate.)

31. $V = \pi \int_0^{2\pi} (\cos^2 x)^2 dx = \pi \int_0^{2\pi} (\frac{1+\cos 2x}{2})^2 dx = \frac{\pi}{4} \int_0^{2\pi} (1 + 2 \cos 2x + \cos^2 2x) dx$

   $= \frac{\pi}{4} \int_0^{2\pi} (1 + 2 \cos 2x + (\frac{1 + \cos 4x}{2})) dx = \frac{\pi}{4} \int_0^{2\pi} (\frac{3}{2} + 2 \cos 2x + \frac{\cos 4x}{4}) dx$

   $= \frac{\pi}{4}[\frac{3}{2}x + \sin 2x + \frac{\sin 4x}{8}]_0^{2\pi} = \frac{\pi}{4} \cdot \frac{3}{2} \cdot 2\pi = \frac{3\pi^2}{4}.$

34. $a(t) = \sin^2 t \cos t \implies v(t) = \frac{1}{3} \sin^3 t + C.$ Using $v(0) = 10$, we get $C = 10.$

   $v(t) = \frac{1}{3}\sin^3 t + 10 \implies s(t) = -\frac{1}{3}\cos t + \frac{1}{9}\cos^3 t + 10t + D$ (using the result

   of Exercise 19). Using $s(0) = 0$, we get $0 = -1/3 + 1/9 + D \implies D = 2/9.$

   Thus $s(t) = \frac{1}{9}(-3 \cos t + \cos^3 t + 90t + 2).$

EXERCISES 9.3

1. Let $x = 2 \sin \theta$, $dx = 2 \cos \theta \, d\theta$, $\sqrt{4-x^2} = 2 \cos \theta.$ $\int \frac{x^2}{\sqrt{4-x^2}} \, dx =$

   $\int \frac{4 \sin^2 \theta (2 \cos \theta) d\theta}{2 \cos \theta} = 2 \int (1 - \cos 2\theta) d\theta =$

   $2(\theta - \frac{\sin 2\theta}{2}) + C = 2(\theta - \sin \theta \cos \theta) + C =$

   $2(\sin^{-1} \frac{x}{2} - \frac{x}{2} \cdot \frac{\sqrt{4-x^2}}{2}) + C.$

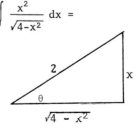

4. Let $x = 3 \tan \theta$, $dx = 3 \sec^2 \theta \, d\theta$, $\sqrt{x^2+9} = 3 \sec \theta.$ $\int \frac{dx}{x^2 \sqrt{x^2+9}} =$

$$\int \frac{3 \sec^2\theta \; d\theta}{9 \tan^2\theta \; (3 \sec \theta)} = \frac{1}{9} \int \frac{\sec \theta}{\tan^2\theta} \, d\theta = \frac{1}{9} \int \frac{(1/\cos \theta)}{(\sin \theta/\cos \theta)^2} \, d\theta =$$

$$\frac{1}{9} \int \frac{\cos \theta}{\sin^2\theta} \, d\theta = \frac{1}{9} \int \cot \theta \; \csc \theta \; d\theta =$$

$$-\frac{1}{9} \csc \theta + C = -\frac{\sqrt{x^2+9}}{9x} + C.$$

7.   You could use $x = 2 \sin \theta$, but the quickest way is the simpler substitution

$u = 4-x^2$, $du = -2x \; dx$ so that $\int \frac{x}{\sqrt{4-x^2}} \, dx = -\frac{1}{2} \int u^{-1/2} \, du = -u^{1/2} + C =$

$-\sqrt{4-x^2} + C.$

10.   Noting that $\sqrt{4x^2-25} = 2\sqrt{x^2 - 25/4}$, we let $x = \frac{5}{2} \sec \theta$, $dx = \frac{5}{2} \sec \theta \tan \theta \; d\theta$,

$\sqrt{x^2 - 25/4} = \frac{5}{2}\tan \theta.$ Then $\int \frac{dx}{\sqrt{4x^2-25}} = \frac{1}{2} \int \frac{dx}{\sqrt{x^2 - 25/4}} =$

$\frac{1}{2} \int \frac{\frac{5}{2}\sec \theta \tan \theta}{\frac{5}{2} \tan \theta} \, d\theta = \frac{1}{2} \int \sec\theta d\theta = \frac{1}{2} \ln|\sec \theta + \tan \theta| + C$

$= \frac{1}{2} \ln\left|\frac{2x}{5} + \frac{\sqrt{4x^2-25}}{5}\right| + C = \frac{1}{2} \ln\left|2x + \sqrt{4x^2-25}\right| + C',$

$C' = C - \frac{1}{2} \ln 5.$

13.   Noting that $\sqrt{9-4x^2} = 2\sqrt{9/4 - x^2}$, we let $x = \frac{3}{2} \sin \theta$, $dx = \frac{3}{2} \cos \theta \; d\theta$,

$\sqrt{9/4 - x^2} = \frac{3}{2} \cos \theta.$ $\int\sqrt{9-4x^2} \, dx = 2 \int\sqrt{9/4 - x^2} \, dx =$

$2 \int (\frac{3}{2} \cos \theta) (\frac{3}{2} \cos \theta) d\theta = \frac{9}{2} \int \cos^2\theta \; d\theta = \frac{9}{4} \int (1 + \cos 2\theta) d\theta$

$= \frac{9}{4}(\theta + \frac{\sin 2\theta}{2}) + C = \frac{9}{4}(\theta + \sin \theta \cos \theta) + C =$

$\frac{9}{4}(\sin^{-1}\frac{2x}{3} + (\frac{2x}{3}) (\frac{\sqrt{9-4x^2}}{3})) + C.$

16.   We could let $x = 3 \sec \theta$, but the quickest way is to let $u = x^2-9$, $du = 2x \; dx$,

and $\int x\sqrt{x^2-9} \, dx = \frac{1}{2} \int u^{1/2} \, du = \frac{1}{3}u^{3/2} + C = \frac{1}{3}(x^2-9)^{3/2} + C.$

19.   Let $x = \sqrt{3} \sec \theta$, $dx = \sqrt{3} \sec \theta \tan \theta \; d\theta$, $\sqrt{x^2-3} = \sqrt{3} \tan \theta.$ $\int \frac{1}{x^4\sqrt{x^2-3}} \, dx$

$= \int \frac{\sqrt{3} \sec \theta \tan \theta}{9 \sec^4\theta(\sqrt{3} \tan \theta)} \, d\theta = \frac{1}{9} \int \frac{1}{\sec^3\theta} \, d\theta = \frac{1}{9}\int \cos^3\theta \; d\theta$

$= \frac{1}{9} \int (1 - \sin^2\theta)\cos \theta \; d\theta = \frac{1}{9} \int (1-u^2) du$ (with $u =$

$\sin \theta)$ $= \frac{1}{9}(u - \frac{u^3}{3}) + C = \frac{1}{9}(\sin \theta - \frac{\sin^3\theta}{3}) + C$

$= \frac{\sin \theta}{27}(3 - \sin^2\theta) + C = \frac{\sqrt{x^2-3}}{27x}(3 - \frac{x^2-3}{x^2}) + C$

$= \frac{\sqrt{x^2-3}(2x^2+3)}{27x^3} + C.$

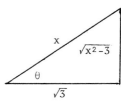

22. Let $x = \sin\theta$, $dx = \cos\theta\,d\theta$, $\sqrt{1-x^2} = \cos\theta$. $\displaystyle\int \frac{3x-5}{\sqrt{1-x^2}}\,dx$

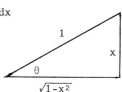

$= \displaystyle\int (3\sin\theta - 5)d\theta = -3\cos\theta - 5\theta + C = -3\sqrt{1-x^2} -$
$5\sin^{-1}x + C.$

25. This is a special case of #28 with $a = 1$.

28. Let $u = a\sec\theta$, $du = a\sec\theta\tan\theta\,d\theta$, $\sqrt{u^2-a^2} = a\tan\theta$. Then $\displaystyle\int \frac{1}{u\sqrt{u^2-a^2}}\,du$

$= \dfrac{1}{a}\displaystyle\int d\theta = \dfrac{1}{a}\theta + C = \dfrac{1}{a}\sec^{-1}\dfrac{u}{a} + C.$

31. $L_0^2 = \displaystyle\int_0^2 \sqrt{1+y'^2}\,dx = \int_0^2 \sqrt{1+x^2}\,dx$. Let $x = \tan\theta$, $dx = \sec^2\theta\,d\theta$, $\sqrt{1+x^2} = \sec\theta$.

Then $\displaystyle\int \sqrt{1+x^2}\,dx = \int \sec^3\theta\,d\theta = \dfrac{1}{2}(\sec\theta\tan\theta +$

$\ln|\sec\theta + \tan\theta|\,) + C$ by Example 6 , Sec. 9.1.

Thus $L_0^2 = \dfrac{1}{2}[x\sqrt{x^2+1} + \ln|\sqrt{x^2+1} + x|\,]\Big|_0^2 =$

$\dfrac{1}{2}[2\sqrt{5} + \ln|\sqrt{5} + 2| - 0] \approx 2.46.$

34. The graph is an ellipse, and by symmetry we can double the area of the upper
     half. $4x^2 + y^2 = 16 \implies y = 2\sqrt{4-x^2}$, $-2 \le x \le 2$ is the upper boundary.

$A = 2\displaystyle\int_{-2}^2 2\sqrt{4-x^2}\,dx$. Let $x = 2\sin\theta$, $dx = 2\cos\theta\,d\theta$, $\sqrt{4-x^2} = 2\cos\theta$.

Since $\theta = \sin^{-1}(x/2)$, $x = \pm 2 \implies \theta = \pm\,\pi/2$. Thus $A = 16\displaystyle\int_{-\pi/2}^{\pi/2} \cos^2\theta\,d\theta =$

$8\displaystyle\int_{-\pi/2}^{\pi/2} (1 + \cos 2\theta)d\theta = 8[\theta + \dfrac{\sin 2\theta}{2}]\Big|_{-\pi/2}^{\pi/2} = 8\pi.$

37. The one identity to remember is $\cosh^2 u = 1 + \sinh^2 u$. Let $x = 5\sinh u$, $dx =$
     $5\cosh u$, $\sqrt{25+x^2} = \sqrt{25(1 + \sinh^2 u)} = 5\cosh u$. Then $\displaystyle\int \frac{1}{x^2\sqrt{25+x^2}}\,dx =$

$\displaystyle\int \frac{5\cosh u}{25\sinh^2 u(5\cosh u)}\,du = \dfrac{1}{25}\int \frac{1}{\sinh^2 u}\,du = \dfrac{1}{25}\int \operatorname{csch}^2 u\,du = -\dfrac{1}{25}\coth u + C$

$= -\dfrac{1}{25}\dfrac{\cosh u}{\sinh u} + C = -\dfrac{1}{25}\dfrac{\sqrt{1+\sinh^2 u}}{\sinh u} + C = -\dfrac{1}{25}\dfrac{\sqrt{1 + x^2/25}}{x/5} + C = -\dfrac{1}{25}\dfrac{\sqrt{25+x^2}}{x} + C.$

40. Let $x = 4\tanh u$, $dx = 4\operatorname{sech}^2 u\,du$, $16-x^2 = 16(1 - \tanh^2 u) = 16\operatorname{sech}^2 u$.

$\displaystyle\int \frac{1}{16-x^2}\,dx = \int \frac{4\operatorname{sech}^2 u}{16\operatorname{sech}^2 u}\,du = \dfrac{1}{4}\int du = \dfrac{1}{4}u + C = \dfrac{1}{4}\tanh^{-1}\dfrac{x}{4} + C.$

43. $u = a \sin \theta$, $du = a \cos \theta$, $\sqrt{a^2-u^2} = a \cos \theta \implies I = \int u^2 \sqrt{a^2-u^2}\,du =$

$a^4 \int \sin^2\theta \cos^2\theta \, d\theta = a^4 \int (\frac{1 - \cos 2\theta}{2})(\frac{1 + \cos 2\theta}{2}) d\theta = \frac{a^4}{4} \int (1 - \cos^2 2\theta) d\theta =$

$\frac{a^4}{4} \int (1 - \frac{1 + \cos 4\theta}{2}) d\theta = \frac{a^4}{8} \int (1 - \cos 4\theta) d\theta =$

$\frac{a^4}{8}(\theta - \frac{\sin 4\theta}{4}) + C$. Now, $\sin 4\theta = 2 \sin 2\theta \cos 2\theta$

$= 4 \sin \theta \cos \theta (\cos^2\theta - \sin^2\theta)$. Since $\theta =$

$\sin^{-1}(u/a)$, $\sin \theta = u/a$ and $\cos \theta = \sqrt{a^2-u^2}/a$.

Combining results, $I = \frac{a^4}{8}(\theta - \frac{u}{a} \frac{\sqrt{a^2-u^2}}{a}(\frac{a^2-u^2}{a^2} - \frac{u^2}{a^2}) + C$

which easily reduces to the formula.

46. $u = a \sec \theta$, $du = a \sec \theta \tan \theta \, d\theta$, $\sqrt{u^2-a^2} = a \tan \theta$

$\implies I = \int \frac{u^2}{\sqrt{u^2-a^2}} \, du = a^2 \int \frac{\sec^2\theta \cdot \sec \theta \tan \theta}{\tan \theta} d\theta =$

$a^2 \int \sec^3\theta \, d\theta$. By #21, Sec. 9.1, $I =$

$\frac{a^2}{2}(\sec \theta \tan \theta + \ln|\sec \theta + \tan \theta|) + C' =$

$\frac{a^2}{2}(\frac{u}{a} \frac{\sqrt{u^2-a^2}}{a} + \ln|\frac{u}{a} + \frac{\sqrt{u^2-a^2}}{a}|) + C' =$

$\frac{u}{2} \sqrt{u^2-a^2} + \frac{a^2}{2} \ln|u + \sqrt{u^2-a^2}| - \frac{a^2}{2} \ln|a| + C'$

which is equivalent to Formula 44 with

$C = c' - \frac{a^2}{2} \ln|a|$.

EXERCISES 9.4

1. $\frac{5x-12}{x(x-4)} = \frac{A}{x} + \frac{B}{x-4}$. Multiplying by $x(x-4)$, we get $5x-12 = A(x-4) + Bx$. Letting

$x = 4$ we find that $20-12 = A(0) + 4B$, or $8 = 4B$, or $B = 2$. Letting $x = 0$, we

get $-12 = -4A$, or $A = 3$. Thus $\int \frac{5x-12}{x(x-4)} dx = \int (\frac{3}{x} + \frac{2}{x-4}) dx = 3 \ln|x| + 2 \ln|x-4|$

$+ C = \ln|x|^3 + \ln|x-4|^2 + C = \ln|x|^3|x-4|^2 + C$.

4. $\frac{4x^2+54x+134}{(x-1)(x+5)(x+3)} = \frac{A}{x-1} + \frac{B}{x+5} + \frac{C}{x+3}$. Multiplying by $(x-1)(x+5)(x+3)$, we get

$4x^2+54x+134 = A(x+5)(x+3) + B(x-1)(x+3) + C(x-1)(x+5)$.

Set $x = 1$ to obtain $192 = 24A$,   or $A = 8$.
Set $x = -5$ to obtain $-36 = 12B$,   or $B = -3$.
Set $x = -3$ to obtain $8 = -8C$,   or $C = -1$.

$$\int \frac{4x^2+54x+134}{(x-1)(x+5)(x+3)} \, dx = \int (\frac{8}{x-1} - \frac{3}{x+5} - \frac{1}{x+3}) \, dx = 8 \ln|x-1| - 3 \ln|x+5| - \ln|x+3|$$

$+ C$, or $\ln \frac{|x-1|^8}{|x+5|^3|x+3|} + C$ where now C is the usual arbitrary constant.

7. $\frac{x+16}{x^2+2x-8} = \frac{x+16}{(x+4)(x-2)} = \frac{A}{x+4} + \frac{B}{x-2}$. Multiplying by $(x+4)(x-2)$ we get $x+16 =$

$A(x-2) + B(x+4)$, and letting $x = 2$ we find $18 = 6B$ or $B = 3$, and letting $x =$

-4, we find $12 = -6A$ or $A = -2$. Thus $\int \frac{x+16}{x^2+2x-8} \, dx = \int (\frac{-2}{x+4} + \frac{3}{x-2}) \, dx =$

$-2 \ln|x+4| + 3 \ln|x-2| + C$.

10. $\frac{4x^2-5x-15}{x^3-4x^2-5x} = \frac{4x^2-5x-15}{x(x-5)(x+1)} = \frac{A}{x} + \frac{B}{x-5} + \frac{C}{x+1}$. $4x^2-5x-15 = A(x-5)(x+1) + Bx(x+1) +$

$Cx(x-5)$.

   Set $x = 0$ to obtain $-15 = -5A$ or $A = 3$.
   Set $x = -1$ to obtain $-6 = 6C$ or $C = -1$.
   Set $x = 5$ to obtain $60 = 30B$ or $B = 2$.

$\int \frac{4x^2-5x-15}{x^3-4x^2-5x} \, dx = \int (\frac{3}{x} + \frac{2}{x-5} - \frac{1}{x+3}) \, dx = 3 \ln|x| + 2 \ln|x-5| - \ln|x+3| + C =$

$\ln \frac{|x|^3|x-5|^2}{|x+1|} + C$ where now C is arbitrary.

13. The integrand is $\frac{9x^4+17x^3+3x^2-8x+3}{x^4(x+3)} = \frac{A}{x} + \frac{B}{x^2} + \frac{C}{x^3} + \frac{D}{x^4} + \frac{E}{x+3}$.

$9x^4 + 17x^3 + 3x^2 - 8x + 3 = Ax^3(x+3) + Bx^2(x+3) + Cx(x+3) + D(x+3) + Ex^4$

Set $x = 0$ to obtain $3 = 3D$ or $D = 1$.

Set $x = -3$ to obtain $324 = 81E$ or $E = 4$.

To obtain the other constants, we set the coefficients of like powers of x
equal to each other:

   x^4: $9 = A + E = A + 4$, or $A = 5$

   x^3: $17 = 3A + B = 15 + B$, or $B = 2$

   x^2: $3 = 3B + C = 6 + C$, or $C = -3$.

$\int \frac{9x^4 + 17x^3 + 3x^2 - 8x + 3}{x^4(x+3)} \, dx = \int (\frac{5}{x} + \frac{2}{x^2} - \frac{3}{x^3} + \frac{1}{x^4} + \frac{4}{x+3}) \, dx =$

$5 \ln|x| - \frac{2}{x} + \frac{3}{2x^2} - \frac{1}{3x^3} + 4 \ln|x+3| + C$.

16. With $u = x-7$, $\int \frac{1}{(x-7)^5} \, dx = \int u^{-5} \, du = \frac{u^{-4}}{-4} + C = -1/4(x-7)^4 + C$.

19. $\frac{x^2+3x+1}{x^4+5x^2+4} = \frac{x^2+3x+1}{(x^2+4)(x^2+1)} = \frac{Ax+B}{x^2+4} + \frac{Cx+D}{x^2+1}$

$x^2+3x+1 = (Ax+B)(x^2+1) + (Cx+D)(x^2+4) = (A+C)x^3 + (B+D)x^2 + (A+4C)x + (B+4D)$

Equating coefficients of like powers of x, we obtain

x^3:   0 = A   + C

x^2:   1 =   B   + D

x^1:   3 = A   +4C

x^0:   1 =   B   +4D

Taking the 1st and 3rd, we get A = -1, C = 1, and from the 2nd and 4th, we get B = 1, D = 0. The integral of the original function reduces to

$$\int (\frac{-x}{x^2+4} + \frac{1}{x^2+4} + \frac{x}{x^2+1}) dx = -\frac{1}{2} \ln(x^2+4) + \frac{1}{2} \tan^{-1}(\frac{x}{2}) + \frac{1}{2} \ln(x^2+1) + C.$$

22. $\frac{x^4 + 2x^2 + 4x + 1}{(x^2+1)^3} = \frac{Ax+B}{x^2+1} + \frac{Cx+D}{(x^2+1)^2} + \frac{Ex+F}{(x^2+1)^3}$

$x^4+2x^2+4x+1 = (Ax+B)(x^2+1)^2 + (Cx+D)(x^2+1) + Ex + F$

$= Ax^5 + Bx^4 + (2A+C)x^3 + (2B+D)x^2 + (A+C+E)x + (B+D+F).$

Equating coefficients as above:

x^5:   0 = A x^4:   1 = B

x^3:   0 = 2A + C x^2:   2 = 2B + D

x^1:   4 = A + C + E x^0:   1 = B + D + F

yielding A = C = D = F = 0, B = 1, E = 4. The integral reduces to

$$\int [\frac{1}{x^2+1} + \frac{4x}{(x^2+1)^3}] dx = \tan^{-1}x - \frac{1}{(x^2+1)^2} + C.$$

25. By long division $\frac{x^6-x^3+1}{x^4+9x^2} = x^2 - 9 + \frac{-x^3+81x^2+1}{x^4 + 9x^2} \cdot \frac{-x^3+81x^2+1}{x^4+ 9x^2} = \frac{-x^3+81x^2+1}{x^2(x^2+9)} =$

$\frac{A}{x} + \frac{B}{x^2} + \frac{Cx+D}{x^2+9}$.

$-x^3+81x^2+1 = Ax(x^2+9) + B(x^2+9) + (Cx+D)x^2 = (A+C)x^3 + (B+D)x^2 + 9Ax + 9B.$

From x^0 and x^1 we get immediately A = 0, B = 1/9. From x^3, -1 = A+C = C;

from x^2, $81 = \frac{1}{9} + D$, or $D = 81 - \frac{1}{9} = \frac{728}{9}$. The integral reduces to

$$\int (x^2 - 9 + \frac{1}{9x^2} - \frac{x}{x^2+9} + \frac{728}{9} \cdot \frac{1}{x^2+9}) dx = \frac{x^3}{3} - 9x - \frac{1}{9x} - \frac{1}{2}\ln(x^2+9)+\frac{728}{27}\tan^{-1}\frac{x}{3}+C.$$

28. $\frac{-2x^4-3x^3-3x^2+3x+1}{x^2(x+1)^3} = \frac{A}{x} + \frac{B}{x^2} + \frac{C}{x+1} + \frac{D}{(x+1)^2} + \frac{E}{(x+1)^3}$.

$-2x^4 - 3x^3 - 3x^2 + 3x + 1 = Ax(x+1)^3 + B(x+1)^3 + Cx^2(x+1)^2 + Dx^2(x+1) + Ex^2.$

Set x = 0 to get 1 = B, and set x = -1 to get -4 = E. To obtain the remaining constants, equate coefficients of three powers of x.

x^4:   -2 = A   + C

x^2:   -3 = 3A + 3B + C + D + E

x :   3 = A + 3B   = A + 3 $\Longrightarrow$ A = 0

Using A = 0 in the first equation, we get C = -2. Using these in the second equation, we get -3 = 0 + 3 - 2 + D - 4, or D = 0. Thus the given integral reduces to $\int (\frac{1}{x^2} - \frac{2}{x+1} - \frac{4}{(x+1)^3}) dx = -\frac{1}{x} - 2 \ln|x+1| + \frac{2}{(x+1)^2} + C.$

31. Note that $x^3-x^2+x-1 = x^2(x-1) + (x-1) = (x-1)(x^2+1)$, and, by long division,
     the integrand can be written as $2x + (4x^2-3x+1)/(x-1)(x^2+1)$.

$$\frac{4x^2-3x+1}{(x-1)(x^2+1)} = \frac{A}{x-1} + \frac{Bx+C}{x^2+1} \,. \quad 4x^2-3x+1 = A(x^2+1) + (Bx+C)(x-1).$$

Set $x = 1$ to obtain $2 = 2A$, or $A = 1$, and set $x = 0$ to obtain $1 = A-C = 1-C$.
Thus $C = 0$. For the third constant we could equate some coefficient or simply
pick another value of x. Let's set $x = -1$ to obtain $8 = 2A + 2B - 2C = 2 +$
$2B$, or $B = 3$. Thus the original integral reduces to $\int (2x + \frac{1}{x-1} + \frac{3x}{x^2+1})dx =$
$x^2 + \ln|x-1| + \frac{3}{2}\ln(x^2+1) + C.$

34. $1/u(a + bu) = A/u + B/(a + bu) \Longrightarrow 1 = A(a + bu) + Bu.$ Set $u = 0$ to obtain
     $1 = Aa$, or $A = 1/a$. Set $u = -a/b$ to obtain $1 = -Ba/b$, or $B = -b/a$. Thus

$$\int \frac{1}{u(a+bu)}\,du = \int (\frac{1/a}{u} - \frac{b/a}{a+bu})du = (1/a)(\ln|u| - \ln|a+bu|) + C = \frac{1}{a}\ln\left|\frac{u}{a+bu}\right| + C.$$

37. $f(x) = x/(x^2-2x-3) = x/(x-3)(x+1)$. Since $x-3 < 0$, $x \geq 0$, and $(x+1) > 0$ on
     $[0,2]$, it follows that $f(x) \leq 0$ there. Therefore, $A = -\int_0^2 f(x)dx$.

Now, $f(x) = \frac{x}{(x-3)(x+1)} = \frac{A}{x-3} + \frac{B}{x+1} \Longrightarrow x = A(x+1) + B(x-3)$. Setting $x = 3$
and -1, we obtain $3 = 4A$, or $A = 3/4$, and $-1 = -4B$, or $B = 1/4$. Thus $A =$
$-\int_0^2 (\frac{3/4}{x-3} + \frac{1/4}{x+1})dx = -[\frac{3}{4}\ln|x-3| + \frac{1}{4}\ln|x+1|]\Big|_0^2 = -\frac{1}{4}[(3\ln|-1| + \ln 3) -$
$(3\ln|-3| + \ln 1)] = -\frac{1}{4}(0 + \ln 3 - 3\ln 3 + 0) = \frac{2\ln 3}{4}$.

40. The distance travelled in the time interval $[1,2]$ is $s(2) - s(1) = \int_1^2 v(t)dt$
$= \int_1^2 (t+3)/(t^3+t)dt$. Now, $\frac{t+3}{t^3+t} = \frac{t+3}{t(t^2+1)} = \frac{A}{t} + \frac{Bt+C}{t^2+1} \Longrightarrow t+3 = A(t^2+1) +$
$(Bt+C)t$ or $t+3 = (A+B)t^2 + Ct + A$. Thus $A = 3$, $C = 1$ and $A+B = 0$, or $B =$
$-A = -3$. Thus $s(2) - s(1) = \int_1^2 (\frac{3}{t} - \frac{3t}{t^2+1} + \frac{1}{t^2+1})dt = 3\ln|t|\Big]_1^2 - \frac{3}{2}\ln(t^2+1)\Big]_1^2 +$
$\tan^{-1}t\Big]_1^2 = 3\ln 2 - (\frac{3}{2}\ln 5 - \frac{3}{2}\ln 2) + \tan^{-1}2 - \tan^{-1}1 = (9/2)\ln 2 -$
$(3/2)\ln 5 + \tan^{-1}2 - \pi/4 \approx 1.027.$

EXERCISES 9.5

1. $x^2-4x+8 = (x-2)^2 + 4$. So, with $u = x-2$, $du = dx$, $\displaystyle\int \frac{dx}{x^2-4x+8} = \int \frac{dx}{(x-2)^2+4}$

$= \displaystyle\int \frac{du}{u^2+4} = \frac{1}{2}\tan^{-1}\frac{u}{2} + C = \frac{1}{2}\tan^{-1}\frac{x-2}{2} + C$. Alternately, after completing the

square, we could have let $x-2 = 2\tan\theta$, $dx = 2\sec^2\theta\, d\theta$ and $(x-2)^2 + 4 =$

$4\sec^2\theta$. Note also that $\tan\theta = (x-2)/2 \implies \theta = \tan^{-1}((x-2)/2)$. The integral

then becomes $\displaystyle\int \frac{2\sec^2\theta}{4\sec^2\theta}\, d\theta = \frac{1}{2}\int d\theta = \frac{1}{2}\theta + C = \frac{1}{2}\tan^{-1}\frac{x-2}{2} + C$ as before. Both

methods are equivalent. Use whichever you prefer.

2. $7 + 6x - x^2 = 16 - (x-3)^2$. So, with $u = x-3$, $du = dx$. $\displaystyle\int \frac{dx}{\sqrt{7+6x-x^2}} = \int \frac{du}{\sqrt{16-u^2}}$

$= \sin^{-1}\frac{u}{4} + C = \sin^{-1}\frac{x-3}{4} + C$.

4. Proceed as in #1 using $x^2-2x+2 = (x-1)^2 + 1$ to get $\tan^{-1}(x-1) + C$.

7. $(x^3-1) = (x-1)(x^2+x+1) = (x-1)[(x + \frac{1}{2})^2 + \frac{3}{4}]$. Let $u = x + \frac{1}{2}$ so that $x-1 =$

$u - \frac{3}{2}$ and $\dfrac{1}{x^3-1} = \dfrac{1}{(u - \frac{3}{2})(u^2+\frac{3}{4})} = \dfrac{A}{u - \frac{3}{2}} + \dfrac{Bu+C}{u^2+\frac{3}{4}}$. As in the previous section,

this yields $1 = A(u^2 + \frac{3}{4}) + (Bu + C)(u - \frac{3}{2})$.

Set $u = 3/2$ to obtain $1 = 3A$ or $A = 1/3$.

Set $u = 0$ to obtain $1 = \frac{3}{4}A - \frac{3}{2}C$ or $C = -1/2$.

Equate u^2 coefficients: $0 = A+B$, or $B = -1/3$. Now we can finally tie this

all together:

$\displaystyle\int \frac{1}{x^3-1}\, dx = \int [\frac{1/3}{u - 3/2} - \frac{1/3\ u}{u^2 + 3/4} - \frac{1/2}{u^2 + 3/4}]\, du = \frac{1}{3}\ln|u - \frac{3}{2}| - \frac{1}{6}\ln(u^2+\frac{3}{4})$

$- \dfrac{1}{\sqrt{3}}\tan^{-1}\dfrac{2u}{\sqrt{3}} + C = \frac{1}{3}\ln|x-1| - \frac{1}{6}\ln(x^2+x+1) - \dfrac{1}{\sqrt{3}}\tan^{-1}\dfrac{2x+1}{\sqrt{3}} + C$.

10. $x^4 - 4x^3 + 13x^2 = x^2(x^2-4x+13) = x^2[(x-2)^2 + 9]$. Let $u = x-2$ so that $x =$

$u+2$ and $\dfrac{1}{x^4-4x^3+13x^2} = \dfrac{1}{(u+2)^2(u^2+9)} = \dfrac{A}{u+2} + \dfrac{B}{(u+2)^2} + \dfrac{Cu+D}{u^2+9}$. Proceeding as in

the previous section, the solution is $A = 4/169$, $B = 1/13$, $C = -4/169$,

$D = -5/169$, and $\displaystyle\int \frac{1}{x^4-4x^3+13x^2}\, dx = \frac{4}{169}\ln|u+2| - \frac{1}{13}\cdot\frac{1}{u+2} - \frac{2}{169}\ln(u^2+9)$

$- \dfrac{5}{3(169)}\tan^{-1}\dfrac{u}{3} + C = \frac{4}{169}\ln|x| - \frac{1}{13x} - \frac{2}{169}\ln(x^2-4x+13) - \frac{5}{507}\tan^{-1}\frac{x-2}{3} + C$.

13. $2x^2 - 3x + 9 = 2(x^2 - \frac{3}{2}x + \frac{9}{16} + (\frac{9}{2} - \frac{9}{16})) = 2((x-3/4)^2 + 63/16)$. So, with

$u = x - \frac{3}{4}$, $\displaystyle\int \frac{1}{2x^2-3x+9}\, dx = \frac{1}{2}\int \frac{1}{(x-3/4)^2+63/16}\, dx = \frac{1}{2}\int \frac{1}{u^2+(3\sqrt{7}/4)^2}\, du =$

$\dfrac{1}{2}\cdot\dfrac{4}{3\sqrt{7}}\tan^{-1}\dfrac{4u}{3\sqrt{7}} + C = \dfrac{2}{3\sqrt{7}}\tan^{-1}\dfrac{4(x-3/4)}{3\sqrt{7}} + C = \dfrac{2}{3\sqrt{7}}\tan^{-1}\dfrac{(4x-3)}{3\sqrt{7}} + C$.

16. By the quadratic formula, $2x^2 + 3x - 4$ has roots $a = (-3 + \sqrt{41})/4$ and $b = (-3 - \sqrt{41})/4$. Thus $2x^2 + 3x - 4 = 2(x-a)(x-b)$. As in the previous section $f(x) = \dfrac{x}{2x^2 + 3x - 4} = \dfrac{x}{2(x-a)(x-b)} = \dfrac{A}{x-a} + \dfrac{B}{x-b}$ where $A = \dfrac{a}{2(a-b)}$, $B = -\dfrac{b}{2(a-b)}$. Thus $f(x) = \dfrac{1}{2(a-b)}(\dfrac{a}{x-a} - \dfrac{b}{x-b})$ and $\int f(x)\,dx = \dfrac{1}{2(a-b)}(a \ln|x-a| - b \ln|x-b|)$, which may be written in many equivalent forms. (Note $2(a-b) = \sqrt{41}$).

19. $A = \displaystyle\int_0^1 \dfrac{1}{x^3+1}\,dx$. An antiderivative of $1/(x^3+1)$ is found in a manner similar to that of problem 7 of this section using $x^3+1 = (x+1)(x^2-x+1) = (x+1)((x-1/2)^2 + 3/4)$. $u = x - 1/2$, $x + 1 = u + 3/2$. In this case, an antiderivative is $F(x) = (1/3)\ln|u+3/2| - (1/6)\ln|u^2+3/4| + (1/\sqrt{3})\tan^{-1}(2u/\sqrt{3})$
$= (1/3)\ln|x+1| - (1/6)\ln(x^2-x+1) + (1/\sqrt{3})\tan^{-1}\dfrac{(2x-1)}{\sqrt{3}}$. Then $A = F(1) - F(0)$
$= [(1/3)\ln 2 - (1/6)\ln 1 + (1/\sqrt{3})\tan^{-1}(1/\sqrt{3})]$
$- [(1/3)\ln 1 - (1/6)\ln 1 + (1/\sqrt{3})\tan^{-1}(-1/\sqrt{3})]$. Using $\tan^{-1}/(\pm 1/\sqrt{3}) = (\pm\pi/6)$, we have $A = (1/3)\ln 2 + (2/\sqrt{3})(\pi/6) \approx 0.83565$.

22. $s(5) - s(0) = \displaystyle\int_0^5 v(t)\,dt = \int_0^5 \dfrac{1}{\sqrt{75 + 10t - t^2}}\,dt$. Now, $75 + 10t - t^2 = 100 - (t-5)^2$, so with $u = t-5$, $du = dt$. $t = 0, 5 \Rightarrow u = -5,0$, and $s(5) - s(0) = \displaystyle\int_{-5}^0 \dfrac{1}{\sqrt{100-u^2}}\,du = \sin^{-1}(u/10)]_{-5}^0 = \sin^{-1}0 - \sin^{-1}(-1/2) = 0 - (-\pi/6) = \pi/6$.

EXERCISES 9.6

1. Let $u = \sqrt[3]{x+9}$. Then $u^3 = x+9$, $x = u^3-9$, $dx = 3u^2\,du$. $\displaystyle\int x\sqrt[3]{x+9}\,dx = \int (u^3-9)u(3u^2)\,du = 3\int (u^6-9u^3)\,du = 3(\dfrac{u^7}{7} - \dfrac{9}{4}u^4) + C$, where $u = (x+9)^{1/3}$.

4. Let $u = (x+3)^{1/3}$. Then $u^3 = x+3$, $x = u^3-3$, $dx = 3u^2\,du$. $\displaystyle\int \dfrac{5x}{(x+3)^{2/3}}\,dx$
$= 5\int \dfrac{(u^3-3)(3u^2)}{u^2}\,du = 15\int (u^3-3)\,du = 15(\dfrac{u^4}{4} - 3u) + C$
$= 15(\dfrac{(x+3)^{4/3}}{4} - 3(x+3)^{1/3}) + C$.

7. As in Example 2, let $x = u^6$. Then $dx = 6u^5\,du$, $\sqrt{x} = u^3$, $\sqrt[3]{x} = u^2$. $\displaystyle\int \dfrac{\sqrt{x}}{1+\sqrt[3]{x}}\,dx$
$= \displaystyle\int \dfrac{u^3(6u^5)}{1+u^2}\,du = 6\int \dfrac{u^8}{1+u^2}\,du = 6\int (u^6-u^4+u^2-1+\dfrac{1}{1+u^2})\,du$ (by long division) =

$$6[\frac{u^7}{7} - \frac{u^5}{5} + \frac{u^3}{3} - u + \tan^{-1}u] + C = 6[\frac{x^{7/6}}{7} - \frac{x^{5/6}}{5} + \frac{x^{1/2}}{3} - x^{1/6} + \tan^{-1}(x^{1/6})]+C.$$

10. Let $u = \sqrt{1+2x}$. Then $u^2 = 1+2x$, $2u\,du = 2\,dx$ or $dx = u\,du$, and $2x+3 =$

$$(u^2-1) + 3 = u^2+2. \quad x = 0,4 \Rightarrow u = 1,3. \quad \text{Thus} \int_0^4 \frac{2x+3}{\sqrt{1+2x}}\,dx = \int_1^3 \frac{(u^2+2)u}{u}\,du$$

$$= \frac{u^3}{3} + 2u]_1^3 = 15 - \frac{7}{3} = \frac{38}{3}.$$

13. Let $u = 1 + e^x$. Then $du = e^x dx$, $e^x = u-1$, and $e^{2x} = (u-1)^2$. $\int e^{3x}\sqrt{e^x+1}\,dx$

$$= \int e^{2x}\sqrt{e^x+1}\ e^x dx = \int (u-1)^2\,u^{1/2}\,du = \int (u^2-2u+1)u^{1/2}\,du =$$

$$\int (u^{5/2} - 2u^{3/2} + u^{1/2})du = \frac{2}{7}u^{7/2} - \frac{4}{5}u^{5/2} + \frac{2}{3}u^{3/2} + C, \text{ where } u = (1+e^x).$$

16. $\int \frac{\sin 2x}{\sqrt{1+\sin x}}\,dx = 2\int \frac{\sin x \cos x}{\sqrt{1+\sin x}}\,dx$. Let $u = 1+\sin x$. Then $du = \cos x\,dx$,

$\sin x = u-1$, and the integral becomes $2\int \frac{u-1}{u^{1/2}}\,du = 2\int (u^{1/2} - u^{-1/2})du =$

$$2(\frac{2}{3}u^{3/2} - 2u^{1/2}) + C = 4(\frac{(1+\sin x)^{3/2}}{3}) - (1+\sin x)^{1/2}) + C.$$

19. Let $u = x-1$, $du = dx$, $x = u+1$. $x = 2,3 \Rightarrow u = 1,2$. Thus $\int_2^3 \frac{x}{(x-1)^6}\,dx =$

$$\int_1^2 \frac{u+1}{u^6}\,du = \int_1^2 (u^{-5}+u^{-6})du = \frac{u^{-4}}{-4} + \frac{u^{-5}}{-5}]_1^2 = -\frac{1}{64} - \frac{1}{160} + \frac{1}{4} + \frac{1}{5} = \frac{-5-2+80+64}{320} = \frac{137}{320}.$$

22. Using $w = \tan(x/2)$, $\cos x = (1-w^2)/(1+w^2)$, $dx = \frac{2}{1+w^2}\,dw$,

$$\int \frac{1}{3 + 2\cos x}\,dx = \int \frac{2/(1+w^2)}{3 + \frac{2(1-w^2)}{1+w^2}}\,dw = \int \frac{2\,dw}{[3(1+w^2) + 2(1-w^2)]} =$$

$$2\int \frac{dw}{5+w^2} = \frac{2}{\sqrt{5}} \tan^{-1}\frac{w}{\sqrt{5}} + C = \frac{2}{\sqrt{5}} \tan^{-1}(\frac{\tan(x/2)}{\sqrt{5}}) + C.$$

25. $I = \int \frac{\sec x}{4 - 3\tan x}\,dx = \int \frac{1/\cos x}{4 - 3\sin x/\cos x}\,dx = \int \frac{1}{4\cos x - 3\sin x}\,dx.$

As above, let $w = \tan x/2$, etc. $I = \int \frac{1}{4(\frac{1-w^2}{1+w^2}) - 3(\frac{2w}{1+w^2})} \cdot \frac{2}{1+w^2}\,dw$

$$= \int \frac{2}{4(1-w^2) - 6w}\,dw = \int \frac{-1}{2w^2+3w-2}\,dw. \quad \text{Now, as in Section 9.4,}$$

$\frac{-1}{2w^2+3w-2} = \frac{-1}{(2w-1)(w+2)} = \frac{A}{2w-1} + \frac{B}{w+2} \Rightarrow -1 = A(w+2) + B(2w-1).$

Setting $w = -2$ and $1/2$ we get $-1 = -5B$, or $B = 1/5$, and $-1 = (5/2)A$, or $A = -2/5$. Thus $I = \int(\frac{-2/5}{2w-1} + \frac{1/5}{w+2})dw = -\frac{1}{5}\ln|2w-1| + \frac{1}{5}\ln|w+2| + C = -\frac{1}{5}\ln|2\tan\frac{x}{2} - 1| + \frac{1}{5}\ln|\tan\frac{x}{2} + 2| + C.$

28.　With $u = \cos x$, $du = -\sin x$, $I = \int \dfrac{\sin x}{5 \cos x + \cos^2 x}\, dx = -\int \dfrac{du}{5u+u^2}$.　Now,

$\dfrac{1}{5u+u^2} = \dfrac{1}{u(5+u)} = \dfrac{A}{u} + \dfrac{B}{5+u} \implies 1 = A(5+u) + Bu$.　Setting $u = 0$ and -5, we

obtain $1 = 5A$, or $A = 1/5$, and $1 = -5B$, or $B = -1/5$.　Thus $I = -\int \dfrac{1}{5}(\dfrac{1}{u} - \dfrac{1}{5+u})du$

$= \dfrac{1}{5}(\ln|5+u| - \ln|u|) + C = \dfrac{1}{5} \ln\left|\dfrac{5+u}{u}\right| + C = \dfrac{1}{5} \ln\left|\dfrac{5}{u} + 1\right| + C$

$= \dfrac{1}{5} \ln\left|\dfrac{5}{\cos x} + 1\right| + C = \dfrac{1}{5} \ln|5 \sec x + 1| + C$.

31.　Hint:　Let $t = 1/u$, $dt = (-1/u^2)du$.　$t = 1, x \implies u = 1, 1/x$.　For the last
　　　part evaluate each integral using $\arctan 1 = \pi/4$.

34.　$I = \int \csc x\, dx = \int \dfrac{1}{\sin x}\, dx = \int \dfrac{1+w^2}{2w}\, \dfrac{2}{1+w^2}\, dw = \int \dfrac{dw}{w} = \ln|w| + C$.

Now, $\cos x = \dfrac{1-w^2}{1+w^2} \implies \cos x + w^2\cos x + w^2 = 1 \implies w^2 = \dfrac{1 - \cos x}{1 + \cos x}$

$\implies w = (\dfrac{1 - \cos x}{1 + \cos x})^{1/2} \implies I = \ln|w| + C = \dfrac{1}{2} \ln\left|\dfrac{1 - \cos x}{1 + \cos x}\right| + C$.

EXERCISES 9.7

1.　The presence of $\sqrt{4+9x^2}$ refers us to the portion of the table involving
　　$\sqrt{a^2+u^2}$.　The integrand, $\sqrt{4+9x^2}/x$ indicates that Formula 23 is the one to use.
　　To convert the given integral into the desired form, we see that $a^2 = 4$,
　　$u^2 = 9x^2$ so that $a = 2$, $u = 3x$, $du = 3\, dx$.　If we multiply and divide by 3,
　　then $u = 3x$ is in the denominator, and $du = 3\, dx$ is in the numerator.　Then
　　$\int \dfrac{\sqrt{4+9x^2}}{x}\, dx = \int \dfrac{\sqrt{2^2 + (3x)^2}}{3x}\, 3\, dx = \sqrt{2^2 + (3x)^2} - 2 \ln\left|\dfrac{2 + \sqrt{2^2 + (3x)^2}}{3x}\right| + C$,
　　which quickly reduces to the given answer.

4.　The integrand, $x^2\sqrt{4x^2-16}$, indicates that Formula 40 is the one to use.　We
　　can make the given integral compatible with that formula as above, or, al-
　　ternately, by changing variables.　Let $u = 2x$, $x = u/2$, $dx = du/2$.　Then
　　$\int x^2\sqrt{4x^2-16}\, dx = \int \dfrac{u^2}{4}\sqrt{u^2-16} \cdot \dfrac{1}{2}\, du = \dfrac{1}{8}\int u^2\sqrt{u^2-16}\, du$, which is in the correct

　　form with $a = 4$.　Using Formula 40, we obtain $\dfrac{1}{8}[\dfrac{u}{8}(2u^2-16)\sqrt{u^2-16} - \dfrac{4^4}{8} \cdot$

　　$\ln|u + \sqrt{u^2-16}| + C = \dfrac{1}{8}[\dfrac{2x}{8}(2(4x^2)-16)\sqrt{4x^2-16} - \dfrac{256}{8} \ln|2x + \sqrt{4x^2-16}| + C$

　　$= \dfrac{x}{4}(x^2-2)\sqrt{4x^2-16} - 4 \ln|2x - \sqrt{4x^2-16}| + C$.

7.　Let $u = 3x$, $dx = (1/3)du$.　Then by Formula 73 with $n = 6$, $I = \dfrac{1}{3}\int \sin^6 u\, du =$

　　$\dfrac{1}{3}[-\dfrac{1}{6} \sin^5 u \cos u + \dfrac{5}{6}\int \sin^4 u\, du]$.　By 73 again, with $n = 4$, $\int \sin^4 u\, du =$

　　$-\dfrac{1}{4} \sin^3 u \cos u + \dfrac{3}{4}\int \sin^2 u\, du$.　Using Formula 63 for the last integral, we obtain: $I = \frac{1}{3}[-\frac{1}{6}\sin^5 u \cos u - \frac{5}{24}\sin^3 u \cos u + \frac{15}{24}(\frac{u}{2} - \frac{\sin 2u}{4})] + C$, where $u = 3x$. (sin 2u can be replaced by 2 sin u cos u if desired.)

10. Directly from Formula 81 with a = 5, b = 3, and u = x, $\int \sin 5x \cos 3x \, dx =$

$-\frac{\cos 2x}{2(2)} - \frac{\cos 8x}{2(8)} + C.$

13. Directly from Formula 98 with a = -3, b = 2, and u = x, $\int e^{-3x} \sin 2x \, dx =$

$\frac{e^{-3x}}{(-3)^2+2^2} (-3 \sin 2x - 2 \cos 2x) + C.$

16. From the integrand, $1/x\sqrt{3x-2x^2}$, the formulas involving $\sqrt{2au-u^2}$ should be consulted, and we see that Formula 120 fits. So, let $u = \sqrt{2}x$ so that $u^2 = 2x^2$, $x = u/\sqrt{2}$, $dx = du/\sqrt{2}$. Then $I = \int \frac{dx}{x\sqrt{3x-2x^2}} = \int \frac{du/\sqrt{2}}{(u/\sqrt{2})\sqrt{3(u/\sqrt{2})-u^2}} =$

$\int \frac{1}{u\sqrt{\frac{3}{\sqrt{2}}u - u^2}} du.$ Thus $2a = \frac{3}{\sqrt{2}}$, or $a = \frac{3}{2\sqrt{2}}$. By the formula,

$I = -\frac{\sqrt{(3/\sqrt{2})u-u^2}}{(3/2\sqrt{2})u} + C = -\frac{\sqrt{3x-2x^2}}{(3/2)x} + C.$

19. Writing $\int e^{2x} \cos^{-1} e^x \, dx = \int e^x \cos^{-1} e^x(e^x)dx$, we let $u = e^x$, $du = e^x dx$ to

$\int u \cos^{-1}u \, du$, which, by Formula 91, equals $\frac{2u^2-1}{4}\cos^{-1}u - \frac{u\sqrt{1-u^2}}{4} + C$, where $u = e^x$.

22. Using Formula 61 with n = 3, u = x, a = 2, b = -1, $I = 7 \int \frac{x^3}{\sqrt{2-x}} dx =$

$7[\frac{2x^3}{-7}\sqrt{2-x} - \frac{12}{-7}\int \frac{x^2}{\sqrt{2-x}} dx]$. Cancel the 7's and use Formula 56 with u = x, a = 2, b = -1 for the last integral to obtain $I = -2x^3\sqrt{2-x} +$

$12[-\frac{2}{15}(32 + 3x^2 + 8x)\sqrt{2-x}] + C.$

25. Combining Formulas 58 and 57 with u = x, a = 9, b = 2, $\int \frac{\sqrt{9+2x}}{x} dx = 2\sqrt{9+2x}$

$+ 9\int \frac{dx}{x\sqrt{9+2x}} = 2\sqrt{9+2x} + \frac{9}{3} \ln\left|\frac{\sqrt{9+2x} - 3}{\sqrt{9+2x} + 3}\right| + C.$ (The first form of 57 was used since a = 9 > 0.).

28. We begin with the rationalizing substitution $u = \sqrt{x}$, $x = u^2$, $dx = 2u\,du$.

$I = \int \frac{dx}{2x^{3/2}+5x^2} = \int \frac{2u}{2u^3+5u^4} du = 2\int \frac{1}{2u^2+5u^3} du = 2\int \frac{1}{u^2(2+5u)} du.$ By

Formula 50 with a = 2, b = 5, $I = 2(-\frac{1}{2u} + \frac{5}{4}\ln\left|\frac{2+5u}{u}\right|) + C = -\frac{1}{\sqrt{x}}$

$+ \frac{5}{2}\ln\left|\frac{2+5\sqrt{x}}{\sqrt{x}}\right| + C.$

EXERCISES 9.8 (Review)

1. $u = \sin^{-1}x$, $dv = x\,dx \Rightarrow du = \dfrac{dx}{\sqrt{1-x^2}}$, $v = \dfrac{x^2}{2}$.

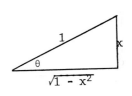

$I = \displaystyle\int x\,\sin^{-1}x\,dx = \dfrac{x^2}{2}\sin^{-1}x - \dfrac{1}{2}\int \dfrac{x^2}{\sqrt{1-x^2}}\,dx$.   With $x =$

$\sin\theta$, $dx = \cos\theta\,d\theta$, $\sqrt{1-x^2} = \cos\theta$, the last integral

is $\displaystyle\int \dfrac{\sin^2\theta\,\cos\theta}{\cos\theta}\,d\theta = \dfrac{1}{2}\int (1 - \cos 2\theta)\,d\theta = \dfrac{1}{2}(\theta - \dfrac{\sin 2\theta}{2})$

$= \dfrac{1}{2}(\theta - \sin\theta\,\cos\theta) = \dfrac{1}{2}(\sin^{-1}x - x\sqrt{1-x^2})$.   Thus

$I = \dfrac{x^2}{2}\sin^{-1}x - \dfrac{1}{4}\sin^{-1}x + \dfrac{1}{4}x\sqrt{1-x^2} + C$.

4.   Let $y^2 = x$, $2y\,dy = dx$, $y = \sqrt{x}$.   Then $\displaystyle\int_0^1 e^{\sqrt{x}}\,dx = 2\int_0^1 y\,e^y\,dy = 2(y-1)e^y\,\Big]_0^1$

$= 2$.   (The y integral was done by parts with $u = y$, $dv = e^y dy$.)

7. $\displaystyle\int \tan x\,\sec^5x\,dx = \int \sec^4x(\sec x\,\tan x)\,dx = \dfrac{\sec^5 x}{5} + C$.

10.   Let $x = 4\sin\theta$, $dx = 4\cos\theta\,d\theta$, $\sqrt{16-x^2} = 4\cos\theta$.   Then $\displaystyle\int \dfrac{1}{x^2\sqrt{16-x^2}}\,dx =$

$\displaystyle\int \dfrac{4\cos\theta\,d\theta}{(16\sin^2\theta)(4\cos\theta)} = \dfrac{1}{16}\int \csc^2\theta\,d\theta = -\dfrac{1}{16}\cot\theta + C$

$= -\dfrac{1}{16}\dfrac{\sqrt{16-x^2}}{x} + C$.

13. $\dfrac{x^3+1}{x(x-1)^3} = \dfrac{A}{x} + \dfrac{B}{x-1} + \dfrac{C}{(x-1)^2} + \dfrac{D}{(x-1)^3}$. $x^3+1 = A(x-1)^3 + Bx(x-1)^2 + Cx(x-1) +$

Dx.   Setting $x = 1$ and then $x = 0$, we see $D = 2$, $A = -1$.   Equating the x^3

coefficients: $1 = A+B = -1 + B \Rightarrow B = 2$; and the x^2 coefficients: $0 = -3A$

$-2B + C = 3-4+C \Rightarrow C = 1$.   Thus $\displaystyle\int \dfrac{x^3+1}{x(x-1)^3}\,dx = -\ln|x| + 2\ln|x-1| -$

$\dfrac{1}{x-1} - \dfrac{1}{(x-1)^2} + C$, where C is now the usual arbitrary constant of integration.

16.   Let $u = x+2$.   Then $du = dx$, and $x-1 = u-3$. $\displaystyle\int \dfrac{x-1}{(x+2)^5}\,dx = \int \dfrac{u-3}{u^5}\,du =$

$\displaystyle\int (u^{-4} - 3u^{-5})\,du = \dfrac{u^{-3}}{-3} - \dfrac{3u^{-4}}{-4} + C = -\dfrac{1}{3(x+2)^3} + \dfrac{3}{4(x+2)^4} + C$.

19.   Let $u^3 = x+8$.   Then $3u^2 = dx$, $x = u^3-8$. $I = \displaystyle\int \dfrac{\sqrt[3]{x+8}}{x}\,dx = \int \dfrac{u(3u^2)}{u^3-8}\,du =$

$3\displaystyle\int (1 + \dfrac{8}{u^3-8})\,du$.   Now, $u^3-8 = (u-2)(u^2+2u+4) = (u-2)[(u+1)^2+3] =$

$(z-3)(z^2+3)$ where $z = u+1$.

$\dfrac{1}{u^3-8} = \dfrac{1}{(z-3)(z^2+3)} = \dfrac{A}{z-3} + \dfrac{Bz+C}{z^2+3}$. $1 = A(z^2+3) + (Bz+C)(z-3)$.

Equating coefficients gives the system

$$z^2: \quad 0 = A + B$$
$$z^1: \quad 0 = -3B + C \} \implies \begin{cases} A = 1/4 \\ B = -1/4 \\ C = -3/4. \end{cases}$$
$$z^0: \quad 1 = A - C$$

Thus $\int \dfrac{1}{u^3-8} du = \dfrac{1}{4} \ln|z-3| - \dfrac{1}{8} \ln(z^2+3) - \dfrac{3}{4} \cdot \dfrac{1}{\sqrt{3}} \tan^{-1} \dfrac{z}{\sqrt{3}}$. Replacing z by

u+1, and combining this with the first calculation for I finally gives

$$I = 3u + 2\ln|u-2| - \ln(u^2+2u+4) - 2\sqrt{3}\, \tan^{-1} \dfrac{u+1}{\sqrt{3}} + C \text{ where } u = \sqrt[3]{x+8}.$$

22. $u = \cos(\ln x)$, $dv = dx \implies du = \dfrac{-\sin(\ln x)}{x} dx$, $v = x$. $\int \cos(\ln x)dx =$

$x \cos(\ln x) + \int \sin(\ln x)dx$. Now, $u = \sin(\ln x)$, $dv = dx \implies du = \dfrac{\cos(\ln x)}{x}$,

$v = x$. $\int \cos(\ln x)dx = x \cos(\ln x) + x \sin(\ln x) - \int \cos(\ln x)dx$.

$\therefore 2 \int \cos(\ln x)dx = x(\cos(\ln x) + \sin(\ln x)) + C.$

25. With $u = 4-x^2$, $du = -2x\, dx$, $\int \dfrac{x}{\sqrt{4-x^2}} dx = -\dfrac{1}{2} \int u^{-1/2} du = -u^{1/2} + C = -\sqrt{4-x^2}+C.$

28. $\dfrac{x^3}{x^3-3x^2+9x-27} = 1 + \dfrac{3x^2 - 9x + 27}{(x-3)(x^2+9)}$ and $\dfrac{3x^2 - 9x + 27}{(x-3)(x^2+9)} = \dfrac{A}{x-3} + \dfrac{Bx+C}{x^2+9}$.

$3x^2-9x+27 = A(x^2+9) + (Bx+C)(x-3)$. Equating coefficients yields

$$x^2: \quad 3 = A + B$$
$$x^1: \quad -9 = \quad - 3B + C \} \implies \begin{cases} A = 3/2 \\ B = 3/2 \\ C = -9/2. \end{cases}$$
$$x^0: \quad 27 = 9A \quad - 3C$$

Putting this all together at last we get

$$\int \dfrac{x^3}{x^3-3x^2+9x-27} dx = x + \dfrac{3}{2}\ln|x-3| + \dfrac{3}{4}\ln(x^2+9) - \dfrac{3}{2}\tan^{-1} \dfrac{x}{3} + C.$$

31. With $u = e^x$, $du = e^x dx$ and $\int e^x \sec e^x dx = \int \sec u\, du = \ln|\sec u + \tan u|$

$+ C = \ln|\sec e^x + \tan e^x| + C.$

34. Recalling that $\sin 2x = 2 \sin x \cos x$, $\int \sin 2x \cos x\, dx = 2 \int \sin x \cos^2 x\, dx$

$= -\dfrac{2}{3} \cos^3 x + C.$

37. With $u = e^x$, $du = e^x dx$, $\int e^x\sqrt{1+e^x}\, dx = \int (1+u)^{1/2} du = \dfrac{2}{3}(1+u)^{3/2} + C$

$= \dfrac{2}{3}(1+e^x)^{3/2} + C.$

40. $x^2 + 8x + 25 = (x+4)^2 + 9$. So, with $u = x+4$, $du = dx$, $3x+2 = 3(u-4) + 2 =$

$3u = 10$, and $\int \dfrac{3x+2}{x^2+8x+25} dx = \int \dfrac{3u-10}{u^2+9} du = \dfrac{3}{2}\ln(u^2+9) - \dfrac{10}{3}\tan^{-1} \dfrac{u}{3} + C$

$= \dfrac{3}{2}\ln(x^2+8x+25) - \dfrac{10}{3}\tan^{-1} \dfrac{x+4}{3} + C.$

43. $u = x$, $dv = \cot x \csc x \, dx \implies du = dx$, $v = -\csc x$. $\int x \cot x \csc x \, dx =$

$-x \csc x + \int \csc x \, dx$. Now use Formula 15.

46. $u = (\ln x)^2$, $dv = x \, dx \implies du = (2 \ln x)/x \, dx$, $v = x^2/2$. $I = \int x(\ln x)^2 dx =$

$(x^2/2)(\ln x)^2 - \int x \ln x \, dx$. In this integral we use $u = \ln x$, $dv = x \, dx \implies$

$du = (1/x)dx$, $v = x^2/2$ to obtain $I = (x^2/2)(\ln x)^2 - [(x^2/2)\ln x - \int x/2 \, dx]$

$= (x^2/2)[(\ln x)^2 - \ln x + 1/2] + C$.

49. First write $\int \dfrac{e^{3x}}{1+e^x} \, dx = \int \dfrac{e^{2x}}{1+e^x} \, e^x dx$. Then let $u = 1+e^x$ so that $du = e^x dx$ and

$e^x = (u-1)$, and we obtain $\int \dfrac{(u-1)^2}{u} \, du = \int \dfrac{u^2-2u+1}{u} \, du = \int (u-2+1/u) \, du =$

$u^2/2 - 2u + \ln|u| + C$, where $u = 1+e^x$.

52. Let $u = 1 + \sin x$. Then $du = \cos x \, dx$ and $\sin x = u-1$. $\int \dfrac{\cos^3 x}{\sqrt{1 + \sin x}} \, dx$

$= \int \dfrac{\cos^2 x}{\sqrt{1 + \sin x}} \cos x \, dx = \int \dfrac{1 - \sin^2 x}{\sqrt{1 + \sin x}} \cos x \, dx = \int \dfrac{1-(u-1)^2}{\sqrt{u}} \, du$

$= \int \dfrac{2u-u^2}{u^{1/2}} \, du = \int (2u^{1/2} - u^{3/2}) \, du = \dfrac{4}{3} u^{3/2} - \dfrac{2}{5} u^{5/2} + C = \dfrac{4}{3}(1 + \sin x)^{3/2} -$

$\dfrac{2}{5}(1 + \sin x)^{5/2} + C$.

55. $\dfrac{1-2x}{x^2+12x+35} = \dfrac{1-2x}{(x+5)(x+7)} = \dfrac{A}{x+5} + \dfrac{B}{x+7} \implies 1-2x = A(x+7) + B(x+5)$. Setting

$x = -7$ and -5 we get $15 = -2B$ and $11 = 2A$, whence $A = 11/2$, $B = -15/2$. The

integral then becomes $\int (\dfrac{11/2}{x+5} - \dfrac{15/2}{x+7}) dx = \dfrac{11}{2} \ln|x+5| - \dfrac{15}{2} \ln|x+7| + C$.

58. First let $u = 3x$, $dx = (1/3)du$. Then use Formulas 73 (with $n = 4$) and 63 to

obtain: $\int \sin^4 3x \, dx = \dfrac{1}{3} \int \sin^4 u \, du = \dfrac{1}{3}[-\dfrac{1}{4} \sin^3 u \cos u + \dfrac{3}{4} \int \sin^2 u \, du]$

$= -\dfrac{1}{12} \sin^3 u \cos u + \dfrac{1}{4}[\dfrac{1}{2}u - \dfrac{1}{4} \sin 2u] + C = -\dfrac{1}{12} \sin^3 3x \cos 3x$

$+ \dfrac{3}{8}x - \dfrac{1}{16} \sin 6x + C$.

61. $\int \dfrac{1}{\sqrt{7+5x^2}} \, dx = \dfrac{1}{\sqrt{5}} \int \dfrac{\sqrt{5} \, dx}{\sqrt{7+(\sqrt{5}x)^2}} = \dfrac{1}{\sqrt{5}} \ln|\sqrt{5} \, x + \sqrt{7+5x^2}| + C$ by Formula 25 with

$u = \sqrt{5} \, x$ and $a^2 = 7$.

64. $\int \cot^5 x \csc x \, dx = \int \cot^4 x(\cot x \csc x) dx = \int (\csc^2 x-1)^2 (\cot x \csc x) dx =$

$-\int (u^2-1)^2 du$ (with $u = \csc x$, $du = -\cot x \csc x \, dx$) $= -\int (u^4-2u^2+1) du =$

$$-\frac{u^5}{5} + \frac{2}{3}u^3 - u + C = \frac{-\csc^5 x}{5} + \frac{2}{3}\csc^3 x - \csc x + C.$$

67. $\int (x^2 - \text{sech}^2 4x)\,dx = \frac{x^3}{3} - \frac{1}{4}\int \text{sech}^2 u\,du = \frac{x^3}{3} - \frac{1}{4}\tanh u + C$, where $u = 4x$.

70. Let $u = x^3 + 1$, $du = 3x^2 dx$ and $x^3 = u - 1$. $\int x^5\sqrt{x^3 + 1}\,dx = (1/3)\int x^3\sqrt{x^3+1}(3x^2)\,dx$

$= (1/3)\int (u-1)\sqrt{u}\,du = \frac{1}{3}\int (u^{3/2}-u^{1/2})\,du = \frac{1}{3}(\frac{2}{5}u^{5/2} - \frac{2}{3}u^{3/2}) + C$

$= \frac{2}{15}(x^3+1)^{5/2} - \frac{2}{9}(x^3+1)^{3/2} + C.$

73. Since $\tan u \cos u = \frac{\sin u}{\cos u}\cos u = \sin u$, $\int \tan 7x \cos 7x\,dx = \int \sin 7x\,dx =$

$-(1/7)\cos 7x + C.$

76. Let $x = 4\sin\theta$, $dx = 4\cos\theta\,d\theta$, $\sqrt{16-x^2} = 4\cos\theta$. Then $I = \int \frac{1}{x^4\sqrt{16-x^2}}\,dx$

$= \int \frac{4\cos\theta}{4^4\sin^4\theta(4\cos\theta)}\,d\theta = \frac{1}{256}\int \frac{1}{\sin^4\theta}\,d\theta = \frac{1}{256}\int \csc^4\theta\,d\theta.$ Now, by Formula

78, with $n = 4$, $I = \frac{1}{256}[-\frac{1}{3}\cot\theta\,\csc^2\theta + \frac{2}{3}\int \csc^2\theta\,d\theta]$

$= \frac{1}{256}[-\frac{1}{3}\cot\theta\,\csc^2\theta - \frac{2}{3}\cot\theta] + C$

$= -\frac{1}{768}\cot\theta\,(\csc^2\theta + 2) + C = -\frac{1}{768}\frac{\sqrt{16-x^2}}{x}(\frac{16}{x^2} + 2) + C$

$= -\frac{\sqrt{16-x^2}(8+x^2)}{384\,x^3} + C.$

79. $\int \frac{\sqrt{9-4x^2}}{x^2}\,dx = 2\int \frac{\sqrt{9-(2x)^2}}{(2x)^2}\,2\,dx = 2[-\frac{\sqrt{9-4x^2}}{2x} - \sin^{-1}\frac{2x}{3}] + C$ by Formula 33

with $u = 2x$, $a^2 = 9$.

82. Let $u = x^2+5$, $du = 2x\,dx$, $x\,dx = (1/2)du$. $\int x(x^2+5)^{3/4}dx = (1/2)\int u^{3/4}du =$

$(1/2)(4/7)u^{7/4} + C = (2/7)(x^2+5)^{7/4} + C.$

85. Let $u = 1 + \cos x$, $du = -\sin x\,dx$. $\int \frac{\sin x}{\sqrt{1+\cos x}}\,dx = -\int u^{-1/2}du = -2u^{1/2}$

$+ C = -2\sqrt{1+\cos x} + C.$

88. $\int \sin^4 x \cos^3 x\,dx = \int \sin^4 x(1 - \sin^2 x)\cos x\,dx = \int u^4(1-u^2)\,du = \int (u^4-u^6)\,du$

$= (1/5)u^5 - (1/7)u^7 + C$, where $u = \sin x$.

91. $x^4+9x^2+20 = (x^2+4)(x^2+5) \implies f(x) = \frac{2x^3+4x^2+10x+13}{(x^2+4)(x^2+5)} = \frac{Ax+B}{x^2+4} + \frac{Cx+D}{x^2+5} \implies$

$2x^3+4x^2+10x+13 = (Ax+B)(x^2+5) + (Cx+D)(x^2+4) = (A+C)x^3 + (B+D)x^2 +$

$(5A+4C)x + (5B+4D)$. Equating the x^3 and x coefficients, we get the system

$A+C = 2$, $5A + 4C = 10$ with solution $A = 2$, $C = 0$. Equating the x^2 and constant

coefficients, we get the system $B+D = 4$, $5B+4D = 13$ with solution $B = -3$,

$D = 7$. Thus $\int f(x)\,dx = \int (\dfrac{2x}{x^2+4} - \dfrac{3}{x^2+4} + \dfrac{7}{x^2+5})\,dx = \ln(x^2+4) - \dfrac{3}{2}\tan^{-1}\dfrac{x}{2}$

$+ \dfrac{7}{\sqrt{5}}\tan^{-1}\dfrac{x}{\sqrt{5}} + C.$

94. $\int \cot^2 x\,\csc x\,dx = \int (\csc^2 x - 1)\csc x\,dx = \int \csc^3 x\,dx - \int \csc x\,dx =$

$(-(1/2)\csc x\,\cot x + (1/2)\ln|\csc x - \cot x|) - \ln|\csc x - \cot x| + C =$

$-(1/2)(\csc x\,\cot x + \ln|\csc x - \cot x|) + C.$ (The integral of $\csc^3 x$ is from #2 or from Formula 72.)

97. Let $u = 2x+3$. Then $x = (1/2)(u-3)$, $dx = (1/2)\,du$. $\int \dfrac{x^2}{\sqrt[3]{2x+3}}\,dx = \int \dfrac{(1/4)(u-3)^2}{\sqrt[3]{u}}.$

$(1/2)\,du = \dfrac{1}{8}\int \dfrac{u^2-6x+9}{u^{1/3}}\,du = \dfrac{1}{8}\int (u^{5/3} - 6u^{2/3} + 9u^{-1/3})\,du =$

$\dfrac{1}{8}[\dfrac{3}{8}u^{8/3} - 6(\dfrac{3}{5})u^{5/3} + 9(\dfrac{3}{2})u^{2/3}] + C$ where $u = 2x+3$.

100. Let $u = x+1$. Then $du = dx$, $u+1 = x+2$. $\int (x+2)^2(x+1)^{10}\,dx = \int (u+1)^2 u^{10}\,du =$ $\int (u^{12}+2u^{11}+u^{10})\,du = (1/13)u^{13} + (2/12)u^{12} + (1/11)u^{11} + C$ where $u = x+1$.

CHAPTER 10

INDETERMINATE FORMS, IMPROPER INTEGRALS,
AND TAYLOR'S FORMULA

EXERCISES 10.1

1. $\lim\limits_{x \to 0} \dfrac{\sin x}{2x} = \lim\limits_{x \to 0} \dfrac{\cos x}{2} = \dfrac{\cos 0}{2} = \dfrac{1}{2}$.

4. $\lim\limits_{x \to 4} \dfrac{x-4}{(x+4)^{1/3} - 2} = \lim\limits_{x \to 4} \dfrac{1}{(x+4)^{-2/3}/3} = \dfrac{3}{8^{-2/3}} = \dfrac{3}{1/4} = 12$.

7. This function is $\underline{NOT}$ indeterminate as $x \to 1$ since the denominator does not approach 0. Thus L'Hôpital's rule cannot be used. However, by the methods of Chapter 2, $\lim\limits_{x \to 1} \dfrac{x^3 - 3x + 2}{x^2 - 2x - 1} = \dfrac{1 - 3 + 2}{1 - 2 - 1} = \dfrac{0}{-2} = 0.$

10. $\lim\limits_{x \to 0} \dfrac{\sin x}{x - \tan x} = \lim\limits_{x \to 0} \dfrac{\cos x}{1 - \sec^2 x} = -\infty$ since $\cos x \to 1$ and $1 - \sec^2 x$ is negative and approaches 0.

13. $\lim\limits_{x \to 0} \dfrac{x - \sin x}{x^3} = \lim\limits_{x \to 0} \dfrac{1 - \cos x}{3x^2} = \lim\limits_{x \to 0} \dfrac{\sin x}{6x} = \lim\limits_{x \to 0} \dfrac{\cos x}{6} = \dfrac{\cos 0}{6} = \dfrac{1}{6}$.

16. $\lim\limits_{x \to 0^+} \dfrac{\cos x}{x} = \infty$ since $\cos x \to 1$ and $x > 0$ as $x \to 0^+$. (Note that L'Hôpital's rule cannot be used.)

19. $\lim\limits_{x \to \infty} \dfrac{x^2}{\ln x} = \lim\limits_{x \to \infty} \dfrac{2x}{1/x} = \lim\limits_{x \to \infty} 2x^2 = \infty.$

22. $\lim\limits_{x \to 0} \dfrac{2x}{\tan^{-1} x} = \lim\limits_{x \to 0} \dfrac{2}{1/(1+x^2)} = \dfrac{2}{1} = 2.$

25. The limit is ∞ since $x \cos x + e^{-x} \to 0(1) + 1 = 1$, and $x^2 \to 0$ and is positive as $x \to 0$. (Again, L'Hôpital's rule cannot be used.)

28. $\lim\limits_{x \to \infty} \dfrac{x^3 + x + 1}{3x^3 + 4} = \lim\limits_{x \to \infty} \dfrac{3x^2 + 1}{9x^2} = \lim\limits_{x \to \infty} \dfrac{6x}{18x} = \dfrac{1}{3}$.

31. If n is an integer, then after n applications of L'Hôpital's rule (possible, since each quotient is the ∞/∞ form), we obtain $\lim\limits_{x \to \infty} \dfrac{x^n}{e^x} = \lim\limits_{x \to \infty} \dfrac{n!}{e^x} = 0$. If n is not an integer, let $k = [n] + 1$ so that $n - k < 0$. After k applications of the rule, $\lim\limits_{x \to \infty} \dfrac{x^n}{e^x} = \lim\limits_{x \to \infty} n(n-1)\ldots(n-k+1) \dfrac{x^{n-k}}{e^x} = 0$, since $x^{n-k} \to 0$ and $e^x \to \infty$ as $x \to \infty$.

34. $\lim\limits_{x \to 0} \dfrac{\sin^2 x + 2 \cos x - 2}{\cos^2 x - x \sin x - 1} = \lim\limits_{x \to 0} \dfrac{2 \sin x \cos x - 2 \sin x}{-2 \cos x \sin x - x \cos x - \sin x}$

$$= \lim_{x \to 0} \frac{\sin 2x - 2 \sin x}{-\sin 2x - x \cos x - \sin x} = \lim_{x \to 0} \frac{2 \cos 2x - 2 \cos x}{-2 \cos 2x - 2 \cos x + x \sin x}$$

$$= \frac{2-2}{-2-2+0} = \frac{0}{-4} = 0.$$

37. Using the identity $\dfrac{\tan x - \sin x}{\tan x} = 1 - \dfrac{\sin x}{\sin x / \cos x} = 1 - \cos x$, we have

$$\lim_{x \to 0} \frac{\tan x - \sin x}{x^3 \tan x} = \lim_{x \to 0} \frac{1 - \cos x}{x^3} = \lim_{x \to 0} \frac{\sin x}{3x^2} = \lim_{x \to 0} \frac{\cos x}{6x} \text{ which does not exist}$$

since $\cos x \to 1$, $6x \to 0$.

40. $\displaystyle \lim_{x \to 0} \frac{2-e^x-e^{-x}}{1-\cos^2 x} = \lim_{x \to 0} \frac{-e^x+e^{-x}}{2 \cos x \sin x} = \lim_{x \to 0} \frac{-e^x-e^{-x}}{2(\cos^2 x - \sin^2 x)} = \frac{-1-1}{2(1-0)} = \frac{-2}{2} = -1.$

43. $\displaystyle \lim_{x \to 0} \frac{x - \tan^{-1} x}{x \sin x} = \lim_{x \to 0} \frac{1 - 1/(1+x^2)}{x \cos x + \sin x} = \lim_{x \to 0} \frac{2x/(1+x^2)^2}{-x \sin x + 2 \cos x} = \frac{0}{0+2} = 0.$

46. $\displaystyle \lim_{x \to 0} \frac{x \sin^{-1} x}{x - \sin x} = \lim_{x \to 0} \frac{x/\sqrt{1-x^2} + \sin^{-1} x}{1 - \cos x}$, which is still of indeterminate form

0/0. Separately, we calculate $D_x \dfrac{x}{\sqrt{1-x^2}} = \dfrac{\sqrt{1-x^2} - x(-2x)/2\sqrt{1-x^2}}{(\sqrt{1-x^2})^2} = \dfrac{(1-x^2)+x^2}{(1-x^2)^{3/2}}$

$= \dfrac{1}{(1+x^2)^{3/2}}$. Continuing the limit calculation from above, using L'Hôpital's

rule, $\displaystyle \lim_{x \to 0} \frac{x/\sqrt{1-x^2} + \sin^{-1} x}{1 - \cos x} = \lim_{x \to 0} \frac{1/(1-x^2)^{3/2} + 1/\sqrt{1-x^2}}{\sin x}$, which does not exist

since the numerator approaches 2 and $\sin x \to 0$ as $x \to 0$.

49. $\displaystyle \lim_{x \to \infty} \frac{2e^{3x} + \ln x}{e^{3x} + x^2} = \lim_{x \to \infty} \frac{6e^{3x} + 1/x}{3e^{3x} + 2x} = \lim_{x \to \infty} \frac{18e^{3x} - 1/x^2}{9e^{3x} + 2} = \lim_{x \to \infty} \frac{54e^{3x} + 2/x^3}{27e^{3x}}$

$$= \lim_{x \to \infty} \left(2 + \frac{2}{27e^{3x} x^3}\right) = 2 + 0 = 2.$$

52. $\displaystyle \lim_{x \to \infty} \frac{x + \cosh x}{x^2 + 1} = \lim_{x \to \infty} \frac{1 + \sinh x}{2x} = \lim_{x \to \infty} \frac{\cosh x}{2} = \infty$

55. All derivatives will be with respect to ω. Thus,

$D_\omega(\cos \omega t) = (-\sin \omega t) D_\omega(\omega t) = (-\sin \omega t)t$. Writing

$s = A\omega^2 \dfrac{\cos \omega t - \cos \omega_0 t}{\omega_0^2 - \omega^2}$, we have

$$\lim_{\omega \to \omega_0} s = A\omega_0^2 \lim_{\omega \to \omega_0} \frac{\cos \omega t - \cos \omega_0 t}{\omega_0^2 - \omega^2} = A\omega_0^2 \lim_{\omega \to \omega_0} \frac{-t \sin \omega t}{-2\omega}$$

$= A\omega_0^2 \dfrac{t \sin \omega_0 t}{2\omega_0} = \dfrac{A\omega_0}{2} t \sin \omega_0 t$. The amplitude, $\dfrac{A\omega_0}{2}t$, increases

as t increases.

58. Note that $\lim\limits_{x\to 0} C(x) = \int_0^0 \cos(u^2)du = 0$. Thus, the functions

in (a) and (b) are of 0/0 form as $x \to 0$. Next, by the Fundamental

Theorem of Calculus, $C'(x) = D_x(\int_0^x \cos(u^2)du) = \cos(x^2)$.

(a) $\lim\limits_{x\to 0} \dfrac{C(x)}{x} = \lim\limits_{x\to 0} \dfrac{C'(x)}{1} = \lim\limits_{x\to 0} \cos(x^2) = 1$.

(b) $\lim\limits_{x\to 0} \dfrac{C(x) - x}{x^5} = \lim\limits_{x\to 0} \dfrac{\cos(x^2) - 1}{5x^4} = \lim\limits_{x\to 0} \dfrac{-2x \sin(x^2)}{20x^3}$

$= -\dfrac{1}{10} \lim\limits_{x\to 0} \dfrac{\sin(x^2)}{x^2} = -\dfrac{1}{10}$ since $\lim\limits_{t\to 0} \dfrac{\sin t}{t} = 1$ and, here $t = x^2$.

EXERCISES 10.2

1. $\lim\limits_{x\to 0^+} x \ln x = \lim\limits_{x\to 0^+} \dfrac{\ln x}{1/x} = \lim\limits_{x\to 0^+} \dfrac{1/x}{-(1/x^2)} = \lim\limits_{x\to 0^+} (-x) = 0$.

4. $\lim\limits_{x\to\infty} x(e^{1/x} - 1) = \lim\limits_{x\to\infty} \dfrac{e^{1/x} - 1}{1/x} = \lim\limits_{x\to\infty} \dfrac{(-1/x^2)e^{1/x}}{(-1/x^2)} = \lim\limits_{x\to\infty} e^{1/x} = e^0 = 1$.

7. $\lim\limits_{x\to 0^+} \sin x \ln \sin x = \lim\limits_{x\to 0^+} \dfrac{\ln \sin x}{\csc x} = \lim\limits_{x\to 0^+} \dfrac{\cos x/\sin x}{-\csc x \cot x} = \lim\limits_{x\to 0^+} \dfrac{\cot x}{-\csc x \cot x}$

$= \lim\limits_{x\to 0^+} \dfrac{1}{-\csc x} = \lim\limits_{x\to 0^+} (-\sin x) = 0$.

10. $\lim\limits_{x\to\infty} e^{-x} \ln x = \lim\limits_{x\to\infty} \dfrac{\ln x}{e^x} = \lim\limits_{x\to\infty} \dfrac{1/x}{e^x} = 0$.

13. This is a 1^∞ indeterminate form. If $y = (1 + \dfrac{1}{x})^{5x}$, then $\ln y = 5x \ln(1 + 1/x)$

$= 5 \dfrac{\ln(1 + 1/x)}{1/x}$. $\lim\limits_{x\to\infty} \ln y = \lim\limits_{x\to\infty} \dfrac{5(-1/x^2)/(1+1/x)}{(-1/x^2)} = \lim\limits_{x\to\infty} \dfrac{5}{1+1/x} = 5$. Thus

$\lim\limits_{x\to\infty} y = e^5$.

16. If $y = x^x$, then $\ln y = x \ln x$ and $\lim\limits_{x\to 0^+} \ln y = \lim\limits_{x\to 0^+} x \ln x = 0$, by #1. Thus,

$\lim\limits_{x\to 0^+} y = \lim\limits_{x\to 0^+} x^x = e^0 = 1$.

19. $(\tan x)^x$ is not indeterminate as $x \to \pi/2^-$ since $\tan x \to \infty$ and $x \to \pi/2 > 1$.

Thus $(\tan x)^x \to \infty$ as $x \to \pi/2^-$.

22. If $y = (1+3x)^{\csc x}$, $\ln y = \csc x \ln(1+3x) = \dfrac{\ln(1+3x)}{\sin x}$.

$$\lim_{x\to 0^+} \ln y = \lim_{x\to 0^+} \frac{\ln(1+3x)}{\sin x} = \lim_{x\to 0^+} \frac{3}{(1+3x)\cos x} = 3. \quad \text{Thus } \lim_{x\to 0^+} y = e^3.$$

25. $\displaystyle\lim_{x\to 0} \left(\frac{1}{x} - \frac{1}{\sin x}\right) = \lim_{x\to 0} \frac{\sin x - x}{x \sin x} = \lim_{x\to 0} \frac{\cos x - 1}{x \cos x + \sin x} = \lim_{x\to 0} \frac{-\sin x}{-x \sin x + 2 \cos x}$

$= \dfrac{0}{0 + 2} = 0.$

28. This is an ∞^0 indeterminate form as $x \to \infty$. $y = (1+e^x)^{e^{-x}} \implies \ln y =$

$e^{-x} \ln(1+e^x) = \dfrac{\ln(1+e^x)}{e^x}$. $\displaystyle\lim_{x\to\infty} \ln y = \lim_{x\to\infty} \frac{e^x/(1+e^x)}{e^x} = \lim_{x\to\infty} \frac{1}{1+e^x} = 0.$ Thus

$\displaystyle\lim_{x\to\infty} y = e^0 = 1.$

31. $\displaystyle\lim_{x\to 0} \cot 2x \tan^{-1}x = \lim_{x\to 0} \frac{\tan^{-1}x}{\tan 2x} = \lim_{x\to 0} \frac{1/(1+x^2)}{2 \sec^2 2x} = \frac{1}{2}$.

34. $\displaystyle\lim_{x\to\infty} (\sqrt{x^2+4} - \tan^{-1}x) = \infty$ since $\sqrt{x^2+4} \to \infty$, but $\tan^{-1}x \to \pi/2$ as $x \to \infty$.

37. $\dfrac{x}{x^2+2x-3} - \dfrac{4}{x+3} = \dfrac{x}{(x+3)(x-1)} - \dfrac{4}{x+3} = \dfrac{1}{x+3}\left(\dfrac{x}{x-1} - 4\right) = \dfrac{1}{x+3}\left(\dfrac{x-4x+4}{x-1}\right) = \dfrac{-3x+4}{(x+3)(x-1)}$.

As $x \to -3$, $(-3x+4) \to 13$ and the denominator $\to 0$. Thus the limit does not exist. (The right-hand limit is $-\infty$, the left-hand limit is ∞.)

40. $\displaystyle\lim_{x\to\pi/2} \sec x \cos 3x = \lim_{x\to\pi/2} \frac{\cos 3x}{\cos x} = \lim_{x\to\pi/2} \frac{-3 \sin 3x}{-\sin x} = \frac{-3(-1)}{-1} = -3.$

43. $y = x^{1/x} \implies \ln y = \dfrac{\ln x}{x}$. $\displaystyle\lim_{x\to 0^+} \ln y = -\infty \implies \lim_{x\to 0^+} x^{1/x} = 0.$ $\displaystyle\lim_{x\to\infty} \ln y =$

$\displaystyle\lim_{x\to\infty} \frac{\ln x}{x} = \lim_{x\to\infty} \frac{1/x}{1} = 0 \implies \lim_{x\to\infty} x^{1/x} = 1.$ Thus $y = 1$ is a horizontal

asymptote. $\ln y = \dfrac{\ln x}{x} \implies \dfrac{y'}{y} = \dfrac{x(1/x) - \ln x}{x^2} \implies y' = \dfrac{x^{1/x}(1 - \ln x)}{x^2} = 0$

at $x = e$. $x < e \implies \ln x < 1 \implies y' > 0$. $x > e \implies \ln x > 1 \implies y' < 0$. Thus $y = f(x)$ is increasing on $(0,e]$, decreasing to the horizontal asymptote ($y = 1$) on $[e,\infty)$ and has a local maximum at $x = e$.

46. Let $y = f(m) = (1 + r/m)^{mt}$, a 1^∞ form as $m \to \infty$. $\ln y = mt \ln(1 + r/m) =$

$t \dfrac{\ln(1 + r/m)}{1/m}$. $\displaystyle\lim_{m\to\infty} \ln y = t \lim_{m\to\infty} \frac{D_m(1+r/m)/(1+r/m)}{D_m(1/m)} = t \lim_{m\to\infty} \frac{(-r/m^2)}{(1+r/m)(-1/m^2)}$

$= t \displaystyle\lim_{m\to\infty} \frac{r}{1+r/m} = \frac{tr}{1+0} = tr.$ Thus $\displaystyle\lim_{m\to\infty} y = e^{rt}$, and the balance (which is the limit of Py) is Pe^{rt}.

EXERCISES 10.3

1. $\displaystyle\int_1^\infty \frac{1}{x^{4/3}}\,dx = \lim_{t\to\infty}\int_1^t x^{-4/3}\,dx = \lim_{t\to\infty} -3x^{-1/3}\,\Big]_1^t = \lim_{t\to\infty}\left(\frac{-3}{\sqrt[3]{t}}+3\right) = 0 + 3 = 3.$

4. $\displaystyle\int_0^\infty \frac{x}{1+x^2}\,dx = \lim_{t\to\infty}\int_0^t \frac{x}{1+x^2}\,dx = \lim_{t\to\infty}\frac{\ln(1+x^2)}{2}\,\Big]_0^t = \lim_{t\to\infty}\left(\frac{\ln(1+t^2)}{2} - \frac{\ln 1}{2}\right) =$

$\displaystyle\lim_{t\to\infty}\frac{\ln(1+t^2)}{2} = \infty.$ Thus the integral diverges.

7. $\displaystyle\int_0^\infty e^{-2x}\,dx = \lim_{t\to\infty}\int_0^t e^{-2x}\,dx = \lim_{t\to\infty}\frac{e^{-2x}}{-2}\,\Big]_0^t = \lim_{t\to\infty}\left(-\frac{1}{2}\right)(e^{-2t}-1) = \frac{1}{2}.$

10. $\displaystyle\int_0^\infty \frac{1}{\sqrt[3]{x+1}}\,dx = \lim_{t\to\infty}\int_0^t (x+1)^{-1/3}\,dx = \lim_{t\to\infty}\left(\frac{3}{2}\right)(x+1)^{2/3}\,\Big]_0^t = \lim_{t\to\infty}\frac{3}{2}[(t+1)^{2/3}-1]$

$= \infty.$ Thus the integral diverges.

13. With $u = \sin x$, $\displaystyle\int \frac{\cos x}{1+\sin^2 x}\,dx = \int \frac{du}{1+u^2} = \tan^{-1}u.$ Thus $\displaystyle\int_0^\infty \frac{\cos x}{1+\sin^2 x}\,dx$

$= \displaystyle\lim_{t\to\infty}\tan^{-1}(\sin x)\Big]_0^t = \lim_{t\to\infty}\tan^{-1}(\sin t) - 0$ which does not exist. As t in-

creases, sin t oscillates between -1 and +1, and $\tan^{-1}(\sin t)$ oscillates

between $-\pi/4$ and $+\pi/4$, approaching no limit. Thus the integral diverges.

15. Using (10.5) with $a = 0$, $\displaystyle\int_{-\infty}^\infty xe^{-x^2}\,dx = \int_{-\infty}^0 xe^{-x^2}\,dx + \int_0^\infty xe^{-x^2}\,dx.$ Now,

$\displaystyle\int_{-\infty}^0 xe^{-x^2}\,dx = \lim_{t\to-\infty}\frac{-e^{-x^2}}{2}\,\Big]_t^0 = \lim_{t\to\infty}\left(\frac{-e^0}{2}+\frac{e^{-t^2}}{2}\right) = -\frac{1}{2}+0 = -\frac{1}{2}.$ The second

integral is $\displaystyle\int_0^\infty xe^{-x^2}\,dx = \lim_{t\to\infty}\frac{-e^{-x^2}}{2}\,\Big]_0^t = \lim_{t\to\infty}\left(\frac{-e^{-t^2}}{2}+\frac{e^0}{2}\right) = 0+\frac{1}{2} = \frac{1}{2}.$ Since

both integrals converge, the original integral converges, and its value is

their sum. Thus $\displaystyle\int_{-\infty}^\infty xe^{-x^2}\,dx = -\frac{1}{2}+\frac{1}{2} = 0.$

16. Using (10.5) with $a = 0$, $\displaystyle\int_{-\infty}^\infty \cos^2 x\,dx = \int_{-\infty}^0 \cos^2 x\,dx + \int_0^\infty \cos^2 x\,dx.$

$\displaystyle\lim_{t\to\infty}\int_0^t \cos^2 x\,dx = \left(\frac{1}{2}\right)\lim_{t\to\infty}\int_0^t (1+\cos 2x)\,dx = \lim_{t\to\infty}\left[t+\frac{\sin 2t}{2}\right] = \infty.$ Since at

least one of the two improper integrals on the right side diverges, the given

integral diverges.

19. $\int_0^\infty \cos x \, dx = \lim_{t\to\infty} \sin x \,]_0^t = \lim_{t\to\infty} \sin t$, which does not exist. Thus, the
     integral diverges.

22. Integrating by parts ($u = x$, $dv = e^{-x}dx$), or using Formula 98, $\int xe^{-x}dx =$
     $-(x+1)e^{-x} + C$. Thus $\int_0^\infty xe^{-x}dx = \lim_{t\to\infty} [-(x+1)e^{-x}]_0^t = \lim_{t\to\infty} [-(t+1)e^{-t} + 1]$
     $= 0 + 1 = 1$. ($\lim_{t\to\infty} (t+1)e^{-t} = \lim_{t\to\infty} \frac{t+1}{e^t} = 0$ by L'Hôpital's Rule.)

25. $1 + x^4 > x^4 \Rightarrow \frac{1}{1 + x^4} < \frac{1}{x^4}$. $\int_1^\infty \frac{1}{x^4} \, dx = \lim_{t\to\infty} \int_1^t x^{-4}dx$
     $= \lim_{t\to\infty} \frac{t^{-3} - 1}{-3} = \frac{1}{3}$. Thus, by Comparison Test (i) with $g(x) = 1/x^4$,
     $f(x) = 1/(1 + x^4)$, the given integral converges.

28. $x \geq 1 \Rightarrow x^2 \geq x \Rightarrow -x^2 \leq -x \Rightarrow e^{-x^2} \leq e^{-x}$.
     $\int_1^\infty e^{-x}dx = \lim_{t\to\infty} \frac{e^{-t} - e^{-1}}{-1} = e^{-1}$. Thus, by comparison test (i),
     with $g(x) = e^{-x}$, $f(x) = e^{-x^2}$, the given integral converges.

31. (a) $A = \int_4^\infty x^{-3/2}dx = \lim_{t\to\infty} \int_4^t x^{-3/2}dx = \lim_{t\to\infty} \frac{t^{-1/2} - 4^{-1/2}}{-1/2} =$
     $\frac{0 - 1/2}{-1/2} = 1$.   (b) $v = \pi\int_4^\infty x^{-3}dx = \pi \lim_{t\to\infty} \frac{t^{-2} - 4^{-2}}{-2} = \frac{\pi}{32}$.

34. $f(x) = e^{-x}$, $f'(x) = -e^{-x} \Rightarrow S = 2\pi\int_0^\infty e^{-x}\sqrt{1 + e^{-2x}}dx$. Let
     $u = e^{-x}$, $du = -e^{-x}dx$. As $x \to \infty$, $u \to 0$ and $x = 0 \Rightarrow u = 1$.
     Thus, $S = 2\pi\int_0^1 \sqrt{u^2 + 1} \, du$. Either by Integral Formula 21 or by
     the trig substitution $u = \tan\theta$, we obtain
     $S = 2\pi(\frac{1}{2})[u\sqrt{u^2 + 1} + \ln(u + \sqrt{u^2 + 1})]_0^1 = \pi(\sqrt{2} + \ln(1 + \sqrt{2}))$.

37. Let x be the distance between the electrons and k, the constant of propor-
     tionality. Then the force function is $f(x) = k/x^2$. Since the electrons
     start 1 cm. apart, $W = \int_1^\infty f(x)dx = \int_1^\infty k/x^2dx = \lim_{t\to\infty} -\frac{k}{x}\,]_1^t = \lim_{t\to\infty} [-\frac{k}{t} + k]$
     $= 0 + k = k$.

40. (a) $A = \int_{20}^\infty 12000e^{-.08t}dt = \lim_{T\to\infty} \frac{12000}{-0.08} e^{-.08t}\,]_{20}^T$
     $-150{,}000 \lim_{T\to\infty} (e^{-.08T} - e^{-.08(20)}) = 150{,}000e^{-1.6} \approx \$30{,}284.50$.
     (b) $A = 12000\int_{20}^\infty e^{.04t}e^{-.08t}dt = 12000\int_{20}^\infty e^{-.04t}dt =$
     $\lim_{T\to\infty} \frac{-12000}{.04}(e^{-.04T} - e^{-.04(20)}) = 300{,}000e^{-.8} \approx \$134{,}800$.

43. $L[1] = \int_0^\infty e^{-sx} \cdot 1\ dx = \lim_{t\to\infty} \frac{e^{-sx}}{-s}\Big]_0^t = \lim_{t\to\infty} \frac{(1 - e^{-st})}{s} = \frac{1}{s}$ provided $s > 0$.

46. $L[\sin x] = \int_0^\infty e^{-sx} \sin x\ dx$. Using Formula 98 from the Table of Integrals

with $a = -s$, $b = 1$, $u = x$ (or 2 integrations by parts), $L[\sin x] =$

$\lim_{t\to\infty} \frac{e^{-sx}}{s^2+1} (-s \sin x - \cos x)\Big]_0^t = \lim_{t\to\infty} \frac{(1 - e^{-ts}(s \sin t + \cos t))}{s^2+1}$. If $s > 0$

then $\lim_{t\to\infty} e^{-ts} (s \sin t + \cos t) = 0$ since $|s \sin t + \cos t| \le s|\sin t| +$

$|\cos t| \le s+1$, independent of t, whereas $\lim_{t\to\infty} e^{-ts} = 0$. The statement then

follows from the Sandwich Theorem. Thus $L[\sin x] = 1/(s^2+1)$, if $s > 0$.

49. Here we will use the formulas (derived by integration by parts)

$\int xe^{-x}\ dx = -(x+1)e^{-x} + C$, $\int x^2 e^{-x}\ dx = -(x^2+2x+2)e^{-x} + C$ and the limit (via

L'Hôpital's rule) $\lim_{t\to\infty} t^n e^{-t} = 0$ for any $n > 0$. (a) $\Gamma(1) = \int_0^\infty e^{-x}\ dx =$

$\lim_{t\to\infty} (1 - e^{-t}) = 1$. $\Gamma(2) = \int_0^\infty xe^{-x}dx = \lim_{t\to\infty} -(x+1)e^{-x}\Big]_0^t = \lim_{t\to\infty} (-(t+1)e^{-t} + 1)$

$= 1$. $\Gamma(3) = \int_0^\infty x^2 e^{-x}dx = \lim_{t\to\infty} -(x^2+2x+2)e^{-x}\Big]_0^t = \lim_{t\to\infty} (-t^2+2t+2)e^{-t} + 2) = 2$.

(b) $\Gamma(n+1) = \int_0^\infty x^n e^{-x}dx$. Let $u = x^n$, $dv = e^{-x}dx$. Then $du = nx^{n-1}\ dx$, $v =$

$-e^{-x}$ and $\Gamma(n+1) = \lim_{t\to\infty} -x^n e^{-x}\Big]_0^t + \int_0^\infty nx^{n-1}e^{-x}dx = \lim_{t\to\infty} (-t^n e^{-t} + 0) + n\Gamma(n)$

$= n\Gamma(n)$.

(c) The truth of the statement $\Gamma(n+1) = n!$ was established in part (a) for

$n = 1$ ($\Gamma(2) = 1$) and $n = 2$ ($\Gamma(3) = 2 = 2!$). So we assume that it is true

for $n = k-1$, i.e. $\Gamma(k) = (k-1)!$ and prove that it is true for $n = k$, i.e. we

must show $\Gamma(k+1) = k!$ By part (b), $\Gamma(k+1) = k\Gamma(k)$ and by the above assump-

tion, this yields $\Gamma(k+1) = k \cdot (k-1)! = k!$, which proves the statement true

for all positive integers.

EXERCISES 10.4

1. Since $1/\sqrt[3]{x}$ is continuous on $(0,8]$ and has an infinite discontinuity at 0, we

get by (10.6), $\int_0^8 1/\sqrt[3]{x}\ dx = \lim_{t\to 0^+} \int_t^8 x^{-1/3}\ dx = \lim_{t\to 0^+} \frac{3}{2} x^{2/3}\Big]_t^8 =$

$\lim_{t\to 0^+} \frac{3}{2}(8^{2/3} - t^{2/3}) = \frac{3}{2}(2^2 - 0) = 6$.

4. Since $1/(x+2)^{5/4}$ is continuous on $(-2,-1]$ and has an infinite discontinuity at

-2, $(10.6) \Rightarrow \int_{-2}^{-1} 1/(x+2)^{5/4} \, dx = \lim_{x \to -2^+} \int_t^{-1} (x+2)^{-5/4} \, dx = \lim_{t \to -2^+} -4(x+2)^{-1/4}]_t^{-1}$

$= \lim_{t \to -2^+} [-4 + \dfrac{4}{(t+2)^{1/4}}] = \infty.$ Thus the integral diverges.

7. Since the integrand is discontinuous at $x = 4$, by (10.6) $\int_0^4 \dfrac{1}{(4-x)^{3/2}} \, dx =$

$\lim_{t \to 4^-} \int_0^t (4-x)^{-3/2} \, dx = \lim_{t \to 4^-} \dfrac{2}{(4-x)^{1/2}}]_0^t = 2 \lim_{t \to 4^-} [\dfrac{1}{(4-t)^{1/2}} - \dfrac{1}{2}] = \infty.$ Thus,

the integral diverges.

10. $\int_1^2 \dfrac{x}{x^2-1} \, dx = \lim_{t \to 1^+} \int_t^2 \dfrac{x}{x^2-1} \, dx = \lim_{t \to 1^+} \dfrac{\ln(x^2-1)}{2}]_t^2 = \lim_{t \to 1^+} \dfrac{\ln 3 - \ln(t^2-1)}{2} = \infty,$

and the integral diverges.

13. $\int_{-2}^0 \dfrac{1}{\sqrt{4-x^2}} \, dx = \lim_{t \to -2^+} \sin^{-1} \dfrac{x}{2}]_t^0 = \lim_{t \to -2^+} (-\sin^{-1} \dfrac{t}{2}) = -\sin^{-1} (-1) = \pi/2.$

16. $x^2 - x - 2 = (x-2)(x+1)$. Thus the integrand has a discontinuity at $x = 2$ in

$[0,4]$, and by (10.7) we use $\int_0^4 = \int_0^2 + \int_2^4$ and examine each of the integrals.

As in Sec. 9.4 , $\dfrac{1}{(x-2)(x+1)} = \dfrac{1/3}{(x-2)} - \dfrac{1/3}{(x+1)}$. Thus $\int_0^2 \dfrac{1}{x^2-x-2} \, dx =$

$\lim_{t \to 2^-} \dfrac{1}{3}[\ln|x-2| - \ln|x+1|]_0^t = \dfrac{1}{3} \lim_{t \to 2^-} (\ln|t-2| - \ln 2 - \ln|t+1|) = -\infty,$ and the

original integral diverges since at least one of the integrals on the right

diverges.

19. $\int_0^{\pi/2} \tan x \, dx = \lim_{t \to \pi/2^-} (-\ln|\cos x|)]_0^t = \lim_{t \to \pi/2^-} (-\ln|\cos t|).$ Now, as

$t \to \pi/2^-$, $|\cos t| \to 0$ and $\ln|\cos t| \to -\infty$. Thus the above limit is ∞, and the

integral diverges.

22. The integrand, $1/x(\ln x)^2$, is continuous on $[1/e, e]$ except at $x = 1$ where

$\ln x = 0$. Thus, by (10.7) we use $\int_{1/e}^e = \int_{1/e}^1 + \int_1^e$. To get the antideriva-

tive, let $u = \ln x$, $du = (1/x) dx$. Then $\int 1/x(\ln x)^2 \, dx = \int 1/u^2 \, du = -1/u + C$

$= -1/\ln x + C$. Then the 2nd integral above is $\int_1^e \dfrac{1}{x(\ln x)^2} \, dx = \lim_{t \to 1^+} [-\dfrac{1}{\ln x}]_t^e$

$= \lim_{t \to 1^+} [-\dfrac{1}{\ln e} + \dfrac{1}{\ln t}] = \infty$ since $\ln t \to 0$ and $1/\ln t \to \infty$ as $t \to 1^+$. Since at at least one of the two integrals above diverges, the original integral diverges.

25. The integrand is continuous on $[0,\pi]$ except at $\pi/2$ since $\sin \pi/2 = 1$.

$$\int_0^\pi = \int_0^{\pi/2} + \int_{\pi/2}^\pi .$$ For the antiderivative, let $u = 1 - \sin x$, $du = -\cos x \, dx$.

$$\int \cos x/\sqrt{1 - \sin x} \, dx = -\int u^{-1/2} \, du = 2u^{1/2} + C = 2\sqrt{1 - \sin x} + C.$$ Thus

$$\int_0^{\pi/2} \frac{\cos x}{\sqrt{1 - \sin x}} \, dx = \lim_{t \to \pi/2^-} [-2\sqrt{1 - \sin x}]_0^t = \lim_{t \to \pi/2^-} (-2\sqrt{1 - \sin t} + 2) =$$

$0 + 2 = 2.$ Also, $\displaystyle\int_{\pi/2}^\pi \frac{\cos x}{\sqrt{1 - \sin x}} \, dx = \lim_{t \to \pi/2^+} [-2\sqrt{1 - \sin x}]_t^\pi =$

$\displaystyle\lim_{t \to \pi/2^-} (-2 + 2\sqrt{1 - \sin t}) = -2 + 0 = -2.$ The given integral is the sum of

the values of these two convergent integrals, namely, $2 + (-2) = 0$.

28. The integrand is discontinuous at $x = \pm 1$. Thus we select any number between

-1 and 1, say, 0, and $I = \displaystyle\int_{-1}^3 = \int_{-1}^0 + \int_0^1 + \int_1^3 = I_1 + I_2 + I_3.$ To get an anti-

derivative, let $u = x^2-1$, $du = 2x \, dx$. Then $\displaystyle\int \frac{x}{\sqrt[3]{x^2-1}} \, dx = \frac{1}{2}\int u^{-1/3} du =$

$\frac{3}{4} u^{2/3} + C = \frac{3}{4}(x^2-1)^{2/3} + C.$ Thus $I_1 = \displaystyle\lim_{t \to -1^+} \frac{3}{4}(x^2-1)^{2/3}]_t^0 =$

$= \displaystyle\lim_{t \to -1^+} (\frac{3}{4} - \frac{3}{4}(t^2-1)^{2/3}) = \frac{3}{4} - 0.$ $I_2 = \displaystyle\lim_{t \to 1^-} (\frac{3}{4}(t^2-1)^{2/3} - \frac{3}{4}) = 0 - \frac{3}{4}.$

$I_3 = \displaystyle\lim_{t \to 1^+} (\frac{3}{4}(3^2-1)^{2/3} - \frac{3}{4}(t^2-1)^{2/3}) = \frac{3}{4}(8)^{2/3} = 3.$ Since all 3 integrals

converge, the given integral converges and its value is $I_1 + I_2 + I_3 =$

$\frac{3}{4} + (-\frac{3}{4}) + 3 = 3.$

31. $1/\sqrt{x} > 0$ and $0 < \sin x \le 1$ on $(0,\pi] \Rightarrow 0 < \sin x/\sqrt{x} \le 1/\sqrt{x}$ on $(0,\pi]$.

$\displaystyle\int_0^\pi \frac{1}{\sqrt{x}} \, dx = \lim_{t \to 0^+} 2(\sqrt{\pi} - \sqrt{t}) = 2\sqrt{\pi}.$ Thus with $f(x) = \sin x/\sqrt{x}$, $g(x) = 1/\sqrt{x}$,

and comparison test (i), $\displaystyle\int_0^\pi f(x) \, dx$ converges.

34. $0 < e^{-x} \le 1$, $1/x^{2/3} > 0 \Rightarrow 0 < e^{-x}/x^{2/3} < 1/x^{2/3}$ on $(0,1]$.

$\displaystyle\int_0^1 1/x^{2/3} dx = \lim_{t \to 0^+} 3(\sqrt[3]{1} - \sqrt[3]{t}) = 3.$ Thus with $f(x) = e^{-x}/x^{2/3}$, $g(x) =$

$1/x^{2/3}$ and comparison test (i), $\displaystyle\int_0^1 f(x) \, dx$ converges.

37.   (a) $A = \int_0^1 1/\sqrt{x}\ dx = \lim_{t\to 0^+} \int_t^1 x^{-1/2}dx = \lim_{t\to 0^+} 2\sqrt{x}\Big]_t^1 = \lim_{t\to 0^+} (2 - 2\sqrt{t}) = 2.$

   (b) $V = \pi\int_0^1 (1/\sqrt{x})^2 dx = \pi \lim_{t\to 0^+} \int_t^1 \frac{1}{x}\ dx = \pi \lim_{t\to 0^+} \ln x\Big]_t^1 = \pi \lim_{t\to 0^+} (\ln 1 - \ln t)$

   $= \infty.$ Thus no value can be assigned.

40.   (a) $A = \int_1^2 \frac{1}{x-1}\ dx = \lim_{t\to 1^+} (\ln 1 - \ln(t-1)) = \infty.$

   (b) $V = \pi\int_1^2 \frac{1}{(x-1)^2}\ dx = \pi \lim_{t\to 1^+} (\frac{1}{t-1} - 1) = \infty.$

   Since both diverge, no value can be assigned in either case.

43.   Solve $y = y_0 e^{-kt}$ for t to obtain $t(y) = -\frac{1}{k} \ln \frac{y}{y_0}$.   (a) $t(y)$ is

   discontinuous at $y = 0$.   Thus, the integral for T is improper.
   (b) Set $t = \tau$, $y = y_0/2$ to obtain $1/2 = e^{-k\tau} \Rightarrow -k\tau = \ln(1/2)$
   $= -\ln 2 \Rightarrow k = (\ln 2)/\tau$, and $t(y) = \frac{-\tau}{\ln 2} \ln \frac{y}{y_0}$.   Then T =

   $\frac{-\tau}{y_0 \ln 2} \int_0^{y_0} \ln \frac{y}{y_0}\ dy.$ Let $u = y/y_0$ so that $du = (1/y_0)dy$,

   $y \to 0 \Rightarrow u \to 0$, $y = y_0 \Rightarrow u = 1.$ Then $T = \frac{-\tau}{\ln 2}\int_0^1 \ln u\ du =$

   $\frac{-\tau}{\ln 2} \lim_{t\to 0^+} [u \ln u - u]_t^1 = \frac{-\tau}{\ln 2} \lim_{t\to 0} [(1 \ln 1 - 1) - (t \ln t - t)] =$

   $\frac{\tau}{\ln 2}.$ (The antiderivative of ln u was found integrating by parts,
   and $\lim_{t\to 0^+} t \ln t = 0$ is by L'Hôpital rule.)

EXERCISES 10.5

1.   With $f(x) = \sin x$, $c = \pi/2$, $n = 3$, we have

   $f(x) = \sin x$ $f(\pi/2) = 1$

   $f'(x) = \cos x$ $f'(\pi/2) = 0$

   $f''(x) = -\sin x$ $f''(\pi/2) = -1$

   $f'''(x) = -\cos x$ $f'''(\pi/2) = 0$

   $f^{(4)}(x) = \sin x$ $f^{(4)}(z) = \sin z.$

   Remembering to divide $f^{(k)}(c)$ by $k!$, we have

   $\sin x = 1 - \frac{1}{2}(x - \frac{\pi}{2})^2 + \frac{\sin z}{4!} (x - \frac{\pi}{2})^4$ for some z between x and $\pi/2$.

4.   With $f(x) = e^{-x}$, $c = 1$, $n = 3$, we have

   $f(x) = f''(x) = f^{(4)}(x) = e^{-x}$ $f(1) = f''(1) = e^{-1}$

                                    $f^{(4)}(z) = e^{-z}$

   $f'(x) = f'''(x) = -e^{-x}$ $f'(1) = f'''(1) = -e^{-1}$

$$e^{-x} = e^{-1} - e^{-1}(x-1) + \frac{e^{-1}}{2!}(x-1)^2 - \frac{e^{-1}}{3!}(x-1)^3 + \frac{e^{-z}}{4!}(x-1)^4$$

for some z between x and 1.

7. With $f(x) = 1/x$, $c = -2$, $n = 5$, we have

$f(x) = 1/x$ $f(-2) = -1/2$

$f'(x) = -1/x^2$ $f'(-2) = -1/4$

$f''(x) = 2/x^3$ $f''(-2) = -2/8$

$f'''(x) = -6/x^4$ $f'''(-2) = -6/16$

$f^{(4)}(x) = 24/x^5$ $f^{(4)}(-2) = -24/32$

$f^{(5)}(x) = -120/x^6$ $f^{(5)}(-2) = -120/64$

$f^{(6)}(x) = 720/x^7$ $f^{(6)}(z) = 720/z^7$

From the entries in the 2nd column, note that $f^{(k)}(-2)/k! = -1/2^{k+1}$ for

$k = 0,1,2,\ldots,5$. Thus $\frac{1}{x} = -\frac{1}{2} - \frac{1}{4}(x+2) - \frac{1}{8}(x+2)^2 - \frac{1}{16}(x+2)^3 - \frac{1}{32}(x+2)^4$

$-\frac{1}{64}(x+2)^5 + \frac{1}{z^7}(x+2)^6$, for some z between -2 and x.

10. We need here, $\sin \pi/6 = 1/2$, $\csc \pi/6 = 2$, $\cot \pi/6 = \sqrt{3}$. So, with $f(x) = \ln \sin x$, $c = \pi/6$, $n = 3$, we obtain

$f(x) = \ln \sin x$ $f(\pi/6) = \ln(1/2)$

$f'(x) = \cos x/\sin x = \cot x$ $f'(\pi/6) = \sqrt{3}$

$f''(x) = -\csc^2 x$ $f''(\pi/6) = -2^2 = -4$

$f'''(x) = 2\csc^2 x \cot x$ $f'''(\pi/6) = 2(4)\sqrt{3} = 8\sqrt{3}$

$f^{(4)}(x) = -2(\csc^4 x + 2\csc^2 x \cot^2 x)$. Thus

$\ln \sin x = \ln(1/2) + \sqrt{3}(x - \pi/6) - \frac{4}{2!}(x - \pi/6)^2 + \frac{8\sqrt{3}}{3!}(x - \pi/6)^3 +$

$\frac{f^{(4)}(z)}{4!}(x - \pi/6)^4$ for some z between $\pi/6$ and x.

13. In Maclaurin's formula $c = 0$. So with $n = 4$, we get

$f(x) = \ln(x+1)$ $f(0) = \ln 1 = 0$

$f'(x) = (x+1)^{-1}$ $f'(0) = 1$

$f''(x) = -(x+1)^{-2}$ $f''(0) = -1$

$f'''(x) = 2(x+1)^{-3}$ $f'''(0) = 2$

$f^{(4)}(x) = -6(x+1)^{-4}$ $f^{(4)}(0) = -6$

$f^{(5)}(x) = 24(x+1)^{-5}$ $f^{(5)}(z) = 24(z+1)^{-5}$

$\ln(x+1) = x - \frac{1}{2!}x^2 + \frac{2}{3!}x^3 - \frac{6}{4!}x^4 + \frac{24x^5}{5!(z+1)^5} = x - \frac{1}{2}x^2 + \frac{1}{3}x^3 - \frac{1}{4}x^4 +$

$\frac{x^5}{5(z+1)^5}$ for some z between x and 0.

16. $f(x) = \tan^{-1}x$ $f(0) = 0$

$f'(x) = (1+x^2)^{-1}$ $f'(0) = 1$

$f''(x) = -2x(1+x^2)^{-2}$ $f''(0) = 0$

$f'''(x) = (6x^2-2)(1+x^2)^{-3}$ $f'''(0) = -2$

$f^{(4)}(x) = 24(x-x^3)(1+x^2)^{-4}$

$\tan^{-1}x = x - \dfrac{2}{3!}x^3 + \dfrac{24(z-z^3)}{4!(1+z^2)^4}x^4 = x - \dfrac{1}{3}x^3 + \dfrac{(z-z^3)}{(1+z^2)^4}x^4$ for some z between

x and 0.

19. $f(x) = (x-1)^{-2}$ $f(0) = (-1)^{-2} = 1$

$f'(x) = -2(x-1)^{-3}$ $f'(0) = -2(-1)^{-3} = 2$

$f''(x) = 6(x-1)^{-4} = 3!(x-1)^{-4}$ $f''(0) = 3!(-1)^{-4} = 3!$

$f'''(x) = -24(x-1)^{-5} = -(4!)(x-1)^{-5}$ $f'''(0) = -(4!)(-1)^{-5} = 4!$

$f^{(4)}(x) = 120(x-1)^{-6} = 5!(x-1)^{-6}$ $f^{(4)}(0) = 5!(-1)^{-6} = 5!$

$f^{(5)}(x) = -720(x-1)^{-7} = -(6!)(x-1)^{-7}$ $f^{(5)}(0) = -6!(-1)^{-7} = 6!$

$f^{(6)}(x) = 5040(x-1)^{-8} = 7!(x-1)^{-8}$ $f^{(6)}(z) = 7!(z-1)^{-8}$

From the entries in the second column, we see that $f^{(k)}(0)/k! = (k+1)!/k! =$
k+1 for k = 0,1,...,5. Thus: $(x-1)^{-2} = 1 + 2x + 3x^2 + 4x^3 + 5x^4 + 6x^5 +$
$7(z-1)^{-8}x^6$ for some z between 0 and x. (The last coefficient is
$f^{(6)}(z)/6!$.)

22. $f(x) = e^{-x^2} \Rightarrow f'(x) = -2xe^{-x^2} \Rightarrow f''(x) = -2x(-2xe^{-x^2})-2e^{-x^2} =$

$(4x^2-2)e^{-x^2} \Rightarrow f'''(x) = (4x^2-2)(-2xe^{-x^2}) + 8xe^{-x^2} = (-8x^3+12x)e^{-x^2} \Rightarrow$

$f^{(4)}(x) = (-8x^3+12x)(-2xe^{-x^2}) + (-24x^2+12)e^{-x^2} = (16x^4-48x^2+12)e^{-x^2}$. Thus
$f(0) = 1$, $f'(0) = 0$, $f''(0)/2! = -2/2 = -1$, and $f'''(0) = 0$. Thus:

$e^{-x^2} = 1 - x^2 + (1/4!)(16z^4-48z^2+12)e^{-z^2}x^4$ for some z between 0 and x.

25. Using Exercise 1 with x = $89° = 90°-1° = \dfrac{\pi}{2} - \dfrac{\pi}{180}$, we have $x - \dfrac{\pi}{2} = \dfrac{-\pi}{180} \approx$

-0.0175 and $\sin 89° = 1 - \dfrac{1}{2}(-\dfrac{\pi}{180})^2 + \dfrac{\sin z}{4!}(-\dfrac{\pi}{180})^4$

$\sin 89° \approx 1 - \dfrac{(.0175)^2}{2} \approx 1 - \dfrac{.0003}{2} = 0.99985$.

$|R_n(x)| \approx \dfrac{|\sin z|(.0175)^4}{4!} < \dfrac{(.02)^4}{24} = \dfrac{16 \times 10^{-8}}{24} < 10^{-8}$. (Compare the answer
obtained above with the value of $\sin 89°$ in a 5-place table. They agree
exactly!)

28. Using Exercise 4 with x = 1.02, x-1 = .02, we have

$e^{-1.02} = e^{-1}(1 - .02 + \dfrac{(.02)^2}{2} - \dfrac{(.02)^3}{6}) + \dfrac{e^{-z}}{24}(.02)^4$

$e^{-1.02} \approx e^{-1}(1 - .02 + .0002 - .000001) = \dfrac{0.980199}{e} \approx 0.3606$. Since 1 < z <

1.02, $e^{-z} < e^{-1} < 1$ and $|R_n(x)| = \dfrac{e^{-z}}{24}(.02)^4 < \dfrac{16 \times 10^{-8}}{24} < 10^{-8}$.

31. From Exercise 13, $\ln(x+1) = x - \frac{x^2}{2} + \frac{x^3}{3} - \frac{x^4}{4} + \frac{x^5}{5(z+1)^5}$ for some z between 0
     and x. Taking $x = 0.25 = 1/4$ and dropping the remainder, we get

     $\ln 1.25 \approx \frac{1}{4} - \frac{1/16}{2} + \frac{1/64}{3} - \frac{1/256}{4} = \frac{1}{4} - \frac{1}{32} + \frac{1}{192} - \frac{1}{1024} = 0.2500 - 0.0312 +$

     $0.0052 - 0.0010$ so that $\ln 1.25 \approx 0.2230$. The error in this approximation is
     the neglected remainder, $R_4(x)$. Since $0 < z < x = 0.25$, we have $z+1 > 1$ and,
     thus, $1/(z+1) < 1$. Now, $R_4(0.25) = \frac{(1/4)^5}{5(z+1)^5}$ so that $0 \le R_4(0.25) < \frac{(1/4)^5}{5(1)^5} =$

     $\frac{1}{5(1024)} = \frac{1}{5120} < 2 \times 10^{-4}$. (3-place accuracy at least.)

34. From Exercise 12, $\log x = 1 + \frac{1}{10 \ln 10}(x-10) - \frac{1}{200 \ln 10}(x-10)^2 +$

     $\frac{1}{3z^3 \ln 10}(x-10)^3$ where z is between 10 and x. Taking $x = 10.01$, so that
     $x-10 = .01$, and dropping the remainder, we obtain $\log 10.01 \approx 1 + \frac{1}{10 \ln 10} \cdot$

     $(.01) - \frac{1}{200 \ln 10}(.01)^2 \approx 1.00043$. The error is the neglected remainder

     $R_3(10.01) = .01^3/3z^3 \ln 10$. Now, $10 < z < 10.01 \implies 1/z < 10 \implies 1/z^3 < 1/10^3$

     $= 10^{-3} \implies 0 \le R_3(10.01) < (.01^3)10^{-3}/3 \ln 10 = 10^{-9}/3 \ln 10$. Since $\ln 10 >$

     2.3, the error is $< 10^{-9}/6.9 \approx 1.4 \times 10^{-10}$. (Thus we'd have 9 place accuracy
     if we knew $\ln 10$ exactly to 9 places.)

35. Maclaurin's formula for $\cos x$ with $n = 3$ is $\cos x = 1 - \frac{x^2}{2} + \frac{(\cos z)x^4}{4!}$

     (Compare #15). Thus if we use $\cos x \approx 1 - \frac{x^2}{2}$, the error is $\frac{|\cos z||x|^4}{24} \le$

     $\frac{1 \cdot (.1)^4}{24} = \frac{10^{-4}}{24} < 5 \times 10^{-6}$. Thus the accuracy is at least 5 places.

37. $f(x) = e^x \implies f^{(k)}(x) = e^x$ and $f^{(k)}(0) = 1$ for all k. Using $n = 2$ in Mac-
     laurin's formula, $e^x = 1 + x + x^2/2 + R_2(x)$ where $R_2(x) = \frac{e^z}{3!}x^3$. Thus if we
     neglect the remainder, $e^x \approx 1 + x + x^2/2$ with error $R_2(x)$. Now, $|x| \le 0.1$
     $\implies -0.1 < x < 0.1$, and since z is between 0 and x, $-0.1 < z < 0.1$ also. Thus
     $e^z < e^{0.1} < e^{\ln 2} = 2$ since $\ln 2 \approx 0.693$ and e^x is an increasing function.
     Thus $|R_2(x)| \le \frac{e^z|x|^3}{6} \le \frac{2(0.1)^3}{6} = \frac{10^{-3}}{3} < 5 \times 10^{-4}$ (i.e., 3 place accuracy).

40. $f(x) = f''(x) = f^{(4)}(x) = \cosh x \implies f(0) = f''(0) = 1$, $f^{(4)}(z) = \cosh z$.
     $f'(x) = f'''(x) = \sinh x \implies f'(0) = f'''(0) = 0$. So, with $n = 3$ in Maclaurin's
     formula, $\cosh x = 1 + \frac{1}{2}x^2 + R_3(x)$ where $R_3(x) = (\cosh z)x^4/4!$ for some z be-
     between 0 and x. Neglecting the remainder yields the approximation formula.

The error is $R_3(x)$. Since $\cosh z = (e^z + e^{-z})/2$ and z is between 0 and x, one exponent in $\cosh z$ is positive and one is negative. The positive exponential is < 2 as in #37 above; the negative exponential is < 1. Thus for such z and x, $0 < \cosh z < 3/2$, and $0 < |R_3(x)| \leq \frac{(\cosh z)|x|^4}{24} \leq \frac{3}{48}(0.1)^4 = \frac{1}{16} \times 10^{-4} =$ 6.25×10^{-6} (at least 4 place accuracy).

EXERCISES 10.6 (Review)

1. $\lim\limits_{x \to 0} \dfrac{\ln(2-x)}{1 + e^{2x}} = \dfrac{\ln 2}{2}$ by the quotient limit theorem, NOT L'Hôpital's rule.

4. $\lim\limits_{x \to 0} \dfrac{\tan^{-1}x}{\sin^{-1}x} = \lim\limits_{x \to 0} \dfrac{1/(1+x^2)}{1/\sqrt{1-x^2}} = \dfrac{1}{1} = 1.$

7. $\lim\limits_{x \to \infty} \dfrac{x^e}{e^x} = \lim\limits_{x \to \infty} \dfrac{ex^{e-1}}{e^x} = \lim\limits_{x \to \infty} \dfrac{e(e-1)x^{e-2}}{e^x} = \lim\limits_{x \to \infty} \dfrac{e(e-1)(e-2)x^{e-3}}{e^x} = 0$ since $x^{e-3} \to 0$
    and $e^x \to \infty$ as $x \to \infty$. (See also #31, Sec. 10.1.)

10. $\lim\limits_{x \to 0} \tan^{-1}x \csc x = \lim\limits_{x \to 0} \dfrac{\tan^{-1}x}{\sin x} = \lim\limits_{x \to 0} \dfrac{1/(1+x^2)}{\cos x} = \dfrac{1}{1} = 1.$

13. $y = (e^x + 1)^{1/x} \implies \ln y = \dfrac{\ln(e^x+1)}{x}$. $\lim\limits_{x \to \infty} \ln y = \lim\limits_{x \to \infty} \dfrac{e^x}{e^x+1} = \lim\limits_{x \to \infty} \dfrac{e^x}{e^x} = 1.$
    Thus $\lim\limits_{x \to \infty} y = \lim\limits_{x \to \infty} e^{\ln y} = e^1 = e.$

16. $\lim\limits_{x \to \infty} \dfrac{3^x + 2x}{x^3 + 1} = \lim\limits_{x \to \infty} \dfrac{3^x \ln 3 + 2}{3x^2} = \lim\limits_{x \to \infty} \dfrac{3^x(\ln 3)^2}{6x} = \lim\limits_{x \to \infty} \dfrac{3^x(\ln 3)^3}{6} = \infty.$

19. Because of the discontinuity at $x = -2$, we write $\int_{-\infty}^{0} = \int_{-\infty}^{-2} + \int_{-2}^{0}$.
    The second integral is: $\int_{-2}^{0} \dfrac{1}{x + 2}\, dx = \lim\limits_{t \to -2^+} \ln|x + 2| \Big]_{t}^{0} =$
    $\lim\limits_{t \to -2^+} (\ln 2 - \ln|t + 2|) = \infty$ since $t+2 \to 0$ as $t \to -2$. Thus
    the original integral diverges.

22. This integral diverges as in the preceding solution by replacing -2 by -4.

25. Multiply numerator and denominator of the integrand so that
    $I = \int_{-\infty}^{\infty} \dfrac{1}{e^x + e^{-x}} dx = \int_{-\infty}^{\infty} \dfrac{e^x}{e^{2x} + 1} dx.$ Now let $u = e^x$, $du = e^x dx$.
    As $x \to \infty$, $u \to \infty$, and as $x \to -\infty$, $u \to 0$. Thus $I = \int_{0}^{\infty} \dfrac{1}{u^2 + 1} du =$
    $\lim\limits_{t \to \infty} \tan^{-1}u \Big]_{0}^{t} = \pi/2.$

28. $\int_0^{\pi/2} \csc x \, dx = \lim_{t \to 0^+} (\ln 1 - \ln|\csc t - \cot t|)$. Note that

 $\csc t - \cot t = \dfrac{1 - \cos t}{\sin t} \to 0$ as $t \to 0^+$ by L'Hôpital. Thus, the integral diverges since $\lim_{t \to 0^+} (-\ln|\csc t - \cot t|) = \infty$.

31. (a) For $f(x) = \ln \cos x$, $c = \pi/6$, and $n = 3$, we have

 $f(x) \quad = \ln \cos x \qquad\qquad\qquad f(\pi/6) \quad = \ln(\sqrt{3}/2)$

 $f'(x) = -\sin x/\cos x = -\tan x \qquad f'(\pi/6) = (-1/2)/(\sqrt{3}/2) = -1/\sqrt{3}$

 $f''(x) = -\sec^2 x \qquad\qquad\qquad f''(\pi/6) = [-1/(\sqrt{3}/2)^2] = -4/3$

 $f'''(x) = -2 \sec x \cdot \sec x \tan x \qquad f'''(\pi/6) = (-2)(4/3)(1/\sqrt{3})$

 $\qquad\qquad\qquad\qquad\qquad\qquad\qquad\qquad = -8/3\sqrt{3}$

 $f^{(4)}(x) = -2(2 \sec x \cdot \sec x \tan x \cdot \tan x + \sec^2 x \cdot \sec^2 x)$

 $\qquad\qquad = -2 \sec^2 x (2 \tan^2 x + \sec^2 x)$.

 $\ln \cos x = \ln \dfrac{\sqrt{3}}{2} - \dfrac{1}{\sqrt{3}}(x - \dfrac{\pi}{6}) - \dfrac{2}{3}(x - \dfrac{\pi}{6})^2 - \dfrac{4}{9\sqrt{3}}(x - \dfrac{\pi}{6})^3$

 $- \dfrac{f^{(4)}(z)}{4!}(x - \dfrac{\pi}{6})^4$ for some z between x and $\dfrac{\pi}{6}$.

(b) For $f(x) = \sqrt{x - 1}$, $c = 2$, $n = 4$, we have

 $f(x) \quad = \sqrt{x - 1} \qquad\qquad\qquad f(2) \qquad = 1$

 $f'(x) \quad = \dfrac{1}{2\sqrt{x - 1}} \qquad\qquad\quad f'(2) \qquad = \dfrac{1}{2}$

 $f''(x) \quad = \dfrac{-1}{4(x - 1)^{3/2}} \qquad\qquad f''(2) \qquad = \dfrac{-1}{4}$

 $f'''(x) \quad = \dfrac{+3}{8(x - 1)^{5/2}} \qquad\qquad f'''(2) \qquad = \dfrac{3}{8}$

 $f^{(4)}(x) = \dfrac{-15}{16(x - 1)^{7/2}} \qquad\qquad f^{(4)}(2) \quad = \dfrac{-15}{16}$

 $f^{(5)}(x) = \dfrac{+105}{32(x - 1)^{9/2}} \qquad\qquad f^{(5)}(z) \quad = \dfrac{105}{32(z - 1)^{9/2}}$.

 $\sqrt{x - 1} = 1 + \dfrac{1}{2}(x - 2) - \dfrac{1}{8}(x - 2)^2 + \dfrac{1}{16}(x - 2)^3 - \dfrac{5}{128}(x - 2)^4$

 $+ \dfrac{7}{256}(z - 1)^{-9/2}(x - 2)^5$ for some z between x and 2.

34. By the calculations of Example 4 of Sec. 10.5, the indicated polynomial is obtained from Maclaurin's Formula for $\sin x$ with $n = 6$. Thus, the error is $|R_6(x)| = \dfrac{|\cos z| |x|^7}{7!} \le \dfrac{|x|^7}{7!} \le \dfrac{(\pi/4)^7}{7!}$

 $\approx \dfrac{(0.7854)^7}{5040} \approx 3.66 \times 10^{-5}$.

EXERCISES 11.1

1. $a_n = \dfrac{n}{3n+2} \implies a_1 = \dfrac{1}{3+2} = \dfrac{1}{5}$, $a_2 = \dfrac{2}{6+2} = \dfrac{1}{4}$, $a_3 = \dfrac{3}{9+2} = \dfrac{3}{11}$, $a_4 = \dfrac{4}{12+2} = \dfrac{2}{7}$.

$\lim\limits_{n \to \infty} a_n = \lim\limits_{n \to \infty} \dfrac{n}{3n+2} = \lim\limits_{n \to \infty} \dfrac{1}{3 + 2/n} = \dfrac{1}{3+0} = \dfrac{1}{3}$.

4. $a_n = \dfrac{4}{8-7n} \implies a_1 = \dfrac{4}{8-7} = 4$, $a_2 = \dfrac{4}{8-14} = -\dfrac{2}{3}$, $a_3 = \dfrac{4}{8-21} = -\dfrac{4}{13}$, $a_4 = \dfrac{4}{8-28} =$

$-\dfrac{1}{5}$. $\lim\limits_{n \to \infty} \dfrac{4}{8-7n} = \lim\limits_{n \to \infty} \dfrac{4/n}{8/n - 7} = \dfrac{0}{-7} = 0.$

7. $a_1 = \dfrac{1 \cdot 4}{2} = 2$, $a_2 = \dfrac{3 \cdot 7}{9} = \dfrac{7}{3}$, $a_3 = \dfrac{5 \cdot 10}{28} = \dfrac{25}{14}$, $a_4 = \dfrac{7 \cdot 13}{65} = \dfrac{7}{5}$.

$\lim\limits_{n \to \infty} \dfrac{(2n-1)(3n+1)}{n^3+1} = \lim\limits_{n \to \infty} \dfrac{(1/n)(2-1/n)(3+1/n)}{1+1/n^3} = \dfrac{0}{1} = 0.$

10. $a_1 = \dfrac{100}{5} = 20$, $a_2 = \dfrac{200}{2\sqrt{2} + 4} = \dfrac{100}{\sqrt{2} + 2}$, $a_3 = \dfrac{300}{3\sqrt{3} + 4}$, $a_4 = \dfrac{400}{8+4} = \dfrac{100}{3}$.

$\lim\limits_{n \to \infty} \dfrac{100n}{n^{3/2} + 4} = \lim\limits_{n \to \infty} \dfrac{100/\sqrt{n}}{1 + 4/n\sqrt{n}} = 0.$

13. $a_1 = 1 + .1 = 1.1$, $a_2 = 1 + (.1)^2 = 1.01$, $a_3 = 1 + (.1)^3 = 1.001$, $a_4 = 1 +$

$(.1)^4 = 1.0001$. $\lim\limits_{n \to \infty} 1 + (.1)^n = 1 + 0 = 1.$

16. $a_1 = 2$, $a_2 = 3/\sqrt{2}$, $a_3 = 4/\sqrt{3}$, $a_4 = 5/2$. $\lim\limits_{n \to \infty} \dfrac{n+1}{\sqrt{n}} = \lim\limits_{n \to \infty} \sqrt{n} + \dfrac{1}{\sqrt{n}} = \infty.$

19. Since $\lim\limits_{x \to \infty} \arctan x = \pi/2$, it follows that $\lim\limits_{n \to \infty} \arctan n = \pi/2$ also.

22. $\lim\limits_{n \to \infty} \dfrac{1.0001^n}{1000} = \infty$ since $1.001 > 1.$

25. Let $f(x) = (4x^4+1)/(2x^2-1)$, an ∞/∞ form as $x \to \infty$. $\lim\limits_{x \to \infty} f(x) = \lim\limits_{x \to \infty} \dfrac{16x^3}{4x} = \infty$

by L'Hôpital's rule. Thus $\lim\limits_{n \to \infty} f(n) = \infty$ also.

28. Let $f(x) = e^{-x} \ln x = \ln x/e^x$. Then $\lim\limits_{x \to \infty} \dfrac{\ln x}{e^x} = \lim\limits_{x \to \infty} \dfrac{1/x}{e^x} = 0$. Thus $\lim\limits_{n \to \infty} e^{-n} \ln n$

$= 0$ also.

31. Since $|\sin n| \leq 1$, $|2^{-n} \sin n| \leq 2^{-n} = (1/2)^n$, and $\lim\limits_{n \to \infty} (1/2)^n = 0$, it follows

that $\lim\limits_{n \to \infty} |2^{-n} \sin n| = 0$. Consequently, $\lim\limits_{n \to \infty} 2^{-n} \sin n = 0.$

34. Let $f(x) = x \sin(1/x)$, an $\infty \cdot 0$ indeterminate form as $x \to \infty$. $\lim\limits_{x \to \infty} f(x) =$

$$\lim_{x \to \infty} \frac{\sin(1/x)}{1/x} = \lim_{x \to \infty} \frac{(-1/x^2)\cos(1/x)}{(-1/x^2)} = \lim_{x \to \infty} \cos\frac{1}{x} = \cos 0 = 1.$$

Thus $\lim_{n \to \infty} n\sin(1/n) = 1.$

37. Let $y = f(x) = x^{1/x}$, an 0 indeterminate form as $x \to \infty$. $\ln y = \dfrac{\ln x}{x} \Longrightarrow$

$\lim_{x \to \infty} \ln y = \lim_{x \to \infty} \dfrac{1/x}{1} = 0 \Longrightarrow \lim_{x \to \infty} y = \lim_{x \to \infty} f(x) = e^0 = 1.$ Thus $\lim_{n \to \infty} n^{1/n} = 1.$

40. This limit does not exist. To see this note that $n^2/(n^2+1) \to 1$ as $n \to \infty$ and that $(-1)^n = +1$ if n is even, -1 if n is odd. Thus if n is large and even, $a_n = (-1)^n n^2/(n^2+1)$ is close to $+1$, but if n is large and odd, then a_n is close to -1. Thus there is no single number which all a_n's are close to when n is large.

43. (a) These equations are quickly derived from the stated conditions. For example, 10% leaving A means that 0.9 remains; 5% of C migrating to A means that 0.05 of C's population is added to A's. Thus $A_{n+1} = 0.9A_n + 0.05C_n$. (b) Let x,y,z denote the limits of A_n, B_n, C_n, respectively, as $n \to \infty$. Since A_{n+1}, B_{n+1}, C_{n+1}, converge to x,y,z also, upon letting $n \to \infty$ in each of the 3 equations in the text, and recalling that the total population is 35,000, we find that x,y,z satisfy the following system:

$$x = 0.9x \qquad\qquad + 0.05z$$
$$y = 0.1x + 0.8y$$
$$z = \qquad\quad 0.2y + 0.95z$$
$$x + y + z = 35,000$$

From the first equation, $.1x = .05z$ or $z = 2x$. The next 2 equations yield $x = 2y$, $z = 4y$. (The equation $z = 2x$ is, therefore, redundant.) Substituting the last 2 into the 4th equation yields $7y = 35000$ so that $y = 5000$, $x = 10,000$ and $z = 20,000$.

46. The sequence is x, 1/x, x, 1/x, ..., which converges only for $x = 1$.

49. (b) Let $n \to \infty$ in $x_{n+1} = \frac{1}{2}(x_n + \frac{N}{x_n})$ to obtain $L = \frac{1}{2}(L + \frac{N}{L}) \Longrightarrow$
$2L = \dfrac{(L^2 + N)}{L} \Longrightarrow 2L^2 = L^2 + N \Longrightarrow L^2 = N \Longrightarrow L = \sqrt{N}.$

EXERCISES 11.2

1. Since a = 3, r = 1/4 < 1, it converges with sum $3/(1 - \frac{1}{4})$ = 4.

4. Since a = 1, r = e/3 < 1, it converges with sum 1/(1-e/3) = 3/(3-e).

7. Writing $\sum_{n=1}^{\infty} 2^{-n}3^{n-1} = \frac{1}{2} + \frac{3}{2^2} + \frac{3^2}{2^3} + \ldots = \frac{1}{2}(1 + \frac{3}{2} + (\frac{3}{2})^2 + \ldots)$ it diverges
    since r = 3/2 > 1.

10. The series diverges since $r = \sqrt{2} > 1$.

13. The terms are obtained by multiplying those in Example 1 (where the sum was
    1) by 5. Thus the series converges by (11.19ii) with c = 5, and its sum is 5.

16. Since $\lim_{n\to\infty} \frac{n}{n+1} = 1 \neq 0$, the series diverges.

19. Since $\sum_{n=1}^{\infty} 1/8^n = \frac{1}{8} + \frac{1}{8^2} + \frac{1}{8^3} + \ldots = \frac{1}{8}(1 + \frac{1}{8} + \frac{1}{8^2} + \ldots)$ is a geometric
    series with a = 1/8, r = 1/8 < 1, it converges and has sum $\frac{1/8}{1-1/8} = \frac{1}{7}$. The 2nd
    series is that of Example 1, convergent with sum 1. Thus the given series
    converges and has sum 1/7 + 1 = 8/7.

22. $\lim_{n\to\infty} \frac{n}{\ln(n+1)} = \lim_{n\to\infty} \frac{1}{1/(n+1)} = \infty$ (L'Hôpital) $\Longrightarrow$ the series diverges.

25. $\lim_{n\to\infty} (\frac{3}{2})^n + (\frac{2}{3})^n = \infty + 0 \neq 0 \Rightarrow$ the series diverges.

28. $\sum 2^{-n} = \sum(1/2)^n$ and $\sum 2^{-3n} = \sum(2^{-3})^n = \sum(1/8)^n$ both converge as geometric
    series with r = 1/2 and 1/8, respectively. Thus the given series converges.

31. Note that $a_n = \ln \frac{n}{n+1} = \ln n - \ln(n+1)$. Thus,
$$S_1 = + a_1 = \ln 1 - \ln 2 = -\ln 2$$
$$S_2 = S_1 + a_2 = -\ln 2 + (\ln 2 - \ln 3) = -\ln 3$$
$$S_3 = S_2 + a_3 = -\ln 3 + (\ln 3 - \ln 4) = -\ln 4$$
    and, in general, $S_n = -\ln(n + 1) \to -\infty$ as $n \to \infty$, and the series
    diverges.

34. The given series, $\sum(-1)^{n+1}$ has terms $(-1)^{n+1}$ and partial sums $S_1 = 1$, $S_2 = 0$,
    $S_3 = 1$, $S_4 = 0$, etc., and the series diverges because $a_n \not\to 0$ and the fact

that $\lim_{n \to \infty} S_n$ does not exist. When grouping is done as shown in the problem,

an entirely new series results. Its terms are $b_1 = a_1 + a_2$, $b_2 = a_3 + a_4$, etc.

so that every term is 0 and, thus, every partial sum of this different series

is 0. Thus, it is not the original series that has sum 0, but rather, it is

the new series, obtained by regrouping, that has sum 0.

37. $3.2\overline{394} = 3.2 + \dfrac{394}{10^4} + \dfrac{394}{10^7} + \dfrac{394}{10^{10}} + \ldots = 3.2 + \dfrac{1}{10^4}(394 + \dfrac{394}{10^3} + \dfrac{394}{10^6} + \ldots)$

$= 3.2 + \dfrac{1}{10^4} \cdot \dfrac{394}{1-10^{-3}}$ by (12.12) with $a = 394$ and $r = 1/10^3 = 10^{-3}$. Thus

$3.2\overline{394} = 3.2 + \dfrac{1}{10^4} \cdot \dfrac{394}{999/1000} = 3.2 + \dfrac{394}{9,990} = \dfrac{3.2(9990)+394}{9990} = \dfrac{32,362}{9,990}$.

40. The total distance is approximately $24 + 24(\frac{5}{6}) + 24(\frac{5}{6})^2 + \ldots = \dfrac{24}{1-5/6} =$
$\dfrac{24}{1/6} = 144$ cm.

43. (a) In the sum, N is the number released today, 0.9N is the
number of survivors from one day ago, 0.9^2N from two days ago,
etc.   (b) If n is large, the number of survivors is almost equal
to the sum of the geometric series $N + .9N + .9^2N + \ldots =$
$\dfrac{N}{1 - .9} = 10N$.   This is 20,000 if $N = 2000$.

46. (a) Let L_n = length of a side of nth square, D_n = diameter of
nth circle. Then $D_n = L_n$, $S_n = L_n^2$, $C_n = \frac{\pi}{4}D_n^2 = \frac{\pi}{4}L_n^2 = \frac{\pi}{4}S_n$. Let
H_n = diagnoal of (n + 1)st square. Then $L_{n+1} = H_n/\sqrt{2}$, and
$S_{n+1} = L_{n+1}^2 = H_n^2/2 = \frac{2}{\pi}C_n = \frac{2}{\pi} \cdot \frac{\pi}{4}S_n = \frac{S_n}{2}$.
(b) A = Shaded area = $(S_1 - C_1) + (S_2 - C_2) + \ldots$.   Note that
$S_2 = S_1/2$, $S_3 = S_2/2 = S_1/2^2$, $\ldots$, $S_n = S_1/2^{n-1}$, and $C_1 = \frac{\pi}{4}S_1$,
$C_2 = \frac{\pi}{4}S_2 = \frac{\pi}{4} \cdot \frac{1}{2}S_1$, $\ldots$, $C_n = \frac{\pi}{4} \cdot \frac{1}{2^{n-1}}S_1$. Thus,
$A = S_1(1 - \frac{\pi}{4}) + S_1(1 - \frac{\pi}{4})\frac{1}{2} + S_1(1 - \frac{\pi}{4}) \cdot \frac{1}{2^2} + \ldots$
$= S_1(1 - \frac{\pi}{4})(1 + \frac{1}{2} + \frac{1}{2^2} + \ldots) = S_1(\frac{4 - \pi}{4})2$.

49. We write the series as: $\dfrac{1}{2} + \dfrac{(x - 3)}{4} + \dfrac{(x - 3)^2}{8} + \ldots + \dfrac{(x - 3)^n}{2^{n+1}}$
$+ \ldots = \frac{1}{2}(1 + \dfrac{(x - 3)}{2} + \dfrac{(x - 3)^2}{2^2} + \ldots + \dfrac{(x - 3)^n}{2^n} + \ldots)$ which
is a geometric series $a + ar + ar^2 + \ldots$, with $a = 1/2$ and
$r = (x - 3)/2$. Its sum is $\dfrac{a}{1 - r} = \dfrac{1/2}{1 - \dfrac{x - 3}{2}} = \dfrac{1}{2 - (x - 3)} =$

$\frac{1}{5 - x}$ if $\left|\frac{x - 3}{2}\right| < 1 \iff |x - 3| < 2 \iff -2 < x - 3 < 2 \iff$

$1 < x < 5.$

EXERCISES 11.3

1. $f(x) = (3+2x)^{-2}$ is > 0, continuous, and decreasing on $[1,\infty)$. (Decreasing

since $f'(x) = -4(3+2x)^{-2} < 0$ there.) Since $\int_1^\infty (3+2x)^{-2} dx = \lim_{t \to \infty} (-\frac{1}{2})\frac{1}{3+2x}\Big]_1^t$

$= \frac{1}{10}$, the given series converges by the integral test.

4. The integral test can apply to a series $\sum\limits_{n=2}^{\infty} f(n)$ if $f(x)$ satisfies the

hypotheses of (11.22) on $[2,\infty)$. The same proof works after a change of

variable $u = \ln x$. Here, $f(x) = 1/x(\ln x)^2$ satisfies all 3 conditions on

$[2,\infty)$. (f is decreasing either by computing $f'(x)$ or by the fact that x and

$\ln x$ are increasing.) Since $\int_2^\infty \frac{1}{x(\ln x)^2} dx = \lim_{t \to \infty} -\frac{1}{\ln x}\Big]_2^t = \frac{1}{\ln 2}$, the given

series converges.

7. $f(x) = (2x+1)^{-1/3}$ is > 0, continuous and decreasing on $[1,\infty)$. $(f'(x) =$

$-(2/3)(2x+1)^{-4/3} < 0$ there.) Since $\int_1^\infty (2x+1)^{-1/3} dx = \lim_{t \to \infty} \frac{3}{4}[(2t+1)^{2/3} - 3^{2/3}]$

$= \infty$, the series diverges.

10. $f(x) = xe^{-x}$ is > 0, continuous and decreasing on $[1,\infty)$. $(f'(x) = (1-x)e^{-x}$

< 0 if $x > 1$.) Since $\int_1^\infty xe^{-x} dx = \lim_{t \to \infty} (-te^{-t} - e^{-t} + 2e^{-1}) = 2e^{-1}$, the

series converges.

13. $f(x) = x2^{-x^2}$ satisfies the hypotheses of (11.22). (It's decreasing on $[1,\infty)$

since $f'(x) = (1-2x^2)(\ln 2)2^{-x^2} < 0$ if $x \geq 1$.) Since $\int_1^\infty x2^{-x^2} dx =$

$\lim_{t \to \infty} \frac{1}{2 \ln 2} (-2^{-t^2}+2^{-1}) = \frac{1}{4 \ln 2}$, the series converges.

16. $f(x) = 1/x\sqrt{x^2-1}$ is > 0, continuous and decreasing on $[2,\infty)$. $(f'(x) =$

$-(2x^2-1)/x^2(x^2-1)^{3/2} < 0$ if $x \geq 2$.) Since $\int_2^\infty (1/x\sqrt{x^2-1})dx = \lim_{t \to \infty}[\sec^{-1}t -$

$\sec^{-1}2] = \frac{\pi}{2} - \frac{\pi}{3} = \frac{\pi}{6}$, the series converges.

19. Since $\frac{1}{n3^n} \le \frac{1}{3^n}$ for $n \ge 1$, and since $\sum(\frac{1}{3})^n$ converges as a geometric series

with $r = 1/3$, the series $\sum \frac{1}{n3^n}$ converges by the Comparison Test.

22. For large n, $\sqrt{n}$ is much larger than 3, and it appears that $2/(3+\sqrt{n})$ behaves very much like $2/\sqrt{n}$. This suggests that we compare the terms of the given series with $1/\sqrt{n}$ (or $2/\sqrt{n}$). Thus with $a_n = 2/(3+\sqrt{n})$ and $b_n = 1/\sqrt{n}$,

$\lim_{n\to\infty} \frac{a_n}{b_n} = \lim_{n\to\infty} \frac{2\sqrt{n}}{3+\sqrt{n}} = 2$. But $\sum b_n = \sum 1/\sqrt{n}$ diverges since it is a p-series with

$p = 1/2$. Thus the given series diverges by the Limit Comparison Test.

25. For large n, neglecting the smaller powers of n, $\sqrt{4n^3-5n} \approx \sqrt{4n^3} = 2n^{3/2}$. So,

with $a_n = 1/\sqrt{4n^3-5n}$, we select $b_n = 1/2n^{3/2}$ and obtain $\lim_{n\to\infty} \frac{a_n}{b_n} =$

$\lim_{n\to\infty} \frac{2n^{3/2}}{\sqrt{4n^3-5n}} = \lim_{n\to\infty} \frac{2}{\sqrt{4-5/n^2}} = \frac{2}{\sqrt{4}} = 1$ (having divided numerator and denominator

by $n^{3/2} = \sqrt{n^3}$). Since $\sum b_n$ is $\frac{1}{2}$ times a p-series with $p = 3/2$, $\sum b_n$ converges,

and, by the Limit Comparison Test, so does $\sum a_n$.

28. $0 \le \sin^2 n \le 1 \implies 0 \le \sin^2 n/2^n \le 1/2^n$. Since $\sum 1/2^n$ converges (geometric series, $r = 1/2$), the given series converges by the Comparison Test.

31. $-1 \le \cos n \le 1 \implies 1 \le 2 + \cos n \le 3 \implies \frac{2 + \cos n}{n^2} \le \frac{3}{n^2}$. Since $\sum 3/n^2$ con-

verges as a p-series with $p = 2$, the given series converges by the Comparison Test.

34. As above, if we retain the largest power of n inside the radical, it appears

that the terms of the series behave like $1/\sqrt{n^3} = 1/n^{3/2}$. Thus with

$a_n = 1/\sqrt{n(n+1)(n+2)}$ and $b_n = 1/n^{3/2}$, $\lim_{n\to\infty} a_n/b_n = 1$. But $\sum b_n$ converges as a

p-series with $p = 3/2$. Thus the given series converges by the Limit

Comparison Test.

37. $n \ge 1 \implies 1/n^n \le 1/n^2$. Since $\sum 1/n^2$ converges $(p = 2)$, $\sum_{n=1}^{\infty} 1/n^n$ converges

by the Comparison Test.

40. Since $n \ge 1 \implies \ln n \ge 0$, we have $a_n = \frac{n + \ln n}{n^2+1} \ge \frac{n}{n^2+1}$. The series

$\sum n/(n^2+1)$ diverges either by the integral test or by the Limit Comparison

Test using $b_n = 1/n$. Thus by the Comparison Test, 2nd conclusion, the given

series diverges.

43. For sufficiently large n, $\ln n < n$ since $\lim_{n \to \infty} \frac{\ln n}{n} = 0$ (by L'Hôpital's Rule).

    (In fact, it can be shown that $\ln n < n$ for all $n \geq 1$.) Thus $\frac{\ln n}{n^3} < \frac{n}{n^3} = \frac{1}{n^2}$.
    Since $\sum 1/n^2$ converges ($p = 2$) so does the given series by the Comparison Test.

46. For $n \geq 1$, $-1 \leq \sin n \leq 1$ and $2 \leq 2^n$. Thus $1 \leq \sin n + 2^n \leq 1 + 2^n$ and
    $\frac{\sin n + 2^n}{n + 5^n} \leq \frac{1 + 2^n}{5^n} = (\frac{1}{5})^n + (\frac{2}{5})^n$. Since both $\sum(1/5)^n$ and $\sum(2/5)^n$ converge
    (geometric series, $r = 1/5, 2/5$), the given series converges by the Comparison Test.

49. (a) Add 1 to each side of $\sum_{k=2}^{n} \frac{1}{k} \leq \int_1^n \frac{1}{x} dx$ for the right inequality
    and use $\int_1^{n+1} \frac{1}{x} dx \leq \sum_{k=1}^{n} \frac{1}{k}$ for the left.

    (b) We seek n such that $S_n > 100$. By the left inequality of (a),
    $S_n > \ln(n + 1)$. Thus we may find n by solving $\ln(n + 1) > 100$
    $\Rightarrow n + 1 > e^{100}$.

55. $E < \int_N^\infty \frac{1}{x^3} dx = \frac{1}{2N^2}$, and we seek N such that $\frac{1}{2N^2} < 0.01 \Rightarrow N^2 > 50$
    $\Rightarrow N = 8$ will work.

EXERCISES 11.4

1. $\lim_{n \to \infty} \frac{a_{n+1}}{a_n} = \lim_{n \to \infty} \frac{3(n + 1) + 1}{2^{n+1}} \cdot \frac{2^n}{3n + 1} = \lim_{n \to \infty} \frac{2^n}{2^{n+1}} \cdot \frac{3n + 4}{3n + 1} =$
   $\lim_{n \to \infty} \frac{1}{2} \cdot \frac{3n + 4}{3n + 1} = \frac{1}{2}$. Since this limit is < 1, the series
   converges.

4. $\lim_{n \to \infty} \frac{a_{n+1}}{a_n} = \lim_{n \to \infty} \frac{2^n}{5^{n+1}(n + 2)} \cdot \frac{5^n(n + 1)}{2^{n-1}} = \lim_{n \to \infty} \frac{2^n}{2^{n-1}} \cdot \frac{5^n}{5^{n+1}} \cdot \frac{n + 1}{n + 2} =$
   $\lim_{n \to \infty} \frac{2}{5} \frac{n + 1}{n + 2} = \frac{2}{5} < 1 \Rightarrow$ series converges.

7. $\lim_{n \to \infty} \frac{a_{n+1}}{a_n} = \lim_{n \to \infty} \frac{(n + 1)!}{e^{n+1}} \cdot \frac{e^n}{n!} = \lim_{n \to \infty} \frac{n + 1}{e} = \infty \Rightarrow$ series diverges.

10. The ratio test is inconclusive since $\lim_{n \to \infty} \frac{a_{n+1}}{a_n} =$
    $\lim_{n \to \infty} \frac{n + 2}{(n + 1)^3 + 1} \cdot \frac{n^3 + 1}{n + 1} = 1$. However, $a_n = \frac{n + 1}{n^3 + 1} \approx \frac{n}{n^3} = \frac{1}{n^2}$
    for large n. Let $b_n = \frac{1}{n^2}$. Then $\lim_{n \to \infty} \frac{a_n}{b_n} = \lim_{n \to \infty} \frac{(n + 1)n^2}{n^3 + 1} = 1$.

    Since Σb_n converges ($p = 2$), the given series converges by the
    Limit Comparison Test.

13. We could use the ratio test, but it is much simpler to observe that $a_n = \dfrac{2}{n^3 + e^n} < \dfrac{2}{n^3} = b_n$. Since Σb_n converges ($p = 3$), the given series converges by the Comparison Test.

16. $\lim\limits_{n \to \infty} \dfrac{a_{n+1}}{a_n} = \lim\limits_{n \to \infty} \dfrac{(n+1)!}{(n+1)^{n+1}} \cdot \dfrac{n^n}{n!} = \lim\limits_{n \to \infty} \dfrac{(n+1)!}{n!} \cdot \dfrac{n^n}{(n+1)^{n+1}} =$

$\lim\limits_{n \to \infty} (n+1) \cdot \dfrac{n^n}{(n+1)^{n+1}} = \lim\limits_{n \to \infty} \dfrac{n^n}{(n+1)^n} = \lim\limits_{n \to \infty} \dfrac{1}{(1+1/n)^n} = \dfrac{1}{e} < 1$

$\Rightarrow$ series converges.

19. $\dfrac{a_{n+1}}{a_n} = \dfrac{[(n+1)!]^2}{[2(n+1)]!} \dfrac{(2n)!}{[n!]^2} = [\dfrac{(n+1)!}{n!}]^2 [\dfrac{(2n)!}{(2n+2)!}] =$

$[(n+1)]^2 \cdot \dfrac{1}{[(2n+2)(2n+1)]} = \dfrac{n^2 + 2n + 1}{4n^2 + 6n + 2} \to \dfrac{1}{4}$ as $n \to \infty$

$\Rightarrow$ series converges.

22. $\lim\limits_{n \to \infty} 3^{1/n} = 3^0 = 1 \neq 0 \Rightarrow$ series diverges by the nth term test.

25. $\lim\limits_{n \to \infty} \dfrac{a_{n+1}}{a_n} = \lim\limits_{n \to \infty} \dfrac{1 \cdot 3 \cdot 5 \ldots (2n-1)(2(n+1) - 1)}{(n+1)!} \cdot \dfrac{n!}{1 \cdot 3 \cdot 5 \ldots (2n-1)}$

$= \lim\limits_{n \to \infty} \dfrac{(2(n+1) - 1)}{n+1} = \lim\limits_{n \to \infty} \dfrac{2n+1}{n+1} = 2 > 1 \Rightarrow$ series diverges.

EXERCISES 11.5

1. The series converges by the Alternating Series Test (AST) since $\lim\limits_{n \to \infty} 1/\sqrt{2n+1} = 0$ and $1/\sqrt{2(n+1)+1} < 1/\sqrt{2n+1}$. Since $|(-1)^{n+1}| = 1$, the series of absolute values is $\Sigma \dfrac{1}{\sqrt{2n+1}}$. With $b_n = \dfrac{1}{\sqrt{n}}$, $\lim\limits_{n \to \infty} \dfrac{a_n}{b_n} = \lim\limits_{n \to \infty} \dfrac{\sqrt{n}}{\sqrt{2n+1}} = \dfrac{1}{\sqrt{2}} > 0$. Since Σb_n diverges ($p = 1/2$) the series of absolute values diverges. Thus the original series converges only conditionally.

4. The series converges by the AST since $\lim\limits_{n \to \infty} \dfrac{n}{n^2+1} = 0$ and the terms decrease. $(f(x) = \dfrac{x}{x^2+1} \Rightarrow f'(x) = \dfrac{1-x^2}{(1+x^2)^2} \leq 0$ for $x \geq 1$.) The series of absolute values is $\Sigma \dfrac{n}{n^2+1}$ which diverges using the Limit Comparison Test with $b_n = \dfrac{1}{n}$, together with $\Sigma \dfrac{1}{n}$ divergent. Thus the original series converges conditionally.

7. $|a_n| = \dfrac{5}{n^3+1} < \dfrac{5}{n^3} \Rightarrow |a_n|$ converges by comparison with the series $\Sigma \dfrac{5}{n^3}$, ($p = 3$). Thus the original series converges absolutely.

10. $\lim\limits_{n\to\infty}\left|\dfrac{a_{n+1}}{a_n}\right| = \lim\limits_{n\to\infty}\dfrac{(n+1)!}{5^{n+1}}\cdot\dfrac{5^n}{n!} = \lim\limits_{n\to\infty}\dfrac{n+1}{5} = \infty \Rightarrow$ series diverges.

13. The series converges by the AST since $\lim\limits_{n\to\infty}\dfrac{\sqrt[3]{n}}{n+1} = 0$ and the terms

decrease. $\left(f(x) = \dfrac{x^{1/3}}{x+1} \Rightarrow f'(x) = \dfrac{1-2x}{3x^{2/3}(x+1)^2} < 0 \text{ if } x \geq 1.\right)$

The series of absolute values is $\sum\dfrac{n^{1/3}}{n+1}$. Since $\dfrac{n^{1/3}}{n+1} \approx \dfrac{n^{1/3}}{n} = $

$\dfrac{1}{n^{2/3}}$ for large n, we choose $b_n = \dfrac{1}{n^{2/3}}$. Then

$\lim\limits_{n\to\infty}\dfrac{a_n}{b_n} = \lim\limits_{n\to\infty}\dfrac{n}{n+1} = 1 > 0$ and, since Σb_n diverges (p = 2/3),

the series of absolute values diverges, and the original series

converges conditionally.

16. $\lim\limits_{n\to\infty}\left|\dfrac{a_{n+1}}{a_n}\right| = \lim\limits_{n\to\infty}\dfrac{\ln(n+1)}{1.5^{n+1}}\cdot\dfrac{1.5^n}{\ln n} = \lim\limits_{n\to\infty}\dfrac{\ln(n+1)}{1.5\ln n} = \dfrac{1}{1.5} = \dfrac{2}{3} < 1$

(by L'Hôpital's rule). Thus the series converges absolutely.

19. The series converges by the AST since $1/n\sqrt{\ln n}$ decreases to zero.

The series of absolute values, $\sum\dfrac{1}{n\sqrt{\ln n}}$, can be handled by the

Integral Test. Let $f(x) = 1/(x\sqrt{\ln x})$, continuous and decreasing

to 0 on $[2,\infty)$. For the antiderivative of $f(x)$, let $u = \ln x$,

$du = \dfrac{1}{x}dx$, and $\int\dfrac{1}{x\sqrt{\ln x}}dx = \int u^{-1/2}du = 2u^{1/2} + C = 2\sqrt{\ln x} + C.$

Thus $\int_2^\infty\dfrac{1}{x\sqrt{\ln x}}dx = \lim\limits_{t\to\infty}2\sqrt{\ln t} - 2\sqrt{\ln 2} = \infty$ and the series of

absolute values diverges since this improper integral diverges.

Thus, the original series converges conditionally.

22. Using the Root Test, $\lim\limits_{n\to\infty}\sqrt[n]{|a_n|} = \lim\limits_{n\to\infty}\dfrac{n^2+1}{n} = \infty \Rightarrow$ series

diverges.

25. If we approximate the sum S by the partial sum S_n, then the error

is less than a_{n+1}, the first term neglected. Since we want 3 place

accuracy, the error is to be less than 5×10^{-4}. Thus we seek n

such that $a_{n+1} < 5\times 10^{-4} \Rightarrow \dfrac{1}{(n+1)!} < 5\times 10^{-4} \Rightarrow$

$(n+1)! > 0.2\times 10^4 = 2000$. Since $6! = 720$ and $7! = 5040$, we

choose $n+1 = 7$, or $n = 6$ terms.

28. $a_{n+1} = \dfrac{1}{(n+1)^5} < 5\times 10^{-4} \Rightarrow (n+1)^5 > 2000$. Since $4^5 = 1024$

and $5^5 = 3125$, we choose $n+1 = 5$ or $n = 4$ terms.

31. $a_{n+1} = \dfrac{1}{(n+1)^2} < 5 \times 10^{-5} \Rightarrow (n+1)^2 > 20{,}000 \Rightarrow n+1 > 141.2$

$\Rightarrow n = 141$ terms.

34. $a_{n+1} = \dfrac{1}{(n+1)^3 + 1} < 5 \times 10^{-5} \Rightarrow (n+1)^3 + 1 > 20{,}000 \Rightarrow$

$n + 1 > 27.14 \Rightarrow n = 27$ terms.

EXERCISES 11.6

1. With $u_n = \dfrac{x^n}{n+4}$, $\lim\limits_{n\to\infty}\left|\dfrac{u_{n+1}}{u_n}\right| = \lim\limits_{n\to\infty}\left|\dfrac{x^{n+1}}{n+5}\cdot\dfrac{n+4}{x^n}\right| = \lim\limits_{n\to\infty}\left|\dfrac{n+4}{n+5}\right||x| = |x|$. So, we

have absolute convergence by the ratio test if $|x| < 1$ or $-1 < x < 1$. If

$x = 1$, the series is $\sum \dfrac{1}{n+4}$, which diverges. If $x = -1$, the series is

$\sum \dfrac{(-1)^n}{n+4}$ which converges by the AST. Thus the interval of convergence is $[-1,1)$.

4. With $u_n = \dfrac{(-3)^n x^{n+1}}{n}$, $\lim\limits_{n\to\infty}\left|\dfrac{u_{n+1}}{u_n}\right| = \lim\limits_{n\to\infty}\dfrac{3n}{n+1}|x| = 3|x|$. So, we have absolute

convergence if $3|x| < 1$ or $-\dfrac{1}{3} < x < \dfrac{1}{3}$. If $x = \dfrac{1}{3}$, the series is $\sum \dfrac{(-3)^n}{n3^{n+1}}$

$= \dfrac{1}{3}\sum\dfrac{(-1)^n}{n}$, which converges by the AST. If $x = -1/3$, the series is

$\sum \dfrac{(-3)^n}{n(-3)^{n+1}} = -\dfrac{1}{3}\sum\dfrac{1}{n}$ which diverges. Thus the interval of convergence is

$(-1/3, 1/3]$.

7. With $u_n = \dfrac{n}{n^2+1} x^n$, $\lim\limits_{n\to\infty}\left|\dfrac{u_{n+1}}{u_n}\right| = \lim\limits_{n\to\infty}\dfrac{n+1}{(n+1)^2+1}\cdot\dfrac{n^2+1}{n}|x| = |x|$. Thus, by the

ratio test we have absolute convergence if $|x| < 1$ or $-1 < x < 1$. If $x = 1$,

the series is $\sum \dfrac{n}{n^2+1}$ which diverges by the limit comparison test $(b_n = 1/n)$,

or the integral test. If $x = -1$, the series is $\sum \dfrac{n}{n^2+1}(-1)^n$, convergent by

the AST. Thus the interval of convergence is $[-1,1)$.

10. $\lim\limits_{n\to\infty}\left|\dfrac{u_{n+1}}{u_n}\right| = \lim\limits_{n\to\infty}\dfrac{10^{n+2}}{3^{2(n+1)}}\cdot\dfrac{3^{2n}}{10^{n+1}}|x| = \lim\limits_{n\to\infty}\dfrac{10}{3^2}|x| = \dfrac{10}{9}|x|$. Thus we have

absolute convergence if $\dfrac{10}{9}|x| < 1$, or $|x| < \dfrac{9}{10}$, or $-\dfrac{9}{10} < x < \dfrac{9}{10}$. If

$x = \pm\dfrac{9}{10}$, $u_n = \dfrac{10^{n+1}}{3^{2n}}(\pm\dfrac{9}{10})^n$. Since $3^{2n} = (3^2)^n = 9^n$, $u_n = \dfrac{10^{n+1}}{9^n}\cdot\dfrac{9^n}{10^n}\cdot(\pm 1)^n$.

Thus, $u_n = 10(\pm 1)^n$, which does not have a limit as $n \to \infty$. (i.e., $u_n \not\to 0$ as $n \to \infty$.) Thus the power series diverges at each end point, and the interval of convergence is $(-\frac{9}{10}, \frac{9}{10})$.

13. $\lim\limits_{n\to\infty} \left| \dfrac{u_{n+1}}{u_n} \right| = \lim\limits_{n\to\infty} \dfrac{(n+1)!}{100^{n+1}} \cdot \dfrac{100^n}{n!} |x| = \lim\limits_{n\to\infty} \dfrac{(n+1)}{100} |x| = \infty$ unless $x = 0$, in which case the limit is 0. Thus the series converges only for $x = 0$.

16. $\lim\limits_{n\to\infty} \left| \dfrac{u_{n+1}}{u_n} \right| = \lim\limits_{n\to\infty} \dfrac{\sqrt[3]{n}}{\sqrt[3]{n+1}} \dfrac{|x|}{3} = \dfrac{|x|}{3} < 1$ if $\dfrac{|x|}{3} < 1$ or $-3 < x < 3$. If $x = 3$, the

series is $\sum \dfrac{(-1)^{n-1}}{\sqrt[3]{n}}$ which converges by the AST. If $x = -3$, the series is

$\sum \dfrac{1}{\sqrt[3]{n}}$ which diverges ($p = 1/3$). Thus the interval of convergence is $(-3,3]$.

19. $\lim\limits_{n\to\infty} \left| \dfrac{u_{n+1}}{u_n} \right| = \lim\limits_{n\to\infty} \dfrac{3^{2(n+1)}}{n+2} \cdot \dfrac{n+1}{3^{2n}} |x-2| = \lim\limits_{n\to\infty} \dfrac{9(n+1)}{n+2} |x-2| = 9|x-2| < 1$ if $|x-2|$

$< 1/9$ or $2 - \dfrac{1}{9} < x < 2 + \dfrac{1}{9}$. If $x = 2 + \dfrac{1}{9}$, $x-2 = \dfrac{1}{9}$, and the series is

$\sum \dfrac{1}{n+1}$, which diverges. (Remember $3^{2n} = (3^2)^n = 9^n$.) If $x = 2 - \dfrac{1}{9}$, $x-2 =$

$-1/9$ and the series is $\sum \dfrac{(-1)^n}{n+1}$ which converges by the AST. Thus the interval

of convergence is $[2 - \dfrac{1}{9}, 2 + \dfrac{1}{9}) = [\dfrac{17}{9}, \dfrac{19}{9})$.

22. $\lim\limits_{n\to\infty} \left| \dfrac{u_{n+1}}{u_n} \right| = \lim\limits_{n\to\infty} \dfrac{2n+1}{2n+3} |x+3| = |x+3| < 1$ if $-1 < x+3 < 1$ or $-4 < x < -2$. If

$x = -2$, $x+3 = 1$ and the series is $\sum \dfrac{1}{2n+1}$ which diverges. If $x = -4$, $x+3 = -1$

and the series is $\sum \dfrac{(-1)^n}{2n+1}$ which converges by the AST. Thus the interval of

convergence is $[-4,-2)$.

25. $\lim\limits_{n\to\infty} \left| \dfrac{u_{n+1}}{u_n} \right| = \lim\limits_{n\to\infty} \dfrac{n}{6(n+1)} |2x-1| = \dfrac{|2x-1|}{6} < 1$ if $|2x-1| < 6$ or $-6 < 2x-1 < 6$ or

$-\dfrac{5}{2} < x < \dfrac{7}{2}$. If $x = \dfrac{7}{2}$, $2x-1 = 6$, and the series is $\sum \dfrac{(-1)^n}{n}$ which converges

by the AST. If $x = -5/2$, $2x-1 = -6$, and the series is $\sum 1/n$ which diverges.

Thus the interval of convergence is $(-5/2, 7/2]$.

28. $u_n = \dfrac{2 \cdot 4 \cdot 6 \cdot \ldots (2n)}{4 \cdot 7 \cdot 10 \cdot \ldots (3n+1)} x^n \implies u_{n+1} = \dfrac{2 \cdot 4 \cdot 6 \cdot \ldots (2n)(2(n+1))}{4 \cdot 7 \cdot 10 \cdot \ldots (3n+1)(3(n+1)+1)} x^{n+1}$. Thus

$$\left|\frac{u_{n+1}}{u_n}\right| = \frac{2\cdot4\cdot6\cdot\ldots(2n)(2n+2)}{4\cdot7\cdot10\cdot\ldots(3n+1)(3n+4)} \cdot \frac{4\cdot7\cdot10\cdot\ldots(3n+1)}{2\cdot4\cdot6\cdot\ldots(2n)} |x| = \frac{2n+2}{3n+4}|x|, \text{ and}$$

$$\lim_{n\to\infty}\left|\frac{u_{n+1}}{u_n}\right| = \lim_{n\to\infty}\frac{2n+2}{3n+4}|x| = \frac{2}{3}|x|. \text{ Thus, by the ratio test, the series con-}$$

verges absolutely if $\frac{2}{3}|x| < 1$ or $|x| < \frac{3}{2}$, and the radius of convergence is $\frac{3}{2}$.

31. $\left|\frac{u_{n+1}}{u_n}\right| = \frac{(n+c+1)!}{(n+1)!(n+d+1)!} \cdot \frac{n!(n+d)!}{(n+c)!}|x| = \frac{n+c+1}{(n+1)(n+d+1)}|x| \to 0$ for all x as

$n \to \infty$. Thus $r = \infty$.

34. The indicated polynomial is the sum of the first 4 terms of the
     series for $J_0(x)$. If $0 \leq x \leq 1$, the terms decrease to zero.
     Thus, by the AST, the error, E, is no greater than the next
     term. Thus $E \leq \frac{x^8}{2^2 4^2 6^2 8^2} \leq \frac{1}{384^2} < 10^{-6}$.

EXERCISES 11.7

1. $\frac{1}{1-x} = 1 + x + x^2 + \ldots = \sum_{n=0}^{\infty} x^n$ for $|x| < 1$. This is just the geometric

series encountered in Sec. 11.2 with $a = 1$ and $r = x$.

4. Replacing x by 4x in #1, $\frac{1}{1-4x} = \sum_{n=0}^{\infty}(4x)^n = \sum_{n=0}^{\infty}4^n x^n$ for $|4x| < 1$ or $|x| < 1/4$.

7. $\frac{x}{2-3x} = \frac{x}{2}\left(\frac{1}{1-\frac{3}{2}x}\right) = \frac{x}{2}\sum_{n=0}^{\infty}\left(\frac{3}{2}x\right)^n = \sum_{n=0}^{\infty}\frac{3^n x^{n+1}}{2^{n+1}}$ for $|\frac{3}{2}x| < 1$ or $|x| < \frac{2}{3}$, where

we replaced x by $\frac{3}{2}x$ in #1.

10. $\frac{3}{2x+5} = \frac{3}{5}\frac{1}{(1+\frac{2}{5}x)} = \frac{3}{5}\sum_{n=0}^{\infty}\left(-\frac{2}{5}x\right)^n = \sum_{n=0}^{\infty}\frac{3(-1)^n 2^n x^n}{5^{n+1}}$ for $|\frac{2}{5}x| < 1$ or $|x| < \frac{5}{2}$,

where we replaced x by $(-\frac{2}{5}x)$ in #1.

13. Since $e^t = 1 + t + \frac{t^2}{2!} + \frac{t^3}{3!} + \ldots + \frac{t^n}{n!} + \ldots$ for all t, replacing t by -x we

obtain $e^{-x} = 1 + (-x) + \frac{(-x)^2}{2!} + \frac{(-x)^3}{3!} + \ldots + \frac{(-x)^n}{n!} + \ldots = 1 - x + \frac{x^2}{2!} - \frac{x^3}{3!}$

$+ \ldots + \frac{(-1)^n x^n}{n!} + \ldots$.

16. $\sinh x = \frac{1}{2}(e^x - e^{-x}) = \frac{1}{2}[(1 + x + \frac{x^2}{2!} + \frac{x^3}{3!} + \ldots) - (1 - x + \frac{x^2}{2!} - \frac{x^3}{3!} + \ldots)]$

$= \frac{1}{2}[2x + \frac{2x^3}{3!} + \frac{2x^5}{5!} + \ldots + \frac{2x^{2n-1}}{(2n-1)!} + \ldots] = x + \frac{x^3}{3!} + \frac{x^5}{5!} + \ldots +$

$\frac{x^{2n-1}}{(2n-1)!} + \ldots$.

19. Replacing x by $(-x^6)$ in #1, $\frac{1}{1+x^6} = 1 - x^6 + x^{12} - \ldots$ for $|x| < 1$. Thus

$\int_0^{1/3} \frac{1}{1+x^6}\,dx = [x - \frac{x^7}{7} + \frac{x^{13}}{13} - \ldots]_0^{1/3} = \frac{1}{3} - \frac{1}{3^7 \cdot 7} + \frac{1}{3^{13} \cdot 13} - \ldots$. This is a

convergent alternating series. Recall that if the sum of such a series is
approximated by the partial sum of the first n terms, the error is less than
the next term, i.e. the first term neglected. Since $\frac{1}{3^7 \cdot 7} \approx 6.532 \times 10^{-5}$ and

$\frac{1}{3^{13} \cdot 13} \approx 4.825 \times 10^{-7}$, we will have 6 place accuracy if we use the first two

terms. Thus the value is $\frac{1}{3} - \frac{1}{3^7 \cdot 7} \approx 0.33333 - 0.00007 = 0.33326$, which

becomes 0.3333 when rounded off to 4 places.

22. $\frac{x^3}{1+x^5} = x^3(1 - x^5 + x^{10} - \ldots)$ for $|x| < 1$. $\int_0^{.2} \frac{x^3}{1+x^5}\,dx = \int_0^{.2} (x^3 - x^8 +$

$x^{13} - \ldots)dx = \frac{(.2)^4}{4} - \frac{(.2)^9}{9} + \ldots$. Since the 2nd term is $\frac{512 \times 10^{-9}}{9} <$

6×10^{-8}, we get 4 (in fact, 6) place accuracy with just the first term.

The value is $\frac{(.2)^4}{4} = .0004$ to 4 places.

25. $f(x) = (1-x^2)^{-1} \implies f'(x) = 2x(1-x^2)^{-2}$. To obtain a series for $2x(1-x^2)^{-2}$,
we can start with the series for f(x) and differentiate it. Replacing x by

x^2 in #1, we obtain $f(x) = \frac{1}{1-x^2} = 1 + x^2 + x^4 + x^6 + \ldots + x^{2n} + \ldots$, for

$|x| < 1$, whence $f'(x) = 2x(1-x^2)^{-2} = 2x + 4x^3 + 6x^5 + \ldots + 2nx^{2n-1}$, for

$|x| < 1$.

28. (a) $\sum_{n=1}^{\infty} P_n = e^{-\lambda} \sum_{n=1}^{\infty} \frac{\lambda^n}{n!} = e^{-\lambda}e^{\lambda} = 1$

(b) The probability of 2 or more photons is $\sum_{n=2}^{\infty} P_n = \sum_{n=0}^{\infty} P_n - P_0 - P_1$

$= 1 - e^{-\lambda} - e^{-\lambda}\lambda$.

30. $\frac{e^t - 1}{t} = [(1 + t + \frac{t^2}{2!} + \frac{t^3}{3!} + \ldots + \frac{t^n}{n!} + \ldots) - 1]/t = 1 + \frac{t}{2!} + \frac{t^2}{3!} + \ldots + \frac{t^{n-1}}{n!} +$

$\ldots$. Thus $f(x) = \int_0^x \frac{e^t - 1}{t}\,dt = \int_0^x (1 + \frac{t}{2!} + \frac{t^2}{3!} + \ldots + \frac{t^{n-1}}{n!} + \ldots)dt =$

$[t + \frac{t^2}{2(2!)} + \frac{t^3}{3(3!)} + \ldots + \frac{t^n}{n(n!)} + \ldots]_0^x = x + \frac{x^2}{2(2!)} + \frac{x^3}{3(3!)} + \ldots + \frac{x^n}{n(n!)} + \ldots$

$= \sum_{n=1}^{\infty} \frac{x^n}{n(n!)}$.

EXERCISES 11.8

1. We calculate:

$$f(x) = \cos x \qquad f(0) = 1$$
$$f'(x) = -\sin x \qquad f'(0) = 0$$
$$f''(x) = -\cos x \qquad f''(0) = -1$$
$$f'''(x) = \sin x \qquad f'''(0) = 0$$

and the subsequent derivatives follow this pattern. Thus $\cos x =$

$(1 - \frac{1}{2!} x^2 + \frac{1}{4!} x^4 - \frac{1}{6!} x^6 + \ldots) = \sum\limits_{n=0}^{\infty} \frac{(-1)^n x^{2n}}{(2n)!}$. The remainder, $R_n(x)$,

is $(\pm \sin z) x^{n+1}/(n+1)!$ if n is even or $(\pm \cos z) x^{n+1}/(n+1)!$ if n is odd. In

either case $|R_n(x)| \le \frac{x^{n+1}}{(n+1)!}$ which has limit 0 as $n \to \infty$ for every x by (11.42).

4. All even-numbered derivatives of $f(x) = \cosh x$ are $\cosh x$ and have value 1 at
x = 0. All odd-numbered derivatives are $\sinh x$ which are all 0 at x = 0. Thus

$\cosh x = (1 + \frac{x^2}{2!} + \frac{x^4}{4!} + \ldots) = \sum\limits_{n=0}^{\infty} \frac{x^{2n}}{(2n)!}$, in agreement with (11.43g). Here,

$R_n(x) = \frac{\sinh z}{(n+1)!} x^{n+1}$ (if n is even) or $R_n(x) = \frac{\cosh z}{(n+1)!} x^{n+1}$ (if n is odd). In

either case $|R_n(x)| \le \frac{e^z + e^{-z}}{2} \frac{|x|^{n+1}}{(n+1)!} \le (\cosh z) \frac{|x|^{n+1}}{(n+1)!}$ since $|\sinh z| \le$

$\cosh z$. Thus $|R_n(x)| \le (\cosh x) \frac{|x|^{n+1}}{(n+1)!}$ since z between 0 and x $\Longrightarrow$ $\cosh z \le$

$\cosh x$. Again by (11.42) the factor $\frac{|x|^{n+1}}{(n+1)!} \to 0$ as $n \to \infty$ and $|R_n(x)| \to 0$ for
all x.

7. Using $e^x = (1 + x + \frac{x^2}{2!} + \ldots)$, $e^{-x} = (1 - x + \frac{x^2}{2!} - \ldots)$, and $\sinh x =$

$(e^x - e^{-x})/2$ we obtain $\sinh x = x + \frac{x^3}{3!} + \frac{x^5}{5!} + \ldots = \sum\limits_{n=0}^{\infty} \frac{x^{2n+1}}{(2n+1)!}$. With $u_n =$

$\frac{x^{2n+1}}{(2n+1)!}$, $\lim\limits_{n\to\infty} \left| \frac{u_{n+1}}{u_n} \right| = \lim\limits_{n\to\infty} \frac{x^2}{(2n+3)(2n+2)} = 0$ for all x. Hence $r = \infty$.

10. Substituting x^2 for x in #1, we obtain $\cos(x^2) = \sum\limits_{n=0}^{\infty} \frac{(-1)^n (x^2)^{2n}}{(2n)!} =$

$\sum\limits_{n=0}^{\infty} \frac{(-1)^n x^{4n}}{(2n)!}$. The ratio test yields $\lim\limits_{n\to\infty} \left| \frac{u_{n+1}}{u_n} \right| = \lim\limits_{n\to\infty} \frac{|x|^4}{(2n+2)(2n+1)} = 0$ for

all x. Hence $r = \infty$.

13. We calculate:

$$f(x) = \sin x \qquad f(\pi/4) = 1/\sqrt{2}$$
$$f'(x) = \cos x \qquad f'(\pi/4) = 1/\sqrt{2}$$
$$f''(x) = -\sin x \qquad f''(\pi/4) = -1/\sqrt{2}$$
$$f'''(x) = -\cos x \qquad f'''(\pi/4) = -1/\sqrt{2},$$

and the higher derivatives follow the same pattern. Thus, the desired

Taylor's series is $\sin x = \dfrac{1}{\sqrt{2}} + \dfrac{1}{\sqrt{2}}(x - \dfrac{\pi}{4}) - \dfrac{1}{2!\sqrt{2}}(x - \dfrac{\pi}{4})^2 - \dfrac{1}{3!\sqrt{2}}(x - \dfrac{\pi}{4})^3 + \ldots$

(Note that the pattern of signs is + + - - + + - -, etc.)

16. $f(x) = e^x \Longrightarrow f^{(n)}(x) = e^x$ and $f^{(n)}(-3) = e^{-3}$ for all integers $n \geq 0$. Thus

$e^x = \sum\limits_{n=0}^{\infty} \dfrac{e^{-3}}{n!}(x-(-3))^n = \sum\limits_{n=0}^{\infty} \dfrac{e^{-3}}{n!}(x+3)^n$.

19. Since powers of $(x+1)$ are required, we want the Taylor series about $c = -1$.
Thus

$\quad f(x) = e^{2x}$ $\qquad\qquad\qquad f(-1) = e^{-2}$

$\quad f'(x) = 2e^{2x}$ $\qquad\qquad\qquad f'(-1) = 2e^{-2}$

$$\vdots$$

$\quad f^{(n)}(x) = 2^n e^{2x}$ $\qquad\qquad f^{(n)}(-1) = 2^n e^{-2}$,

and $e^{2x} = \sum\limits_{n=0}^{\infty} \dfrac{2^n e^{-2}}{n!}(x+1)^n$.

22. $f(x) = \tan x$ $\qquad\qquad\qquad f(\pi/4) = 1$

$\quad f'(x) = \sec^2 x$ $\qquad\qquad\qquad f'(\pi/4) = 2$

$\quad f''(x) = 2\sec^2 x \tan x$ $\qquad\quad f''(\pi/4) = 4$

$\quad f'''(x) = 2\sec^4 x + 4\sec^2 x \tan^2 x$ $\quad f'''(\pi/4) = 16$,

and the first four terms are

$$1 + 2(x - \dfrac{\pi}{4}) + \dfrac{4}{2!}(x - \dfrac{\pi}{4})^2 + \dfrac{16}{3!}(x - \dfrac{\pi}{4})^3 + \ldots$$

25. $f(x) = xe^x$ $\qquad\qquad\qquad\qquad f(-1) = -e^{-1}$

$\quad f'(x) = (x+1)e^x$ $\qquad\qquad\qquad f'(-1) = 0$

$\quad f''(x) = (x+2)e^x$ $\qquad\qquad\qquad f''(-1) = e^{-1}$

$\quad f'''(x) = (x+3)e^x$ $\qquad\qquad\qquad f'''(-1) = 2e^{-1}$

$\quad f^{(4)}(x) = (x+4)e^x$ $\qquad\qquad\qquad f^{(4)}(-1) = 3e^{-1}$,

and the first four terms of the Taylor series are

$$xe^x = -e^{-1} + \dfrac{e^{-1}}{2!}(x+1)^2 + \dfrac{2e^{-1}}{3!}(x+1)^3 + \dfrac{3e^{-1}}{4!}(x+1)^4 + \ldots$$

28. $e^x = \sum\limits_{n=0}^{\infty} \dfrac{x^n}{n!} \Longrightarrow e^{-1} = \sum\limits_{n=0}^{\infty} \dfrac{(-1)^n}{n!}$. This series satisfies the conditions of

the AST. Thus, if we approximate the sum, $S = e^{-1}$, by the nth partial sum,

S_n, the error is $< a_{n+1}$, the 1st term neglected. Thus, for 4 decimal place accuracy we seek n such that $\frac{1}{(n+1)!} < 5 \times 10^{-5}$ or $(n+1)! > 20,000$. With

n = 6, 7! = 5040, but with n = 7, 8! = 40,320. Thus, to 4 places, $e^{-1} \approx S_7$

$$= \sum_{n=0}^{7} (-1)^n/n! = \frac{1}{0!} - \frac{1}{1!} + \frac{1}{2!} - \frac{1}{3!} + \frac{1}{4!} - \frac{1}{5!} + \frac{1}{6!} - \frac{1}{7!} = 1 - 1 + \frac{1}{2} - \frac{1}{6} + \frac{1}{24}$$

$$- \frac{1}{120} + \frac{1}{720} - \frac{1}{5040} \approx 0.3679.$$

31. By (11.43c) with x = 0.1, $\tan^{-1} 0.1 = \sum_{n=0}^{\infty} \frac{(-1)^n (0.1)^{2n+1}}{2n+1}$. Again using the

alternating series criterion, we seek n such that $a_{n+1} = \frac{(0.1)^{2n+3}}{2n+3} <$

5×10^{-5} or $(2n+3) 10^{2n+3} > 2 \times 10^4$. With n = 1, the left side is 5×10^5.

Thus, to 4 places, $\tan^{-1} 0.1 \approx S_1 = \sum_{n=0}^{1} \frac{(-1)^n (0.1)^{2n+1}}{2n+1} = 0.1 - \frac{0.1^3}{3} \approx 0.0997.$

34. Since, by (11.43c), the series for cosh x is a positive term series for all
x, we may NOT use the alternating series criterion as in #28 and #31 above.
Rather, we must determine the number of terms needed by looking at the re-
mainder in Maclaurin's formula. It was shown in #4 above that $|R_n(x)| \le$

$\frac{(\cosh x) x^{n+1}}{(n+1)!}$. Since x = 0.1, $|R_n(0.1)| \le \frac{(\cosh 0.1)(.1)^{n+1}}{(n+1)!}$. Now we have

a problem! We are trying to estimate cosh 0.1, and this very number occurs
in the upper bound for the remainder (which is, of course, the error if we
use the Maclaurin polynomial of degree n to approximate cosh 0.1). We must
replace cosh 0.1 by some larger number for estimation purposes. cosh 0.1 =

$(e^{0.1} + e^{-0.1})/2 < (e^{0.1} + e^{0.1})/2 = e^{0.1} < e^{\ln 2} = 2$ since $0.1 < \ln 2 \approx$

0.693. Thus $|R_n(0.1)| < \frac{2(0.1)^{n+1}}{(n+1)!}$ and this is $< 5 \times 10^{-5}$ if $\frac{(n+1)! 10^{n+1}}{2} >$

2×10^4. With n = 3, the left side is 12×10^4. Thus we need the terms
out to x^3 (n = 3), of the Maclaurin series of cosh x. Since the coefficients

of x and x^3 are 0, $\cosh 0.1 \approx 1 + \frac{(0.1)^2}{2!} = 1.0050$ to 4 places.

37. $\cos t = 1 - \frac{t^2}{2!} + \frac{t^4}{4!} - \dots \implies \cos x^2 = 1 - \frac{x^4}{2!} + \frac{x^8}{4!} - \dots$

$$\int_0^{.5} \cos x^2 \, dx = \int_0^{.5} (1 - \frac{x^4}{2} + \frac{x^8}{24} - \dots) dx = 0.5 - \frac{(0.5)^5}{10} + \frac{(0.5)^9}{9(24)} - \dots .$$

The 3rd term = $1/2^9(216) = 1/(512)(216) < 1/(500)(200) = 10^{-5}$. Thus the
first two terms are sufficient, and the value is 0.5000 - 0.00312 = 0.49688
$\approx 0.4969.$

40. $\sin x = x - \dfrac{x^3}{3!} + \dfrac{x^5}{5!} - \dfrac{x^7}{7!} + \cdots \Rightarrow \dfrac{\sin x}{x} = 1 - \dfrac{x^2}{3!} + \dfrac{x^4}{5!} - \dfrac{x^6}{7!} + \cdots$

$\Rightarrow \int_0^1 \dfrac{\sin x}{x}dx = 1 - \dfrac{1}{3(3!)} + \dfrac{1}{5(5!)} - \dfrac{1}{7(7!)} + \cdots$

Using the alternating series criterion, we see that the 4th term is $1/7(5040) < 5 \times 10^{-5}$. Thus, the 1st 3 terms suffice for 4 place accuracy, and the value is $1 - \dfrac{1}{18} + \dfrac{1}{600} \simeq 0.9461$.

43. The series for $\ln(1 - x)$ is obtained by replacing x by $-x$ in the series for $\ln(1 + x)$. Then $\ln(\dfrac{1 + x}{1 - x}) = \ln(1 + x) - \ln(1 - x) =$

$(x - \dfrac{x^2}{2} + \dfrac{x^3}{3} - \dfrac{x^4}{4} + \cdots) - (-x - \dfrac{x^2}{2} - \dfrac{x^3}{3} - \dfrac{x^4}{4} - \cdots) =$

$2(x + \dfrac{x^3}{3} + \dfrac{x^5}{5} + \cdots).$

46. $\pi = 4[\tan^{-1}(1/2) + \tan^{-1}(1/3)]$

$= 4[\displaystyle\sum_{n=0}^{\infty} (-1)^n \dfrac{(1/2)^{2n+1}}{2n + 1} + \displaystyle\sum_{n=0}^{\infty} (-1)^n \dfrac{(1/3)^{2n+1}}{2n + 1}]$

$\simeq 4[(\dfrac{1}{2} - \dfrac{1}{24} + \dfrac{1}{160} - \dfrac{1}{896} + \dfrac{1}{4608}) + (\dfrac{1}{3} - \dfrac{1}{81} + \dfrac{1}{1215} - \dfrac{1}{15,309} + \dfrac{1}{177,147})]$

$= 3.1417.$

49. Let $f(t) = (1 - \cos t)/\cos t = \sec t - 1$ so that $\delta = xf(kL)$. We compute the first few terms of the Maclaurin series of $f(t)$:

$f(0) = 0$, $f'(t) = \sec t \tan t \Rightarrow f'(0) = 0$,

$f''(t) = \sec^3 t + \sec t \tan^2 t \Rightarrow f''(0) = 1$. Thus, if t is small,

$f(t) \simeq \dfrac{f''(0)}{2}t^2 = \dfrac{t^2}{2}$, and $\delta = xf(kL) \simeq x\dfrac{(kL)^2}{2} = \dfrac{xL^2 P}{2E}$.

EXERCISES 11.9

1. (a) Using (11.44) with $k = 1/2$, $\sqrt{1+x} = (1+x)^{1/2} = 1 + \dfrac{1}{2}x + \dfrac{(1/2)(-1/2)}{2!}x^2$

$+ \dfrac{1/2(-1/2)(-3/2)}{3!}x^3 + \cdots + \dfrac{(1/2)(-1/2)\cdots(1/2-n+1)}{n!}x^n + \cdots$ which reduces

to the answer given. By (11.44), $r = 1$.

(b) Using part (a) and replacing x by $-x^3$, we have $\sqrt{1-x^3} = 1 - \dfrac{x^3}{2} +$

$\displaystyle\sum_{n=2}^{\infty} (-1)^{n-1} \dfrac{1 \cdot 3 \cdots (2n-3)}{2^n n!} (-x^3)^n$ which again reduces to the answer given.

4. For $|x| < 1$, $(1 + x)^{1/4} = 1 + \dfrac{1}{4}x + \dfrac{((1/4)(-3/4)}{2!}x^2 +$

$\dfrac{(1/4)(-3/4)(-7/4)}{3!}x^3 + \cdots + \dfrac{(1/4)(-3/4)\cdots(1/4 - n + 1)}{n!}x^n + \cdots =$

$1 + \dfrac{1}{4}x - \dfrac{3}{4^2 2!}x^2 + \dfrac{7 \cdot 3}{4^3 3!}x^3 + \cdots + (-1)^{n-1}\dfrac{3 \cdot 7 \cdot 11 \ \cdots \ (4n - 1)}{4^n n!}x^n$

$+ \cdots$

7. For $|x| < 1$, $(1 + x)^{-2} = 1 - 2x + \dfrac{(-2)(-3)}{2!}x^2 + \dfrac{(-2)(-3)(-4)}{3!}x^3$

+ $\ldots = 1 - 2x + 3x^2 - 4x^3 + \ldots + (-1)^n(n + 1)x^n$.

10. Using (11.44) with $k = -2$, $x(1+2x)^{-2} = x[1 + (-2)(2x) + \dfrac{(-2)(-3)(2x)^2}{2!} + \ldots$

+ $\dfrac{(-2)(-3)\ldots(-2-n+1)(2x)^n}{n!} + \ldots] = x[1 + \sum\limits_{n=1}^{\infty} (-1)^n \dfrac{2\cdot3\ldots(n+1)2^n}{n!} x^n]$. By

(11.44), we have convergence if $|2x| < 1$ or $|x| < 1/2$. Thus $r = 1/2$.

13. We begin by computing $(1+x)^{-1/2} = 1 - \dfrac{1}{2}x + \dfrac{(-1/2)(-3/2)}{2!} x^2 + \ldots +$

$\dfrac{(-1/2)(-3/2)\ldots(-1/2 - n + 1)}{n!} x^n + \ldots = 1 + \sum\limits_{n=1}^{\infty} (-1)^n \dfrac{1\cdot3\cdot5\ldots(2n-1)}{2^n n!} x^n$.

Setting $x = -t^2$, we have $\dfrac{1}{\sqrt{1-t^2}} = 1 + \sum\limits_{n=1}^{\infty} (-1)^n \dfrac{1\cdot3\cdot5\ldots(2n-1)}{2^n n!} (-1)^n t^{2n}$.

Using $(-1)^n \cdot (-1)^n = (-1)^{2n} = 1$, we obtain $\sin^{-1} x =$

$\int_0^x \dfrac{1}{\sqrt{1-t^2}} dt = x + \sum\limits_{n=1}^{\infty} \dfrac{1\cdot3\cdot5\ldots(2n-1)x^{2n+1}}{(2n+1) 2^n n!}$. $r = 1$ since the series expansion

was valid for $|x| = |t^2| < 1 \iff |t| < 1$, and integrating does not alter r.

15. (This will be done rather than #16 since it relates to #1(b) above.)

By #1(b), $\sqrt{1-x^3} = 1 - \dfrac{x^3}{2} - \dfrac{x^6}{8} - \dfrac{3x^9}{48} + \ldots$ and, replacing x by -x, we have

$\sqrt{1+x^3} = 1 + \dfrac{x^3}{2} - \dfrac{x^6}{8} + \dfrac{x^9}{16} - \ldots$ and $\int_0^{1/2} \sqrt{1+x^3}\, dx = [x + \dfrac{x^4}{8} - \dfrac{x^7}{56} + \dfrac{x^{10}}{160} - \ldots]_0^{1/2}$

$= [\dfrac{1}{2} + \dfrac{1}{2^4\cdot8} - \dfrac{1}{2^7\cdot56} + \dfrac{1}{2^{10}\cdot160} - \ldots]$. Since this is an alternating series,

the error in using only the 1st 2 terms is less than the 3rd term. Since

$\dfrac{1}{2^7\cdot56} = \dfrac{1}{7168} \approx 0.00014$, this will yield 3 place accuracy. Thus the value of

the integral to 3 places is $\dfrac{1}{2} + \dfrac{1}{2^4\cdot8} = \dfrac{1}{2} + \dfrac{1}{128} = 0.5000 + 0.0078 \approx 0.508$.

19. Replacing x by x^3 in #7, $(1 + x^3)^{-2} = 1 - 2x^3 + 3x^6 - 4x^9 + \ldots$,

and $\int_0^{0.3} (1 + x^3)^{-2} dx = x - \dfrac{1}{2}x^4 + \dfrac{3}{7}x^7 - \dfrac{4}{10}x^{10} + \ldots]_0^{0.3} =$

$0.3 - \dfrac{(0.3)^4}{2} + \dfrac{4}{7}(0.3)^7 - \dfrac{4}{10}(0.3)^{10} + \ldots$. The 3rd term of this

alternating series is $< 5 \times 10^{-4}$. Thus we obtain 3 place accuracy

with only the 1st 2 terms: $0.3 - \dfrac{0.3^4}{2} = 0.3 - 0.00405 \approx 0.296$.

EXERCISES 11.10 (Review)

1. $f(x) = \dfrac{\ln(x^2+1)}{x} \implies \lim\limits_{x\to\infty} f(x) = \lim\limits_{x\to\infty} \dfrac{\ln(x^2+1)}{x} = \lim\limits_{x\to\infty} \dfrac{2x/(x^2+1)}{1} = 0$ by

L'Hôpital's rule. Thus $\lim\limits_{n\to\infty} \dfrac{\ln(n^2+1)}{n} = 0$ also.

4. If n is even, $(-2)^n$ is positive and becomes larger as n does. If n is odd,

$(-2)^n$ is negative and becomes more so as n increases. So, even though $1/n \to$

0, the sequence behaves essentially like $(-2)^n$ which oscillates wildly, and

no limit exists.

7. A preliminary analysis suggests that the terms behave like $1/\sqrt[3]{n^3} = 1/n$. Thus

with $a_n = 1/\sqrt[3]{n(n+1)(n+2)}$ and $b_n = 1/n$, $\lim\limits_{n\to\infty} \dfrac{a_n}{b_n} = \lim\limits_{n\to\infty} \dfrac{n}{\sqrt[3]{n(n+1)(n+2)}} = 1$. Since

$\sum b_n$ diverges (p = 1, the harmonic series), the given series diverges also.

10. $\lim\limits_{n\to\infty} \dfrac{1}{2+(1/2)^n} = \dfrac{1}{2} \neq 0 \implies$ the series diverges.

13. Since $\dfrac{n!}{\ln(n+1)} \geq \dfrac{n}{\ln(n+1)}$ for all $n \geq 1$, and since $\lim\limits_{n\to\infty} \dfrac{n}{\ln(n+1)} = \lim\limits_{n\to\infty} \dfrac{1}{1/(n+1)}$

$= \infty$ (by L'Hôpital's rule), the series diverges.

16. A preliminary analysis suggests we choose $b_n = 1/n^2$. So, with $a_n = \dfrac{n + \cos n}{n^3 + 1}$,

$\lim\limits_{n\to\infty} \dfrac{a_n}{b_n} = \lim\limits_{n\to\infty} \dfrac{n^3 + n^2\cos n}{n^3 + 1} = \lim\limits_{n\to\infty} \dfrac{1 + (\cos n)/n}{1 + 1/n^3} = 1$. Since $\sum b_n$ converges (p = 2),

the given series converges.

19. The series diverges since $\lim\limits_{n\to\infty} \sqrt[n]{n} = 1$, and thus, $\lim\limits_{n\to\infty} 1/\sqrt[n]{n} = 1 \neq 0$. To obtain

this limit, let $y = x^{1/x}$, an ∞^0 indeterminate form as $x \to \infty$. Then $\ln y =$

$\dfrac{1}{x} \ln x$, and $\lim\limits_{x\to\infty} \ln y = \lim\limits_{x\to\infty} \dfrac{\ln x}{x} = \lim\limits_{x\to\infty} \dfrac{1/x}{1} = 0$. Thus $\lim\limits_{x\to\infty} y = \lim\limits_{x\to\infty} x^{1/x} = e^0 = $

1, and $\lim\limits_{n\to\infty} n^{1/n} = \lim\limits_{n\to\infty} \sqrt[n]{n} = 1$ also.

22. If u_n denotes the nth term of the series, then $|u_n| = \dfrac{\sqrt[3]{n-1}}{n^2-1} \approx \dfrac{\sqrt[3]{n}}{n^2} = \dfrac{1}{n^{5/3}}$.

This preliminary analysis suggests we use $b_n = 1/n^{5/3}$ in the limit comparison

test. $\lim\limits_{n\to\infty} \dfrac{|u_n|}{b_n} = \lim\limits_{n\to\infty} \dfrac{n^{5/3}\sqrt[3]{n-1}}{n^2-1} = \lim\limits_{n\to\infty} \dfrac{\sqrt[3]{1-1/n}}{1-1/n^2} = 1$, having divided numerator

and denominator by n^2. Since $\sum b_n$ converges (p = 5/3), so does $\sum |u_n|$, and the

given series converges absolutely.

25. $\left|\dfrac{1 - \cos n}{n^2}\right| = \dfrac{|1 - \cos n|}{n^2} \le \dfrac{1 + |\cos n|}{n^2} \le \dfrac{2}{n^2}$. Since $\sum \dfrac{2}{n^2}$ converges (p = 2), the given series converges absolutely.

28. $\lim\limits_{n\to\infty} \left|\dfrac{u_{n+1}}{u_n}\right| = \lim\limits_{n\to\infty} \dfrac{3^n}{(n+1)^2+9} \cdot \dfrac{n^2+9}{3^{n-1}} = \lim\limits_{n\to\infty} 3 \dfrac{n^2+9}{(n+1)^2+9} = 3(1) = 3 > 1$. Thus the series diverges by the ratio test.

31. With $f(x) = \sqrt{\ln x}/x$ we have $f'(x) = (1-2 \ln x)/2x^2\sqrt{\ln x} < 0$ for $x \ge 2$.

Moreover, $\lim\limits_{x\to\infty} f(x) = \lim\limits_{x\to\infty} \dfrac{\sqrt{\ln x}}{x} = \lim\limits_{x\to\infty} \dfrac{1/2x\sqrt{\ln x}}{1} = 0$. Thus, the given series,

$\sum (-1)^n \dfrac{\sqrt{\ln n}}{n}$ converges by the AST. Now, if $n \ge 3$, $\dfrac{\sqrt{\ln n}}{n} > \dfrac{1}{n}$ (ln x is an in-

increasing function $\Rightarrow$ ln n $\ge$ ln 3 > ln e = 1). Since $\sum \dfrac{1}{n}$ diverges (harmonic

series), $\sum \left|(-1)^n \dfrac{\sqrt{\ln n}}{n}\right| = \sum \dfrac{\sqrt{\ln n}}{n}$ diverges by the comparison test. Thus the

original series converges conditionally.

34. $f(x) = \dfrac{x}{\sqrt{x^2-1}}$ is continuous, positive and decreasing on $[2,\infty)$. ($f'(x) =$

$-1/(x^2-1)^{3/2} < 0$.) $\displaystyle\int_2^\infty f(x)\,dx = \lim\limits_{t\to\infty}\int_2^t x(x^2-1)^{-1/2}\,dx = \lim\limits_{t\to\infty} (x^2-1)^{1/2}\Big]_2^t =$

$\lim\limits_{t\to\infty} (\sqrt{t^2-1} - \sqrt{3}) = \infty$, and the series diverges. (Note: divergence can be

obtained immediately by observing $n/\sqrt{n^2-1} \to 1$ as $n \to \infty$.)

37. $f(x) = 10/\sqrt[3]{x+8}$ is positive, continuous and decreasing on $[1,\infty)$. $\displaystyle\int_1^\infty f(x)\,dx =$

$\lim\limits_{t\to\infty}\int_1^t 10(x+8)^{-1/3}\,dx = \lim\limits_{t\to\infty} (15(t+8)^{2/3}-15(9)^{2/3}) = \infty$, and the series

diverges.

40. The series converges by the AST. Since the 10th term, $\dfrac{1}{100(101)} < 10^{-4}$, we

will get at least 3 place accuracy if we take only the 1st 9 terms as the

approximation to the sum, A. Thus $A \approx \dfrac{1}{1(2)} - \dfrac{1}{4(5)} + \dfrac{1}{9(10)} - \dfrac{1}{16(17)} + \dfrac{1}{25(26)}$

$- \dfrac{1}{36(37)} + \dfrac{1}{49(50)} - \dfrac{1}{64(65)} + \dfrac{1}{81(82)}$. $A \approx .5000 - .05000 + .0111 - .0037 +$

$.0015 - .0008 + .0004 - .0002 + .0002 \approx .4585 \approx .458.$

43. With $u_n = \dfrac{(x+10)^n}{n2^n}$, $\lim\limits_{n\to\infty}\left|\dfrac{u_{n+1}}{u_n}\right| = \lim\limits_{n\to\infty} \dfrac{n}{2(n+1)}\,|x+10| = \dfrac{|x+10|}{2} < 1$ if $|x+10| <$

2 or $-2 < x+10 < 2$ or $-12 < x < -8$. If $x = -8$, $x+10 = 2$, and the series is

$\sum \dfrac{1}{n}$ which diverges. If $x = -12$, the series is $\sum \dfrac{(-1)^n}{n}$ which converges by the

AST. Thus the interval of convergence is $[-12,8)$.

46. With $u_n = \frac{(x+5)^n}{(n+5)!}$, $\lim\limits_{n\to\infty} \left|\frac{u_{n+1}}{u_n}\right| = \lim\limits_{n\to\infty} \frac{|x+5|}{n+6} = 0$ for all x. Thus the interval

of convergence is $(-\infty, \infty)$.

49. The quickest way for this problem is to recognize that $\sin x \cos x = \frac{1}{2}\sin 2x$

$= \frac{1}{2}\sum\limits_{n=0}^{\infty} \frac{(-1)^n(2x)^{2n+1}}{(2n+1)!} = \sum\limits_{n=0}^{\infty} \frac{(-1)^n 2^{2n} x^{2n+1}}{(2n+1)!}$. $r = \infty$ by the ratio test or from

the nature of the sine series. Another method for this problem would be to multiply the Maclaurin series for sin x and cos x together.

52. See the solution to #13, Sec. 11.9.

55. One way is to write $\sqrt{x} = \sqrt{4+(x-4)} = 2\sqrt{1 +(x-4)/4}$ and use #1(a), Sec. 11.9 replacing x by $(x-4)/4$ and multiplying by 2. Another way is to compute the

Taylor series of $f(x) = x^{1/2}$ about c = 4. This yields $f(x) = x^{1/2}$, $f'(x) = \frac{1}{2}x^{-1/2}$, $f''(x) = \frac{1}{2}(-\frac{1}{2})x^{-3/2}$, $f'''(x) = \frac{1}{2}(-\frac{1}{2})(-\frac{3}{2})x^{-5/2}$, ..., $f^{(n)}(x) = (\frac{1}{2})(-\frac{1}{2})\ldots(\frac{-(2n-3)}{2})$. $x^{-(2n-1)/2}$ for $n \geq 2$. Evaluating these derivatives

at x = 4 and using the Taylor series formula, the given answer results.

58. $\sin x = x - \frac{x^3}{6} + \frac{x^5}{120} - \frac{x^7}{5040} + \cdots$

$\frac{\sin x}{\sqrt{x}} = x^{1/2} - \frac{x^{5/2}}{6} + \frac{x^{9/2}}{120} - \frac{x^{13/2}}{5040} + \cdots$

$\int_0^1 \frac{\sin x}{\sqrt{x}} dx = \frac{2}{3}x^{3/2} - \frac{2}{42}x^{7/2} + \frac{2}{1320}x^{11/2} -$

$\frac{2}{75,600}x^{15/2} + \cdots \Big]_0^1$

$= \frac{2}{3} - \frac{1}{21} + \frac{1}{660} - \frac{1}{37,800} + \cdots$

Since the 4th term of this alternating series is $< 10^{-4}$, the sum of the 1st 3 yields 3 place accuracy at least. This sum is .6667 - .0476 + .0015 = .6206 $\approx$ .621.